I0213840

# THE TURNING WHEEL

## The Story Of General Motors Through Twenty-Five Years 1908-1933

**By**

**Arthur Pound**

**Drawings by WILLIAM HEYER**

Cover Photograph: Pets4Dawn

ISBN: 978-1-78139-182-2

# Contents

# List of Illustrations

# Publisher's Note

IT is probable that no invention of such far-reaching importance was ever diffused with such rapidity or so quickly exerted influences that ramified through the national culture, transforming even habits of thought and language." This quotation from the report of the Hoover Research Com- mittee on Social Trends refers to the motor vehicle.

The commonplaceness of motor cars in our daily lives makes us unaware of their significance. It is almost im- possible to realize a present-day world without automobiles, and yet motor cars are little more than a generation old.

This book, then, not only helps to make us conscious of the marvelously rapid development of a new art, a new convenience, a new means of transportation, but also, in giving the history of one of our important industries, it provides a view of the vast social consequences of inven- tion and enterprise. And yet General Motors is but twenty- five years old.

Innumerable histories of nations, rulers, wars, and peoples have been published of much less significance than this story of a great industry. Our leading business groups will find here many instances and examples of enterprising public service. Here is a broad yet carefully written history of an industrial enterprise which directly or indirectly affects intimately the lives of our people.

In this book will be found illuminating accounts of in- ventors, financial geniuses, scientists, business statesmen. Their accomplishments have altered our lives and will affect those of our grandchildren and great-grandchildren. While making motor cars, they have also been making history.

## Foreword

GENERAL MOTORS in 1933 reached its twenty-fifth milestone. Since the founding of General Motors Company of New Jersey in 1908, the growth of the organization has contributed a unique chapter to American industrial his- tory. From beginnings so small that its birth escaped notice in financial centers, General Motors has worked its way steadily forward to a place where its leadership in many of the most exacting branches of production and distribution is taken for granted and where it meets the public of many lands with a wide variety of merchandise and services. Scientific research, close attention to dealer and consumer needs, and constructive public policies are among the fac- tors accounting for General Motors' present strength.

The older companies of General Motors, now known as divisions, go back to the early days of the automotive indus- try, and some of them far beyond. Their taproots reach down to carriage- and wagon-building, to firearms, station- ary and marine gasoline engines, milling machinery, roller bearings, bicycle gears, lathes, and even to door-bells. Their branch roots stretch back to the beginnings of scientific experiment, since the self-propelled vehicle is the child of physics and chemistry. Chapters n and in trace that long evolution. As one follows the rise of General Motors against the broad background of latter-day industry and science, he comprehends that the flowering of large-scale production in our day is the inevitable result of generations of inventiveness, organizing ability, and the willingness of capital and labor to pull together toward common objec- tives. Among those objectives are the lifting of the stand- ards of living, the satisfaction of old wants with less labor, and the creation of new wants on a higher level of comfort, convenience, and culture. Modern industry has conquered the old-time dearth of goods, and more and more it searches for a balance wheel through whose steadying influ- ence its products can remain available to all industrious men and women at all times. To steady economic life is perhaps as real an industrial need today as mass production was fifteen years ago, as real a need as the automobile was forty years ago, when men traveled at the pace of the horse over wretched roads.

My acquaintance with General Motors began at its birth in 1908, and as a somewhat impartial observer of social trends I have watched its progress with keen interest ever since. After observing General Motors employees as work- men and citizens, I began, more than thirteen years ago, to write in Flint, Michigan, my Iron Man papers, noting some of the social effects of modern industrialism. The Cor- poration seemed then to foreshadow many of the develop- ments on the social side which have since come to pass. I welcomed the opportunity to complete a full-length study with access to records, believing that the story of a great corporation's growth through twenty-five years would be of more than passing value, since corporations are the most efficient of modern groupings and probably also the most meaningful from the standpoint of basic social relations: work and wages, production and distribution, consumption and investment. If it appears that approval is voiced here more freely than the reverse, that is because the record is clean and clear.

As I review the history of General Motors in my mind, I think of the many thousands of men and women who made its present competence come to pass by their labors in factories, laboratories and offices, and in the field; of workers in all branches of production; of craftsmen and designers striving to combine beauty with serviceability; of scientists patiently attacking problems in chemistry, electricity, metallurgy, and engineering; of foremen, superin- tendents, and inspectors; of dealers and salesmen in every land searching for the sales by which the Corporation lives and by means of which it pays wages and dividends. This book is the history of a joint effort which succeeded be- cause, when a long, strong pull was needed, team-play tri- umphed over the frictions which tend to dissipate human efforts and destroy institutions.

Those who recognize in General Motors a force of the first magnitude in America's economic life will find here several references to the Corporation's policies which have contributed to its present standing. Some of these policies apply to interdivisional operations, others to employees, dealers, suppliers, and the public. Beyond the equities in- volved in strictly commercial contacts, any large corpora- tion which touches the lives of millions of persons here and abroad can scarcely escape being rated and appraised by public opinion. General Motors has been a leader in pro- viding full and detailed accounting to stockholders and in giving the general public accurate news of what it is doing and why it is doing it.

The history of General Motors records scientific and commercial achievements of a high order, but this is true of many corporations. What makes this corporation most interesting is the fact that its expansion was rapid and yet it was marked by relatively few of the discords usually connected with swift industrial growth.

In the appendices will be found historical notes on many subsidiaries and affiliates not treated in the text.

The author acknowledges gratefully the assistance of persons too numerous to mention who have contributed in- formation to this work, including many formerly active in Corporation affairs but now retired, and others whose busi- ness relationship with General Motors in its early days qualifies them to testify on the events of that period.

Special thanks are due to Dr. Dixon Ryan Fox, Professor of American History at Columbia University, for his in- terest and encouragement, to Dr. John S. Worley, Pro- fessor of Transportation at the University of Michigan, and to Dr. M. M. Quaife, secretary-editor, the Burton Historical Collection, Detroit Public Library, for their aid in research.

ARTHUR POUND

# Chapter I

## AMERICA ON WHEELS

In a single picture are caught and recorded centuries of history. In the distance, ready to vanish over the hill, is an Indian family departing with its poor goods and beaten gods. A tiny pony strains between two poles, across which is a laden platform. The poles drag on the earth. In the foreground is the settler's covered wagon, drawn by strong horses and ready to roll westward as long as its tall, ironshod wheels hang together.

The wagon holds more wheels and the produce of wheels: a spinning wheel and the cloths which wheels have fabricated; a plowshare which some wheel has helped to smooth and point. There is a rifle fashioned on a lathe to which the principle of the wheel has been adapted. A continent is being surrendered to those who come on wheels. The conqueror sets his wheels in motion and moves on with calm assurance to occupy the empire which wheels have enabled him and his kind to possess.[1]

It is one of the paradoxes of history that America, where modern civilization runs on wheels, was a wheel-less country before the white man came to its shores. Other peoples, no

---

[1] Bernard de Voto, Mark Twain's America, p. 24: "That winter he [Marcus Whitman, the missionary] gathered a large mission. Also he found a wife; and she and another woman joined the caravan that took the trail the following spring [1836]. The caravan possessed, too, wagons and a light cart that were to go further into transalpine America than wheels had ever gone before. The western filtrate had now no boundary to its passage but the Pacific sea."

more primitive than the American Indian, had discovered and applied the wheel in the dawn of civilization; indeed, it is likely that without the wheel Western civilization would never have emerged from its birth throes. The two-wheeled chariots of Nineveh broke down the barriers of space and the boundaries of empires built on self-contained river valleys. The mergings of peoples, exchange of ideas, tools, and goods over wide distances in short, the early education of the race in commerce, mechanics, language, and thought were destined henceforth to proceed through and by, and to a large extent directly on, wheels.

Yet prehistoric America somehow missed this potent application of the wheel to earth. Their grave lack was early noted by scholars. The learned Dr. Robert Hooke, in the 1726 edition of his Philosophic Experiments and Observations, says that ignorance of the wheel in aboriginal America indicates that its inhabitants could not have come hither from Europe, Africa, or Asia, since the wheel "is an invention of so great use, that it seems impossible to be lost by mankind, after it be once known."

Ancient Egyptian Wheel

The American Indians possessed marked capacities in many directions. They were excellent workers in stone, as well as hardy hunters and bold warriors. They traversed wide areas in both war and chase; they were essentially moral and lived a primitive religion of exalted concepts and close harmonies. In their intertribal relations they were capable of acute political sagacity; The Great Peace Pact of the Five Nations of the Iroquois Confederacy, which endured at least three centuries, is one of the most subtle adjustments of conflicting national political interests ever devised, more intricate and better balanced than the League of Nations. Yet the Indian gave way before the whites because the newcomers had better tools, especially better weapons and vehicles.

At one time or another the American aborigines had used the other five primary machines: the lever, wedge, screw, pulley, and the inclined plane. With these the Mayas of Yucatan and Guatemala and the Incas of Peru built massive edifices. But there is no evidence that the Red Man ever had command of the wheel. In the meantime, through the centuries of recorded time and before, the Asiatics, Europeans, and Egyptians had not only evolved wheeled carts and chariots, but also they had made so many efficient combinations of the wheel with the other five primary machines that they had evolved the compass to guide them to America, domestic tools of many sorts, the well-ground sword blade, and the musket which spat fire. The conquerors had inherited a superior technology as well as superior form of carriage on land and sea. Hence the aborigines gave way. One reason and perhaps the root reason for the failure of the aboriginal inhabitants effectively to occupy and defend North America may well have been their failure to master the wheel.

But merely to put two wheels on the ground, with an axletree between them, is not enough to satisfy an artful folk with pressing problems to solve problems of sustenance and distribution, problems of state and war. The fabled wheel of Ceres, the goddess of Agriculture, might be turned into a country cart, the scythed chariot might do for war, but not for long could either satisfy the eternal demand of a questing people for more speed, more thrills, more wealth. The ancients applied more power to wheels by harnessing eight and twelve horses to their chariots and as many oxen to their wagons, increasing their teams of draft animals up to the full limit of the wit and muscle of men in controlling them. Within the limits prescribed by flesh and blood these ambitious men of ancient days did what they could to increase their power in transport. They built gorgeous equipages of state, rode furiously in their hunts, developed strains of horseflesh suitable to various purposes chargers for war, draft horses for tillage, light palfreys for milady to ride, high-stepping carriage horses. They developed wheeled vehicles in great variety, from the simple cart to impressive and beautiful coaches of royalty. But ever and always each generation was restricted in its burning desire to conquer space and time by the fact that the horse represented at once the strongest and most flexible power plant he could apply to the wheels of his vehicles. Other animals might outdraw the horse, but no other animal could combine pulling power with as prompt acceleration. The unique conformation of the horse's leg from hoof to hip-joint gives him a leverage out of all proportion to his weight, and for centuries he was the best motive force man had at his disposal for transportation purposes.

The Horse Age lasted from the dawn of history to recent times. In those slow-moving centuries enormous advances were made in other directions, but land transport remained keyed to draft animals. Man charted the solar system, discovered the mass of the earth by laying out a geographical degree on the plains of Greece, applied the arch, tenon, and pillar in architecture, constructed huge cathedrals of surpassing beauty, worked out the basic laws of physics and mathematics, began experiments in chemistry, developed marine transportation from the oar to the sail, applied the compass to the discovery of America and the circumnavigation of the earth, built large and cunningly contrived vessels for the mastery of wind and water, and harnessed water power to turn looms. Laboriously he brought water transport by means of canals to assist in his problems of land carriage. In all the practical arts the advance of knowledge had been tremendous, but from the dawn of history down to modern times mankind had available in quantity for land transport no better motive power than the horse.

Yet the challenge toward improvement persisted. Dimly the more progressive peoples have understood the truth laid down in the report of a Select Committee of the British House of Commons on Highways in 1808, that "next to the general influence of the seasons upon which the regular supply of our wants and a great proportion of our comfort so much depend, there is perhaps no circumstance more interesting to men in a civilized state than the perfection of the means of interior communication."

A philosophical investigator of transportation has said, "with the exception of land and ruins there are few things of any material value to man which do not derive that value, in part at least, from transport from their original position." One might go further and say that life, shorn of the prospect of ever improving transportation, would be dull, flat, and unprofitable. Truly civilization has run on wheels from the stone discs of the primitive cart to the rubber tires of the automobile.

# Chapter II

## THE EVOLUTION OF SELF-PROPELLED VEHICLES

THE evolution of transport has been worked out by John Brisben Walker and others somewhat as follows:

| | |
|---|---|
| Floating log | Sailing chariot |
| Animal's back | Coach and carriage |
| Sledge down hill | Velocipede |
| Horse-drawn sledge cart | Free balloon |
| Canoe | Steam carriage |
| Ox cart | Steamship |
| Chariot | Railroad |
| Oared galley | Bicycle |
| Sedan chair | Cable car |
| Sailing vessel | Electric trolley car |
| Canal | Automobile |
| Man-powered carriage | Airplane and airship |

Observe, in the sequence, the progress from the simplest of motive powers, water running under the pull of gravity, to the explosions of rarefied gases in the internal combustion engine. Those in the first half of the list are shrewd adaptations of forces found in nature. Of special interest is the sailing chariot, known in China, and brought to its peak in Holland, where one designed by Simon Stevin, about 1600, covered forty-two miles in two hours, carrying twenty-eight persons, and was used quite regularly.

Even before man tried to harness the wind to land transport, he endeavored to "beat the horse" by various combinations of muscle and machinery. Heliodorus notes a triumphal chariot at Athens moved by slaves who worked machinery. A Jesuit missionary in China, Matthieu Ricci (1552-1610), declared that he had traveled there inside a great wheel, propelled by two fellow

Simon Stevin's sailing chariot, Holland, about A. D. 1600

passengers who used levers to give the wheel forward motion. A beautiful series of drawings comes down from the Middle Ages, showing men propelling heavy and ornate carriages by worm and other gears. The last supreme effort in that direction seems to be Sir John Anderson's 1831 patent, in which twenty-four oarsmen provided the energy.

We are accustomed to think of the steam engine as a world-shaking invention, but the fact is that for centuries after its first application the world wagged on without giving the steam engine a second thought. Hero of Alexandria, living in the second century before Christ (130 B.C.), described in his Pneumatica his famous self-propelled apparatus known as the Aeolipile, and also a fountain, both operated by steam. The principle used in the engine which worked the fountain became highly important centuries later, but the scientific Hero appears to have had no discoverable successor until the forgetful Middle Ages had come and gone, although engines like his probably were used in unimportant ways and more to amuse than to perform useful work. After the lapse of 1300 years, that bold monk, Roger Bacon (1214-94) uttered a prophecy to which no one listened: "It will be possible to construct chariots so that without animals they may be moved with incalculable speed." The prophecy is less remarkable than the revelation it gives of the complete loss of interest in science between Hero's time and that of Roger Bacon; so the latter could do no more than prophesy in a field where the former actually had experimented.

With the Renaissance, interest revived in science as well as in art and letters. Giovanni Baptista della Porta in 1601 began where Hero had left off sixteen centuries earlier, thus anticipating Savery T s first practical and productive steam engine by nearly a century.

Ramsay and Wildgoose were not thinking of steam when they applied for an English patent, in 1619, for "drawing-carts without horses." However, the edge of the curtain was being drawn aside a little about this time for the introduction of the modern Steam Age. In 1629 Giovanni Branca, an Italian, came forward with a steam turbine, now rated superior to Hero's steam engine and also to that of the same type generally accredited to the great English physicist, Sir Isaac Newton (1642-1727), which was along the same line. It is likely that the Newton engine was made after Newton's suggestions rather than by Newton himself. Both Branca and Newton suggested certain uses for their simple engines, among them the propulsion of a vehicle by a jet of steam, a means still used in cheap toys. A Jesuit missionary in Pekin, China Father Verbiest, about 1630 used an aeolipile, says Lavergne, "with jets of steam playing on a revolving winged wheel geared to the wheels of a car." In 1633, Edward Somerset, Marquis of Worcester, fathered a double-action steam engine, with displacement chambers, which was the first useful engine, though not commercially successful.

While Englishmen and Italians were pioneering in steam, a Dutch scientist, Christian Huyghens, in 1680 described the first step in the evolution of the modern internal combustion engine, the invention of a first explosion engine. In this ancestor of all internal combustion engines the ignition of gunpowder in a cylinder produced a vacuum into which outside atmospheric pressure pushed the piston. Ten years later, in 1690, Denis Papin, a young French doctor acquainted with Huyghens, substituted steam for gunpowder, thereby earning himself a place in history as the inventor of the earliest piston-and-cylinder steam engine, although his efforts fell short of practical success.

In 1698 Thomas Savery (1650-1715) obtained a patent for a steam engine designed to raise water, and applied to it a safety valve by J. T. Desaguliers, thus realizing on the suggestion of della Porta, put forth nearly a century earlier. Raising water from coal mines seems to have been more of a problem at this time than swift transportation, and for nearly a hundred years that field was cultivated by inventors to the practical exclusion of others. Thomas Newcomen, an associate of Savery, produced various atmospheric steam engines, from 1705 to 1711, which found immediate and high favor with the mine operators. Other engineers contributed to this development, yet none of them seems to have followed seriously Papin's lead toward locomotion. But witty Bishop Berkeley could see further than the men who knew tools; in 1740 he said: "Mark me, ere long we shall see a pan of coals brought to use in place of a feed of oats."

Soon men again began to think about power transport. Spring-driven carriages were made in Germany in the Middle Ages. The spring method of propulsion lingered for a long time and developed some fairly successful vehicles, but of course they did not solve at all the problems of speed and power. A step in the latter direction came a little later, in 1753, when Daniel Bournoulli demonstrated to the French Academy the point at which steam power could be applied to navigation, and was given a prize by that body which even then exercised an inspiring influence on the new science. The stationary steam engine had proved its usefulness, and was at work daily in hundreds of places. The question was: how to make the steam engine portable, gear it to wheels, and get it on the road?

This question tormented the learned Dr. Erasmus Darwin (1731-1802), who urged Boulton and Watt, the steam-engine manufacturers, to produce a "fiery chariot" which would fulfill the startling prophecy contained in Darwin's poem, The Botanic Garden:

> Soon shall thy arm, unconquered steam, afar
> Drag the slow barge, or drive the rapid car;
> Or on the waving wings, expanded bear
> The flying chariot through the field of air;
> Fair crews triumphant, leaning from above,
> Shall wave their fluttering kerchiefs as they move,
> Or warrior bands alarm the gaping crowds,
> And armies shrink beneath the shadowy clouds.

James Watt was a canny Scot who is generally regarded as the inventor of the steam engine, though we have seen that he had several predecessors. He made, however, a good many telling improvements on Newcomen's engines from 1769 on, bringing the steam engine to the point where others more daring than himself were willing to place it in carriages and vessels.

So the scene shifts to France for the great experiment in which a full-sized steam carriage, designed for a special task, took the road. It will hardly do to say flatly that this was the first self-propelled vehicle, but it was certainly the first definitely constructed for a concrete and practical purpose in contrast to those strictly experimental and of small scale. Under the stern call of war Captain Nicholas Joseph Cugnot designed and directed the construction of an artillery gun tractor which, had it been successful, would have given Royalist France undisputed control of Europe. Seldom have the destinies of the world waited so directly upon the success or failure of a machine.

It was a three-wheeler, with the single front wheel carrying a tremendous load in the boiler and engine, all of which had to be moved on a pivot in order to change direction. But it also had the failing, repeated over and over during the next century, of being underpowered for the work to be done. It traveled at about three miles an hour, had to be refueled at fifteen miles, and on the second test, upset. The fickle Ministry, which had

Cugnot's artillery tractor, 1769,
first self-propelled vehicle built for road work

financed construction at great expense, turned upon the inventor, and he went into exile. When Napoleon came to power twenty years later, he recalled Cugnot, put him on a pension, and sought to revive the project, but by failing to follow up this idea of artillery transport, he lost a telling chance to increase the mobility of his artillery, just as his failure to adopt Fulton's submarine lost him his chance to invade England.

Three years after Cugnot's daring effort, Oliver Evans (1755-1819) of Delaware dreamed, in 1772, of using steam to propel carriages, but, by his own statement, did not work out the plans for his engine till 1784. He was certainly the first American to apply steam to a road vehicle. It was he who received the first United States patent covering a "self-propelled carriage," dated October 17, I789.[1]

'Orukter Amphibolis" Oliver Evans' amphibian, driven in Philadelphia, 1804

This extraordinary man, whose genius was better recognized abroad in his day, and still is, than at home, anticipated Trevithick in applying the high-pressure principle to steam engines. He sent to England the drawings of his 1787 engine; these are said to have been seen by Trevithick in 1794-95 and applied by him. In 1804 Evans astonished the mechanical world by moving in Philadelphia, partly by water and partly by land, a combination steam-wagon and flatboat, the first authentic amphibian of record. A picture of Evans guiding his 21-ton Orukter Amphibolis hangs in the Automobile Wing of the Arts and Manufactures Section of the Smithsonian Institution at Washington, D. C., facing the display of ancient motor vehicles which are in direct line of descent from Evans[2] first American self-propelled vehicle. He is the great American name in the early stages of motor vehicle development.

Another candidate for inventor of the first American steam carriage is Dr. Apollos Kinsley of Hartford, who is described in

---

[1] Encyclopaedia Britannica, 11th ed. vol. 10, p. 2.

[2] The Young Steam Engineer's Guide, by Oliver Evans. First edition, Philadelphia, 1805.

Forest Morgan's Connecticut as a Colony and a State as "driving the highways of Hartford in one of the first steam carriages ever conceived, of which he was the inventor at the close of the 18th century."

Many an idea which events have shown to be sound was unduly delayed by the inability of inventors to find backing. A strange example of this misfortune is William Murdock (1754-1839), a subordinate of Boulton and Watt, who made a model of a steam carriage which was far ahead of its time in the simplicity of the application of power, but he never patented it because his employers discouraged him.

There is something dynamic and challenging in the turn of a century which stirs the soul of man to fresh creative effort. It was so in 1800. Inventions flooded into the British patent office. Then Richard Trevithick (1771-1833), the "Captain Dick" who was destined to taste the triumph of placing the first locomotive on wheels, to make and lose fortunes, and to die at last in poverty, stepped into the field of the steam carriage. Wherever Dick Trevithick took his stand in his prime he made history. He brought forward in 1801 a steam carriage whose boiler was of American descent, since Trevithick was directly influenced by the Americans, Oliver Evans and Nathan Read. It produced sixty pounds pressure. Trevithick used a long-stroke engine coupled directly to the driving wheels. He had been at work on this design five years. It was put together in a blacksmith shop. One of his friends left this lively picture of its initial trial:

> In the year, 1801, upon Christmas Eve, Captain Dick got up steam, out in the highroad, just outside the shop. When we see'd that Captain Dick was a-goin' to turn on steam, we jumped as many as could, maybe seven or eight of us. 'Twas a stiffish hill but she was off like a little bird... . They turned her and came back to the shop.

Merry Christmas for Captain Dick that of 1801! He ran this car in and out of London several times at an average speed of five miles an hour. But there was then no market for such an innovation, and Dick and his partner

Vivian approached the end of their slim resources. In May, 1803, Trevithick thought he could sell one quarter interest in his patent for 10,000, but the deal fell through. His carriage tore down a fence and was denied the road by purblind authority. The carriage was dismantled; the engine sold to operate a mill. Alas for dreams! But within another year Captain Dick was ready to place upon the rails the first steam locomotive, and by so doing earn beyond cavil a secure fame.

England in the first decade of the nineteenth century was still a rural and rustic land, full of conservatives, rich and poor. The roads were not generally good, but the main roads were kept passable by toll-roads associations in which considerable capital was invested. The coaching interests and the toll-road interests coincided on one point at least, that newfangled vehicles should be discouraged.

First self-propelled vehicle to attain speed of ten miles per hour, 1810

These interests were sure to rise against a new form of highway transport which drew nearer practicability year by year. In 1811 Blenkinsop of Leeds made a practical application of steam power in transport by drawing thirty coal cars from Middletown to Leeds, three and a half miles, in one hour. Interest rose until hardly a technical magazine appeared without mention of some new vehicle or experiment. The Monthly Magazine (London) for November 1, 1819, contains five items on self-propelled vehicles, mostly hand or foot-propelled in ingenious ways, and all probably bone-shakers. But the extent of the interest in self-propelled vehicles in London can be seen in this sentence: "A steam carriage has been invented in Kentucky, of which word is eagerly awaited."

In 1821 Julius Griffiths led off, with a well-made steam carriage, a line of these vehicles destined to become so numerous in the next few years that the highway problem they created caused a Parliamentary investigation.

In the primitive state of the art, inventors had not yet learned even the primary truth that driving wheels would give them enough traction, and that passengers could be safely carried on the same chassis as the engine. Like the locomotive, which from the first was a "drag" pulling other vehicles, many of the early steam carriages made no provision for passengers, and these were accommodated in stagecoaches attached as trailers. Of the latter type was the Patent Steam Carriage of 1824, designed by W. H. James and produced by him with the aid of Sir James Anderson, Bart. It contained several features of novelty and long-continuing interest. James used four cylinders coupled to two crankshafts. Each pair of cylinders and each driving wheel was independent of the other, to avoid a compensating gear in turning corners. In a second vehicle, 1829, James forsook his original principles of design, "mistaking failure in detail for failure in principle," and produced a steam tractor more in accord with common practice.

While England was entering into a fury of experimentation, there appeared in France, in 1828, a steam carriage by Pecqueur of Paris which Lavergne says contained the germ of all the vital mechanisms of the modern automobile, chief among these being a driving wheel geared to the rear axle, planet gearing, spring suspension, and a competent steering arrangement. What English invention lacked in these respects it made up for in aroused feeling among inventors, whose search for capital and public acclaim lashed them into furies of controversy against each other. It is impossible, at this distance in time, to sift the wheat from the chaff in the printed evidence of these broils. Sir Goldsworthy Gurney occupies the chief place in this literature of controversy. As a boy he had seen Trevithick's steam carriage and followed that lead by diligently producing a succession of coaches known as Gurney's steam carriages. He is said to have made, in 1829, the first long sustained journey by mechanical means, from London to Bath and return, a distance of 200 miles, at a speed of 15 miles per hour. This record is

Sir Goldsworthy Gurney's steam carriage, London to Bath, 1829

challenged by a rival, however: the picturesque soldier of fortune, Colonel Francis Maceroni. The Colonel, with a rival machine in hand, had fish of his own to fry.

This and many other jealousies were brought into the open at and immediately following the hearings of the Select Committee of the House of Commons in 1831, one of the important milestones in the history of self-propelled vehicles, because England then led the world as an industrial nation, and the state of the art was presented to the Committee by the leading inventors and engineers. This Parliamentary inquiry seems to have been an honest attempt to discover how far the highways were likely to be injured by steam vehicles, and what promise mechanical transport held out to the nation. We will understand its importance if we recall that the new departure in transportation was under fire from several substantial interests: the alliance between turnpike associations and stagecoach operators, the widespread antipathy of a public always conservative and in those years of Chartist discontent bitterly opposed to the introduction of machinery in all lines of work, and the newly arisen railroad interest. The first railroad for regular passenger and freight traffic, the Stockport and Darlington, had begun operations six years earlier. Other railroads had been opened; some of them were actually earning money, and one of them was in a position to pay a first dividend. With this record of earnings, capital generally was for "clearing the track" of competition.

Almost from the moment when Trevithick put his first steam engine on the rails, British railroading had marched ahead, until, after only twenty years of trial, railroading had begun to pay. Against this notable progress, steam travel on the highways could produce nothing comparable. Steam on the highways had never earned a shilling, had killed some persons, and thrown whole countrysides into terror, and both the influential and the ignorant set their weight to hold the highways for horses and pedestrians. Sir Goldsworthy Gurney, for instance, was mobbed and his engineer injured at a country fair.

In the ten years from 1828 to 1838 Walter Hancock built nine large steam carriages, of which six were used in carrying passengers. His buses were safe, dependable, and handsome. The public liked them, and Hancock took in large revenues but was always under heavy expense from pioneering experiments and exorbitant tolls. He testified that one of his coaches, in three months service on the Paddington road, covered 4,200 miles and carried 12,761 persons, without mishap or serious delay. He offered to carry mails at the then high speed of 20 miles an hour, the usual mail speed, attained through relays of horses, being 12 miles an hour on the fastest routes. He is credited with being the first power-vehicle designer to use chain transmission and the first to make tight metallic joints. His wedged drive wheels are also considered important.

Not far behind Hancock in the point of impressing the Committee with reliability of operation was Sir Charles Dance. Against Dance it was alleged that his steamer was noisy and dangerous, dropping coals and driving horsemen and teams into the fields. Another interesting development of the Committee hearings was the reappearance of Captain Dick Trevithick, to say that he had a new boiler with a condenser attached, so that the same water could be used over and over.

With full information before it, the Select Committee of Commons reported that it was convinced that:

1. Carriages can be propelled by steam on common roads at 10 miles per hour.
2. That at this rate they have conveyed upwards of 14 persons per vehicle.
3. That their weight, including engine, fuel, water and attendants, may be under three tons.
4. That they can ascend and descend hills of considerable inclination with safety and facility.
5. That they are perfectly safe for passengers.
6. That they are not (or need not be, if properly constructed) nuisances to the public.
7. That they will become a speedier and cheaper mode of conveyance than carriages drawn by horses.
8. That as they admit of greater breadth of tire than other carriages, and as the roads are not acted on so injuriously as by the feet of horses in common draught, such carnages will cause less wear of roads than coaches drawn by horses.
9. That rates of toll have been imposed on steam carriages which would prohibit their being used on several lines of road, were such charges to remain unaltered.

A clear victory, on paper, for the road steamer men! If Parliament had seen fit to follow the findings of the 1831 report with appropriate legislation, Great Britain would have led the world in motor transport, adding greatly to her already dominant industrial leadership. England had the inventors, the roads, the capital, and a clear lead in the practical operation of self-propelled vehicles, but she surrendered this advantage by placing the new industry under handicaps which became more and more severe until at last, by the Act of 1865, it was decreed that no power vehicle could use a highway unless it was preceded by a man on foot carrying a red flag. Sometimes the rule was relaxed to let the red flag be carried by a man on horseback, as is still done on Tenth Avenue, New York, in advance of

New York Central freight trains. This is a stock example of the restraint of speed and flexibility of transportation by insisting that the old methods shall control the new.

Arbitrary discrimination in tolls was in itself almost enough to drive steam carriages from English highways; their use decreased greatly after 1836. Britain deliberately had turned her back on progress in an art where she was once ascendant. It has been said that an English Rip Van Winkle, falling asleep in 1831, would have awakened sixty years later in a world little further advanced, as respects highway transport, than that in which he fell asleep. The price Britain paid for this unjust interference which exerted state influence in favor of stagecoaches and railroads has been enormous in loss of trade and prestige. Because of this interference France and America, where government and public opinion both favored the new idea in transport, forged ahead. The French Academy honored inventors, and the French public was favorable, though still limiting highway speed for self-propelled vehicles to four miles an hour in the country and two miles in town.

America at the very beginning of power transport encouraged transport improvements in the highway field, falling into the errors of repressive control much later than England. Maryland early gave Oliver Evans the right to use its roads in his experiments. One of the first American steam carriages of record, which Thomas Blanchard brought out on the streets of Springfield, Massachusetts, in 1825, so far commended itself to the Massachusetts legislature that Blanchard secured an official endorsement; but he could find no buyers. The Kentucky vehicle which England was eager to hear about in 1819 seems to have disappeared from history, and we have but a faint trace of another, which one T. W. Walker of Edgar County, Illinois, late of Vincennes, Indiana, is said to have made and operated in 1824 or 1825. William T. James of New York City produced one steam carriage in 1829 and another in 1830. But whatever these vehicles were, we can be sure that they were applauded and went their way without official interference, because such interference would be of record. These American vehicles, like Evans' early efforts, encountered another sort of handicap which was effective in discouragement bad roads. America was still in the pioneer stage of development; few city streets were paved; in the country the roads were wretched, and even in the long-settled East they remained almost impassable in wet weather. This country of magnificent distances turned to railroads as an easy escape from mud.

The race for public favor and use between the steam railroaders and the steam highway school of transport ended in victory for the former in both Great Britain and America. There remained France, blessed with a splendid system of highways, a fertile inventive genius, and a sporting element both in the aristocracy with means to finance experiments and in a public which relished thrills.

In England mechanical transport had been pushed into the fields. From Worby's tractor in 1841 clear through the 'seventies there is a steady development of the steam tractor under the encouragement of certain great English landlords. There were also some notable developments in the highway field, but since use of the highways was so restricted, these were restricted to private courses or back roads, or else their owners were compelled to a "bootleg" traffic, eluding the authorities. As might have been expected, various enthusiasts outwitted the authorities by using the roads illegally. A vehicle which has received a good deal of attention because of its "bootleg" operations was the Fly-by-Night, a steam carriage weighing nearly seven tons, built by W. O. Carrett of Carrett & Marshall, Leeds. After it was publicly exhibited and found of no legal use, Carrett gave it to Frederick Hodges of the London Distillery, under whose ownership it acquired its gay reputation and name. Filled with sporting characters, it was run at night in defiance of the authorities and in this way is said to have traveled about eight hundred miles in Kent and Surrey. After being summoned for excessive speeds, the owners equipped it with fire hose and gave their passengers brass helmets to wear, escaping in this way arrest at the hands of village constables. At length it was necessary to convert the Fly-by-Night into a slower-moving vehicle. The advance of the mechanical arts in England during these trying years may be briefly indicated by listing four quite startling innovations. In 1833 Richard Roberts of Manchester applied differential gearing to a steam carriage, which solved, in accordance with modern practice, the turning of corners, thereby escaping the necessity of having one wheel running loose on the turn, which could hardly be considered safe.

Far ahead of its time was William Barnett's contribution to the internal combustion engine. In 1838 he invented a double-acting gas engine. Its single cylinder was vertically placed, and explosions occurred on either side of the piston.

Another astonishing invention, considering the fact that the self-exciting dynamo was not invented until 1871, was the electric vehicle produced by Robert Anderson of Aberdeen, Scotland, in 1839, apparently the first electric vehicle in the world. His carriage was driven by a primitive electric motor consisting of bars of iron on a drum. These were drawn around by electro-magnets, "probably on the principle of some old toys which had a star wheel to produce the necessary make-and-break." Gibson thinks it was dependent upon intermittent primary cells.

England pioneered also, in the development of pneumatic tires; at least, Robert William Thompson was the first to make a clear patent specification of the pneumatic principle.

In the field of the explosive engine France assumed the lead after 1860, when Lenoir of Paris made a gas engine more practical than Barnett's English effort of twenty-two years earlier. Lenoir made history by igniting his explosive mixture with a spark produced through the use of an electric battery and an inductive coil. His engine started easily and was fairly quiet. Used to drive lathes, printing presses, and water pumps, it was the first gas engine to come into commercial use. In 1862 Lenoir placed one of his gas engines in a vehicle and, using street gas for fuel, successfully drove from Paris to Joinville-le-Pont but was always handicapped by low speed and a narrow range of action.

The British, clinging manfully to steam in spite of the legal restrictions which pressed upon them more and more, created some notable machines in the decade of the 'sixties, being encouraged somewhat by the hope of export trade and the fond expectation that the road restrictions would be eased in their favor. Instead, however, the restrictions were tightened.

American interest lagged through the middle period, but Richard Dudgeon of New York built a steam carriage

Richard Dudgeon's steam carriage, New York, 1860

which richly deserves notice because of its unusual span of life. Put into storage in 1866, it was resurrected in 1903 and even then ran at ten miles an hour.

The early evolution of the steam car has been traced, and notice taken of the creation, almost stillborn as far as its influence on inventive activity is concerned, of the first electric vehicle. To these two grand divisions a third was destined to be added and eventually to triumph over the others in popular approval and use. This was the car using petroleum or one of its derivatives which, after many trials, developed into the gasoline car of the present.

Various steps in the evolution of the internal combustion engine have been cited. In 1866 Otto and Langen of Germany opened wide the door to the evolution of the modern high-speed engine by inventing their famous gas engine, a comprehensive improvement on Lenoir's, since it used only one half the fuel of its predecessor. In 1867 Otto and Langen were granted United States patents on improvements on three kinds of combustion engines: a two-cycle non-compression engine, a two-cycle compression engine for which the mixture was compressed outside the cylinders, and a four-cycle compression engine. The last named is the one of historic importance, as its cycle of charging, compression, explosion, and expulsion is still adopted. It was carried along from gas to oil by Gottlieb Daimler, an associate of Otto and Langen, whose petrol motor was the first to be manufactured in quantity.[1]

Nearly twenty years were required for Daimler to evolve his 1885 petrol engine from Otto's gas engine of 1866 but in the meantime inventors elsewhere were busy in other directions. Todd of England built a small steam carriage for two passengers. In the same year Charles Ravel of Paris attempted to propel vehicles with gas motors, and in this country Austin constructed a small car in Massachusetts which was exhibited along the Atlantic seaboard.

A year later, in 1871, Dr. J. N. Carhart of Racine, Wisconsin, built a steam buggy, certified to by J. D. Donald, Secretary of State of Wisconsin, under date of December 14, 1914. Dr. Carhart was assisted by his brother, H. S. Carhart, Professor of Physics at Northwestern University and later at the University of Michigan. Compared with the foreign machines of the same period the Carhart vehicle must be rated as a rather crude product. In particular, it revealed the curse destined to beset American inventors for many years to come that of viewing the "horseless carriage" as the only escape from the horse itself. For at least twenty years more, American inventors in the self-propelled vehicle field would be trying to fit engines into carriages of the general type and style to which their prospective buyers were accustomed, instead of evolving designs better suited to carry their power plants.

---

[1] See Chap.III for a further discussion of the evolution of gas and oil engines.

The Carhart experiments seem to have aroused great local interest, for in 1875 the Wisconsin Legislature passed an act appropriating $10,000 as a bounty for the invention of a steam road wagon which would meet certain tests as to performance and durability. This act was amended several times, but $5,000 was actually awarded in 1879.

In the 'seventies France began to move definitely toward assured leadership. In Paris Amadee Bollee built a number of steam omnibuses which enjoyed a long and profitable life. Other French achievements of the late 'sixties and early 'seventies are Tellier's ammonia engine, and the compression (not internal compression, however) motors of Brothier and Carre Tellier received a United States patent on an ammonia engine for road work in 1871.

Siegfried Markus of Vienna, at an isolated point as respects the art in general, constructed a petrol car, but some doubts are expressed by Lavergne and others that it ever ran. Markus is sometimes credited with building the first automobile in 1869; elsewhere the date is given as 1877. Like so many other claims to first position, this one would seem to rest on a narrow definition of the word "automobile," yet even that word had not yet been invented. His experiments, however, no doubt stimulated the activities of men in more favorable locations.

In 1876 George B. Brayton, an Englishman living in Boston, exhibited a petrol engine at the Philadelphia Centennial which attracted a good deal of attention, as it seemed to represent a distinct advance in American eyes. Described in the Encyclopaedia Britannica as the first engine to use "fuel oil under compression instead of gas," in the end it proved to be inferior to the four-cycle engine then being developed by Daimler in Germany. Its chief significance for us lies in the fact that George Baldwin Selden, a patent attorney and inventor of Rochester, New York, whose keen mind foresaw a demand for small self-propelled vehicles using oil fuel, also perceived in what way the basic difficulties involved could be solved. Selden sent an agent to Philadelphia to study Brayton's engine and, on the basis of using a similar engine as motive power, drew up the specifications which accompanied his application May 8, 1879, for the famous Selden patent which was finally granted to him in 1895.

The evolution of self-propelled vehicles has been traced from the first mention in history down to the year 1879, a dramatic date in the American automobile industry. Steam, the first motive power of artificial origin, came close to successful application on highways in England before being blocked by adverse legislation. The distant beginnings of the internal combustion engine and the beginnings of the evolution of the oil or gasoline engine by descent from the gas engine, have been noted. While it is impossible to allocate credit accurately, the names of the chief contributors are here recorded.[1]

These individuals and many others worked in an atmosphere beset by difficulties to an extent which moderns can scarcely comprehend. Under the circumstances, the wonder is not that it took centuries to master the art of propelling wheeled vehicles by power plants, but rather that the inventive spirit persevered in these men so strongly that the evolution of the motor car proceeded as swiftly as it did. Many of them sacrificed their fortunes as well as their lives to the great task of advancing in some degree man's mastery over space and time.

The inventors of the 'seventies could see that the railroads, despite their enormous gains in mileage, speed, and dependability, had not solved the transportation problem. Tied to tracks of steel, the railway lacked the flexibility necessary to make rapid transit available on the highways leading everywhere. Away from the rails, land transport was still geared to the horse, as it had been from the days of the Pharaohs. Goods moved swiftly by rail from depot to depot, but away from depots to consumer they were subject to the ancient lag of the Dark Ages. In particular, rural America lay in a quagmire of mud at certain seasons, and even under good weather conditions its beauties could be explored by few city dwellers. Vast regions where railroading could not hope to pay a dividend remained in a backward condition, almost untouched by the forces of pro-

---

[1] For other significant dates and names in the evolution of self-propelled vehicles, see Appendix I.

gress which steam locomotion had brought to the more favored areas.

The railroads also had created a serious social problem not foreseen in the early stages of their development. The advantages which they offered for residence and manufacturing could be reaped only near their stations and particularly at junction points. As a result, cities grew to enormous size, were densely overcrowded, and slum life became an appalling reality. To relieve the congestion which rail transport had created, other rail systems first cable cars and later electric trolleys were placed in operation on city streets, but these were never entirely effective because they reproduced, in a small way, the same influences which the railroads brought into the situation in a larger way. Suburbs so served were strung out thinly along lines of transit with large open areas between them.

The man of 1879 could see, as the man of 1829 could not see, that humanity needed more than ever a mode of transportation more flexible than anything tied to rails and faster than the horse. The railroad had extended tremendously the range of the individual and the productivity of society, but clearly it did not satisfy completely the instinctive demand of mankind for greater freedom of action and a wider radius of effective toil, trade, travel, acquaintanceship, and understanding.

In the Lynds's Middletown, the most complete picture of the evolution of an American community, an old man who had entered Indiana on an ox cart is quoted as saying that he can put the cause of social change in his day in four letters "A-U-T-O." The gasoline automobile in 1879 made its first appearance in American history, not as a practical vehicle, but as a definite, compelling, and driving idea under the spell of which many men would labor shrewdly and devotedly to compass its reality.

# Chapter III

## THE FORMATIVE PERIOD: 1879-1899

ALTHOUGH American inventors hopefully built steam carnages of strange design at intervals from 1800 on, it remained for George Baldwin Selden in 1879 to catch the first confused vision, on this side of the water, of the modern gasoline motor car. A shrewd lawyer, Selden kept his application alive from 1879 until 1895, when he secured a sweeping patent under which millions were collected from automobile manufacturers until, in 1911, its broad scope was limited and held not infringed by the actual types of motor vehicles then in use. (Columbia Motor Car Company vs. C. A. Duerr & Co., 184 Fed. 493.) All General Motors cars until then were manufactured under licenses from the Association of Licensed Automobile Manufacturers having been formed to exploit the Selden Patent. The litigation ending this monopoly is the most famous in American automobile history.

The period from Selden's application of 1879 to the founding of Olds Motor Works in 1899 measures the birth pangs of the gasoline automobile in America. At the beginning of that period the French were leading the van of automotive progress; at its close America had caught up and was poised for adventure in quantity production.

Economic needs have a way of permeating society, so that it is a commonplace of the history of invention to find minds far distant from one another in time and space, wrestling with the same problems and independently producing results roughly identical. In the long and intricate evolution of the automobile this was many times the case. Advances in other mechanical arts had to be tried and tested by time before they could be adapted to transportation. The older industrial countries necessarily provided the background for the beginning of the automobile's march to assured success.

Since the internal combustion engine is the heart of the modern automobile, its development is of prime interest. The chief American name is that of Drake, who in 1855 introduced incandescent metal as an ignition agent for gaseous mixtures. Lenoir's double-acting gas engine, fired by an electric spark, came into practical use. Four hundred of these engines were in use in Paris in 1865. They were quiet and smooth in action but expensive in fuel consumption. With the year 1867, when Otto and Langen introduced their free-piston engine at the Paris Exposition, the gas engine approached full utility. By 1876 Dr. Otto had applied the cycle of operations proposed earlier by de Rochas; this cycle, now known as the Otto cycle, was worked out independently because de Rochas never brought his ideas to execution, and in the meantime practical difficulties in the way of completion along his lines had been overcome.

The affinity between gas and oil engines is, of course, close, the vapor being produced from oil in the latter instead of being present as constant gas as in the former. The first practical oil engine is credited to Hock of Vienna in 1870, but his product was not a commercial success. Brayton's oil engine, working on the constant pressure system without explosion, is said to have been the first of this type to use oil instead of gas.

It seems impossible to determine precisely who first used an oil fuel to drive a motor vehicle. Gibson, a considerable authority, says that Lawson of England invented an engine in 1880 which was driven by the explosion of gas prepared from gasoline stored in a receptacle carried on the vehicle, which was a tricycle. Worby-Beaumont is inclined to credit the priority, at least as far as actual propulsion of a vehicle by an oil motor is concerned, to Edward Butler, who invented a three-wheeled motor tricycle, the Petrocycle, using benzine or benzoline. This machine was invented in 1883, proved in 1884, and exhibited in 1885. The motor had double-acting cylinders coupled direct to a single power wheel, and burned vapor of benzoline, which was exploded electrically. A three-wheeled oil motor carriage by Knight is said by some English authorities to have been the

First motorcycle oil motor

first oil motor carriage actually to run. At first benzoline was used as the fuel, but later the engine was adapted to ordinary lamp petroleum.

In France the early 'eighties saw the earnest efforts of a few petrol enthusiasts to enter a transportation field where steam-power vehicles were at their highest stage of development under the leadership of Comte de Dion and his associates, Bouton and Trepardoux, later manufacturing as De Dion, Bouton & Cie. The further progress of steam was assured when Leon Serpollet, in 1888-89, invented his "flash generator," the capillary water boiler known by his name. This improvement gave steam a new lease of life on highways, and led to the use of smaller steam cars of which, during the next decade, the White and Stanley were the first American examples.

In 1883, Delamare-DeBouteville, borrowing an idea from Lenoir, constructed a gas tricycle which used illuminating gas. A year later DeBouteville joined with Malandin in building a petrol car which the French claim was the first to operate.

The year 1885 furnishes material for what is still a bitter controversy, with traces of nationalist feeling as between France and Germany. In that year Gottlieb Daimler invented his famous petrol engine in which the vapor from oil was burned in the same cycle worked out by Otto for his gas engine.

With Gottlieb Daimler's appearance in the internal combustion field we approach the significant application of the internal combustion engine to the differentialized wheel-and-axle, which has given us the modern motor car. Until 1883, when Daimler conceived the construction of small high-speed engines with light moving parts, the various oil and gas engines were of heavy construction, rotating at 150 to 200 revolutions a minute. He attained 800 to 1,000 revolutions a minute without great sacrifice of durability and smoothness.

The French, however, have put forward the claim of Fernand Forest, a humble but excellent mechanic, who built a four-cylinder motor in 1885. The 1885 motors of both Daimler and Forest seem to have done excellent work, but Daimler's came into commercial use in 1890 when 350 were manufactured and, throughout, Daimler was far more influential than Forest. The latter deserves, however, an honorable niche in the annals of automobile invention, not only for the work he did on motors, but also for the improvements which he made in carbureters. Apparently without question he should be credited with the first water jacket constructed to warm the carbureter.

In Germany, Daimler's chief rival was Carl Benz, who put out a gasoline tricycle in 1885 which, like all of Benz's work, was strongly built and satisfactory in use. He seems always to have been a little ahead of Daimler in putting the gasoline engine into road work, for in 1886 he matched Daimler's "bone-shaker" with a car containing several highly progressive features a device allowing variable speeds and an automatic control of the gas supply through a clutch lever operating a stop-cock. This car developed fifteen miles per hour. The Benz car of 1886 is sometimes spoken of as the first gasoline motor car, but this can hardly be taken literally, although it was certainly a great advance on its predecessors in design and dependability. Daimler soon followed with a four-wheel motor car, but his first efforts in that direction were not as successful as those of Benz. There seems to be no clear warrant for the statement of the Encyclopaedia Britannica that Daimler "ran for the first time a motorcar propelled by a petrol engine." Daimler certainly had the correct motive means well in hand by 1883, even if he did not apply it to a motor car until 1886.

Daimler's light and fast engine thrust itself through the field of invention as an earthquake heaves a new mountain into view. M. Levassor, of Panhard & Levassor, secured the French patent right from Daimler and, in adapting it to highway use, fixed many of the trends which have made the automobile what it is today. The influence of cycle design on the French inventive mind freed French inventors from the obsession of carriage design which dominated early American and English efforts the stagecoach in England, and the buggy in America. Levassor placed his engine in front, with the axis of the crankshaft parallel to the side members of the frame.

> The drive was taken through a clutch to a set of reduction gears and thence to a differential gear on a countershaft from which the road wheels were driven by chains. With all recent modification of details, the combination of clutch, gear-box and transmission remains unaltered, so that to France, in the person of M. Levassor, must be given the honor of having led in the development of the motorcar.

This solemn verdict of Encyclopaedia Britannica seems entirely justified. America's early effort to use the Daimler engine in automobiles was made by William Steinway on Long Island in 1888. Steinway secured the American rights and spent large sums on mechanical equipment, but had little success.

To this preeminence of France in the history of effective motor-car design certain conditions which America at that period lacked were contributing factors:

(1) A magnificent system of highways which encouraged travel, especially cycling, which in turn led to an efficient machine-shop industry.

(2) An open and friendly attitude on the part of the French population and authorities which permitted those highways to be used without legal restrictions such as discouraged experiment and adaptation in England.

(3) The approval by wealthy French sportsmen and aristocrats, who from the beginning gave the automobile both financial and moral support.

In England the development of self-propelled vehicles had been greatly retarded by adverse legislation and the hostility of the population. Here in America it was retarded by wretched highways, the indifference of financiers, and to some extent by the hostility of the public, particularly of the farmers who were later to become the chief beneficiaries of the flexible transportation. Also, the hard times of the early 'nineties introduced a discouraging economic factor into the American situation.

For these reasons and others, America lagged behind Europe during the important ten years following the French application of Daimler's engine to road transport, but nevertheless there had been substantial American progress. Through the growing use of stationary internal combustion engines, and the design and manufacture of marine engines, both American inventors and a portion of the public worked slowly toward the predestined goal. The American vehicles which took the road were chiefly propelled by steam, and down to the late 'eighties they showed little if any advance over English "steamers" of a much earlier day.

A direct result of English influence would seem to be the four-wheeled steam car produced by John Clegg, an excellent mechanic, English born and trained, and his son, Thomas J., in the village machine shop at Memphis, Michigan, which the younger Clegg still operates. Thomas Clegg describes this vehicle as driven by a single cylinder, steam being produced in a tubular boiler carried in the rear of the car. It had seating capacity for four persons, including driver and stoker. Cannel coal was the fuel. Leather belts were used to transmit power, and spring adjustments on them provided enough play to let the car negotiate corners.

R. E. Olds's first horseless carriage (steam), constructed 1886-87

This machine is significant as the first self-propelled vehicle on record as being built in Michigan, now and for many years the leading state in the Union in the manufacture of automobiles. The reasons for Michigan's rise in the automobile world will be examined later. The great drive of the Wolverine State to leadership in the industry presently would begin, but the Clegg "steamer" nevertheless created hardly a ripple of excitement beyond a twenty-mile circle of rural countryside which it disturbed with its journeys through its short life of six months. Built in the winter of 1884-85, it ran perhaps five hundred miles in some thirty tests during the succeeding summer, its longest trip being to Emmet and return, a distance of fourteen miles. In 1887 Ransom E. Olds, who with his father Pliny Olds was engaged in manufacturing gasoline engines for farm use, drove on the streets of Lansing, Michigan, a three-wheeled steam vehicle. Two years later he brought out another "steamer" with a vertical boiler. The Olds steamers will be discussed later.

The three-cornered struggle between oil-and-gasoline, electricity, and steam was being waged all over the world and would not be settled for another fifteen years. Nowhere was it waged more hotly than in America. Each of the three types had its advantages and disadvantages. Steam gave smooth acceleration, but it carried the handicap of delay required for generation beside the difficulty of carrying enough fuel for a long trip. Electric propulsion meant quiet and ease of operation but had the disadvantages of extreme weight, limited radius of operation, and long waits for battery charging. Oil or gasoline as fuel meant a quick-starting vehicle and large reserve power in comparison to weight, yet the early vehicles of this type were noisier and more complicated in their operation than the others. While they possessed a wider range of action and more adaptability to road conditions, these could not always be realized on because of mechanical difficulties. They did possess, however, two supreme advantages which tended, as design improved, to give them supremacy. The fuel they used was rather widely distributed from the start, and this distribution could be and was easily expanded, so that nearly every country store could become a supply station for motor cars as soon as demand increased. And, secondly, they could be built at a price to fit a wide range of pocketbooks, a fact of supreme moment in those early years when the measure of transportation investment was the cost of buying and maintaining a horse and carriage.

In 1890 the Olds Gasoline Engine Works, parent company of the oldest unit of General Motors, was incorporated at Lansing, Michigan, for $30,000. The experience of R. E. Olds in producing gasoline engines there would turn him soon toward automobiles so driven. For the present, however, Mr. Olds continued his experiments with steam power. In 1891 he produced the steam horseless carriage with a float boiler, which machine he sold to India, thereby consummating the first recorded sale of an American self-propelled vehicle.

In spite of "hard times," or perhaps because of the adversities of those times, inventors in transport everywhere were busily at work bringing out novelties which soon were

R. E. Olds's first factory, River Street, Lansing, Michigan

recognized as significant. In 1891 Thomas B. Jeffrey patented the clincher rim. In the following year Charles E. Duryea and his brother Frank built and ran at Springfield, Massachusetts, the first successfully operated American gasoline car.

This car is preserved in the Smithsonian Institution. The date given is 1892-93; apparently the car received preliminary road trials in 1892 but was not definitely introduced to the public until 1893. It weighs 700 pounds, is propelled by a 4 horsepower motor weighing 120 pounds and is fittingly described as a "horseless buggy." It set a style, as well as inaugurating an industry. America was so thoroughly a horse country, back from the railroads, that for years American manufacturers favored carriage styles. It is said that the first body delivered to Packard was equipped with a whip-socket.

Acutely aware that the American market did not want power cycles, American producers followed Duryea's example of the "horseless buggy," paying little heed to the lead given by Panhard & Levassor in Paris in creating a design fundamentally different from both cycle and carriage. Levassor revealed the full possibilities of present-day automobile design in which a style carriage is moved by a power plant placed ahead of it on a chassis, but American manufacturers for years to come would still be reproducing buggies as closely as possible and hiding their power plants in narrow spaces beneath the seats. Of course, Panhards were expensive; perhaps no country lacking a rich and sporting aristocracy such as that of France could have provided a broad enough market for them. In America the inventors strove from the first for cheaper cars, and while their initial designs were faulty, their goal was one which has been realized so fully that the automobile is in possession of the common people here to an extent matched nowhere else in the world.

Elwood Haynes in 1894 and R. E. Olds in 1895 produced gasoline-powered cars, the latter the forerunner of the famous curved-dash runabout which in a few years was destined to become the first American car produced in quantity. The two-seater Oldsmobile in the National Museum, while highly interesting as a specimen car of the times (1897), lacks commercial significance, as it was the only one of its kind produced.

Automobile interest the world over was tremendously stimulated by the first Paris Rouen race for motor cars in June, 1894, and the Paris Bordeaux race the following year. One result of these races was the filing of 500 applications for patents on all varieties of self-propelled vehicles at the United States Patent Office. In the 1898 race, the winning Panhard covered 726 miles at an average speed of fifteen miles an hour.

America owes its initiation into automobile racing and red-hot automobile news to H. H. Kohlsaat, publisher of the Chicago Times-Herald, who offered prizes for the first motor vehicle contest ever held in America. After being postponed, the race was finally run at Chicago on Thanksgiving Day, 1895, under road conditions which provided a stern test for the entrants. Snow and slush filled the streets, which were soon churned into "a slough of mud" by the narrow tires of the competing vehicles. The storm gave the public excellent proof of the superiority of gasoline-power over horses, steam, or electricity. The winning Duryea "horseless buggy" covered the 54.36 miles of the muddy course in seven and one half hours, despite stops and accidents which caused sixteen miles of extra travel.

The horseless buggy, one of Charles E. Duryea s first vehicles, 1893

Noteworthy is the fact that although all three motive powers steam, electricity, and gasoline contested, the gasoline cars were most numerous and finished one, two, and three. Morris & Salom's famous Electrobat, though given a prize for the best showing in. preliminary tests, could not cope with the harsh weather and severe going. Duryea's victory and the good, showing of other gasoline-type cars helped to fix in the American mind the truth that the most flexible and dependable automobile, in all weathers and on all roads, was the gasoline car.

So impressed was the race's sponsor by the achievements of the day that he declared his faith that, in five years more, Chicago streets would show five automobiles for every horse. Thanks to the Chicago race the American automobile had been launched as "news," and it has continued to be news ever since.

In both the Times-Herald Chicago competition, and the French races of the 'nineties, gasoline cars had proved their superiority over both steam and electric automobiles in those qualities most suitable to the American scene. They started more quickly than steam cars and could deliver more power per pound than the electrics. Since the fuel they consumed could be found at almost any country store, their cruising radius was limited only by the condition of the roads, and the difficulty of making repairs when wear and tear proved too much for their mechanisms. All in all, the "horseless buggies" of the 'nineties were seen to be rough-and-ready performers capable of meeting the severe tests of wretched American roads well enough to augur their future ascendency.

In England, Wallis-Tayler might say that steam would eventually carry the day against gasoline and electricity[1], but America even then was engaged in a fury of development destined to overturn his solemn verdict. Between 1895 and 1900 top speed for gasoline cars rose from fifteen miles an hour to nearly fifty, and problems of supply and repair were being solved. Many companies had sprung up, searching for the key to market success.

Looking backward, it is easy to see the main outlines of their merchandising problem, but to the pioneer companies the situation no doubt had its puzzles. There was an avid public interest in the new means of transport, but the country was still depressed as a result of the bitterly hard times of the early 'nineties. As cars one by one appeared, they were bought by rich folk of a sporting and adventurous turn of mind who sometimes used them ruthlessly on the highways, rousing the opposition of pedestrians and horse drivers, whose steeds were affrighted by the noisy novelty of the few cars

[1] There can be little doubt that the vast majority of people would prefer a smooth-running, reliable steam engine for use as the propelling medium of a pleasure or light business carriage, to the evil smelling, dangerous, wasteful, and at best uncertain and unreliable engine heretofore chiefly employed for that purpose in motors of recent construction." A. J. Wallis-Tayler in Motor Cars or Power Carriages for Common Roads, 1897.

they met. Unless automobiles could come into common use, so that an individual of average means could look forward to possessing one in the future, there was danger that the small boy's "Get a horse" would be translated into restrictions of the sort which had blasted English enterprise two generations before.

"Horsey Horseless Carriage" designed by Uriah Smith of Battle Creek, Michigan, to keep Dobbin from shying on the road

The extent of this hostility against automobiles in the late 'nineties, especially marked in the rural sections, can be indicated by relating the fate of the first motorcar introduced into South Dakota. This was a "home-made" horseless wagon planned and assembled by Louis Greenough and Harry Adams of Pierre, using a two-cylinder Wolverine gasoline motor and a special Elkhart wagon, in which the engine was housed under the rear seat and power transmitted by chains to the rear axle. It could carry eight persons and altogether was a competent vehicle, which that then frontier state might well have been proud to welcome as a home product. But such was the public opposition that Greenough and Adams were refused the right to carry passengers for hire at county fairs, which was their only prospect of securing prompt returns on their investment. At Mitchell they were refused permission even to bring their machine inside the town limits. The Press and Dakotan voiced the public verdict when it said: "It is a dead moral certainty that that infernal machine will frighten horses and endanger the lives of men, women, and children."[1] So the checkmated pioneers of motoring gave up their efforts at a time when it was painfully clear to them that motor cars would be an unmixed blessing in that state of vast distances. Like hostility manifested itself in many other parts of the country.

Obviously, this popular distrust of the motor car had to be overcome, and the way to break it down was to make automobiles so common that thousands could drive them and more thousands ride in them daily; then horses and humans alike would grow accustomed to their passage; then the farmer and the working-man alike would learn to look upon the automobile, not as a rich man's toy, but as a convenience which he might hope some day to possess. The conversion of the populace would begin as soon as any considerable number of car owners started taking their neighbors for rides. After even one ride the small boy would be on fire to own a car when he grew to be a man.

Quantity production, it is now clear, was the key not only to the financial success of the industry but also to winning the public mind away from its traditional enmity. Only through quantity production of a single model would costs be reduced sufficiently to bring the automobile within reach of the average American. A car so produced had to be small and strong and simple. The first manufacturer who could bring a car of that kind to the public "at a price" would score an immediate financial advantage and at the same time clear the way for the whole industry to surge toward large proportions.

The first company to take that most important step was the Olds Motor Works, the oldest unit of the General Motors Corporation, with its curved-dash runabout. The evolution of that car and of the company which produced it appears in the next chapter.

---

[1] Encyclopedia of South Dakota.

# Chapter IV

## OLDSMOBILE: FIRST "QUANTITY" CAR

As THE twentieth century drew near, the prospect of high fortune for those manufacturers who could build automobiles in quantity rode scores of ambitious men like a witch, stirring them to extraordinary efforts, sleepless nights, ceaseless planning. Their customers, the devoted "auto-mobiliers" as the phrase of the day ran, were scarcely less excited. They had endured the jibes of a prejudiced populace; many of them had been denied the use of streets and highways and had cheerfully braved arrest to win freedom of movement for their fellows, as Dave Hennen Morris and Augustus Post had done in Central Park, New York City. Bold citizens who caught a glimpse of the future not only braved the clutches of the law and the wrath of mobs; they also did volunteer publicity work for the cause of motordom. Mr. Post was one of the leaders in. this field, assisting in organizing and publicizing many historic contests, including the famous Glidden tours, carefully planned long-distance endurance runs, annually featured from 1905 to 1911, for a trophy presented by C. J. Glidden of Boston. Foreign observers returned to praise American initiative. In London, John Munro describes the encouragement given to motor-car manufacturers in France and concludes that the French army will soon be motorized. With clairvoyant sense, he foresees the "tank": "We are in a measurable distance of the ironclad on shore." England was awakening from lethargy caused by earlier legal restraints; London, he observes, has an Automobile Club, to match those of the United States, France, Germany, Austria, Belgium and Italy; while an American, James Gordon Bennett, has presented a cup to the Automobile Club of France, to be competed for yearly by cars representing clubs from both sides

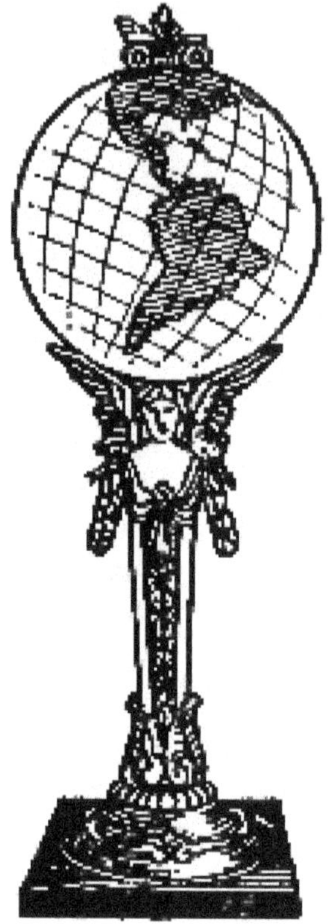

The Glidden Trophy

of the ocean, the first race to be held in France in midsummer, 1900, over a course of 550 to 650 kilometers, in 150 kilometer stages. The French Academy had even invented an enduring name for the self-propelled vehicle "automobile."

American developments might be praised by foreign writers but the American reporters on the state of the art were a little apologetic. Writes Ray Stannard Baker in McClure's for July, 1899:

> Never before has Yankee genius and enterprise created an important business interest in so short a time. And yet the motor vehicle in America is in its babyhood... . Here it has hardly passed the stage of promotion and promise.

Notwithstanding this modesty, the achievements recorded in the same article might be considered excuse for letting the eagle scream:

> Five years ago there were not thirty self-propelled carriages in practical use anywhere in the world. A year ago there were not thirty in America. And yet between the first of January and the first of May, 1899, companies with the enormous aggregate capital of $388,000,000 have been organized in New York, Boston, Chicago and Philadelphia for the sole purpose of manufacturing these new vehicles.
>
> At least eighty establishments ... 200 types of vehicles, with nearly half as many methods of propulsion.
>
> A motor ambulance is in operation in Chicago; motor trucks are at work in several cities; a motor gun carriage will be ready for army use in the summer.
>
> The Santa Fe railroad has ordered a number of horseless coaches for an Arizona mountain route.
>
> A trip of 720 miles has actually been made in a gasoline carriage (Cleveland to New York), and an enthusiastic automobile traveler is now on his way from New England to San Francisco.

Through most of the articles of the time runs the horse motif. Some might lament his passing, as Chauncey M. Depew did in Horseless Age, 1899: "As to the ordinary, everyday horse, hc is certainly doomed." Of course Mr. Depew put no time limit on his prophecy and it may yet come to pass, but on the record Dobbin has outlived the magazine containing his sentence of doom. Other writers rejoiced in the prospect of clean, horseless streets. The statisticians of the new industry worked out the comparative costs of hay and gasoline, harness and tires. Their general conclusions, that motors were as economical as horses, seem a little over-optimistic when one observes their neglect of depreciation and replacement, which was then even more of an item than at present.

The truth that here was a new transportation tool and not merely an improvement on the horse, had not dawned even on the industry itself. Its leaders thought of the automobile as making headway only through putting horses out of work, whereas the fact is that the number of automobiles increased many times before the horse population showed signs of falling off. What happened was this: a more efficient use of human time lifted the productivity of the country, bettered man's economic life, and increased the wealth of the nation, so that the new means of transport more than paid its way without prematurely displacing the equipment already available.

Even the motor poets were obsessed by the horse; a popular verse in praise of the motor-car, declared:

> It doesn't shy at papers as they blow along the street; It cuts no silly capers on the dashboard with its feet; It doesn't paw the sod up all around the hitching post; It doesn't scare at shadows as a man would at a ghost; It doesn't gnaw the manger and it doesn't waste the hay. Nor put you into danger when the brass bands play.

These are virtues, but after all they are negative virtues. The qualities which gave the automobile a chance to enter quantity production, after its long experimental evolution, were positive virtues power to carry persons and goods faster than they had ever been carried on highways since the dawn of time, ability to overcome bad highways until better ones could be built, resistance to shock and stress, high mobility in rural areas where speed had never been applied before, saving precious time, widening the individual radius of action, making groups more effective and cohesive, life more various, thrilling, and productive. These advantages were destined not only to substitute the automobile for the horse on highways, but also to change America into a nation of travelers moving on ever improving highways at an ever increasing tempo toward wider and wider horizons, higher and higher standards of living.

Public interest was rising: how could it be maintained, exploited, brought to bank? Through lower prices, obviously. Yet costs and prices could only be reduced by increasing volume.

But who dared to commit himself to a quantity production program which would absorb all one's capital and wreck its originator if the goods could not be sold? There seemed no escape from that circle except gradual enlargement of volume, an opinion expressed by one of the most ardent automobile advocates, W. H. Maxwell, Jr., in Metropolitan Magazine for

R. E. OLDS Father of the Oldsmobile

November, 1900. To grasp the opportunity required high courage; to win the leadership required shrewdness as well. There was only one man in America who, in 1899, refused to wait any longer for the "gradual enlargement of the volume of business." While others pushed on bit by bit, Olds Motor Works made the leap into the dark.

Their take-off for the great leap was Detroit. Various explanations have been given for Detroit's position in the automobile world, but the one that seems conclusive is this: In Detroit, the Oldsmobile early proved a money-maker; hence, the pioneer manufacturer could find capital support there easier than elsewhere.

To accomplish quantity production under reasonably good management was to reap profits. Nothing else serves to commend an industry to a community as swiftly as a good earning record. Detroit early became automobile-minded, and for that the City of the Straits has the Olds Motor Works to thank, since it was the Oldsmobile which demonstrated before any other American car that automobiles could be made and sold in quantity, and fortunes reaped from their manufacture. Once that demonstration had been made, Detroit put its money on the automobile, and in so doing began its population climb from the eleventh city in the United States in 1900 to the fourth city in 1930.

Another factor in Michigan's development as an automobile center was its location on navigable water. Olds, Buick, and Leland three pioneers of the industry made gasoline engines for marine use before turning to production of motor cars.

Bringing Oldsmobile to Detroit was in itself the result of capital investment built upon an astonishing faith in the new vehicle of transportation, and in a wealthy old man's confidence in a young man, poor in purse, but rich in ambition and practical ideas.

The young man was Ransom E. Olds, born in Geneva, Ohio, in 1864. With his father, Pliny S. Olds, a competent mechanic, he came to Lansing, Michigan, and though little more than a lad worked in his father's machine shop in River Street, where among other products was developed a gasoline engine for farm and marine use. By the time he was twenty-one R. E. Olds had saved $300, and putting that into the business, along with a note for $800 at 8 percent interest, he became a half-owner of his father's business. In 1892 he bought out the balance of his father's interest and incorporated the Olds Gasoline Engine Works for $30,000, S. L. Smith being a considerable stockholder.

Two ideas possessed Olds. One was to improve the gasoline engine: to that end he invented a new type which drew gasoline directly into the cylinder. It has since been discarded in favor of engines using air-and-gasoline mixtures. No matter: Olds was on his way. The other idea was that of putting power behind wheels. In 1887 he had produced and driven down the streets of Lansing a three-wheeled steam-power horseless carriage, arising early to avoid shocking the citizens and scaring the horses. Lansing was then the quiet capital of an agricultural state. The city's population was only 2,000; there were no paved streets, and there were those in the community who objected to having their roads used by horseless carriages.

Olds went on from his three-wheeler to other steamers of the more practical four-wheel type, reaching dependability in that field in 1893. One of his steamers, equipped with a flash boiler of Olds's designing, became known all over the world through an article in the Scientific American, so that its fame reached even to India, resulting in its sale to the Francis Times Company of Bombay, India certainly the first sale of an American self-propelled vehicle for export and perhaps the first American-made passenger car u sold for value." This transaction antedates by at least five years the Winton sale usually listed as the first American sale of an American motor car, and in the meantime, Duryea had probably sold some of his "horseless buggies." As we have seen, a number of steam cars had been produced in the United States earlier, but no authentic record of a bona fide earlier sale than this one is available.

A Lansing lad, by name Roy Chapin, later a notable figure in the automobile world and Secretary of the United States Department of Commerce, running after one of Olds's vehicles then and there decided that he wanted to be an automobile man. The same decision was being made by thousands of the most progressive youths of his generation. A little later they would be found energizing the young industry with their superb vitality, until the "automobile game" became the phrase describing the business in its most energetic and resourceful stride.

But after all, young R. E. Olds's experiences with steam were a side issue. His livelihood and his dearest dreams were centered in gasoline engines. After four years of struggle with debt, he finally stood clear with enough working capital to feel safe. The business prospered enough to afford a margin to finance

First Oldsmobile, 1897, now on display at Smithsonian Institution, Washington, D. C.

experiments. While these ran chiefly to steam, it was inevitable that a manufacturer of gasoline engines would eventually abandon steam cars for gasoline cars. In actually selling a steam car to India, Olds foresaw market possibilities; but concluded that the future belonged to gasoline rather than steam. As he said years later, "The gasoline engines were our bread-and-butter business, and most people thought the car was just a toy, but I knew that the car was the big venture."

This reasoning was built on hard experience with steam cars. He knew that boiler troubles would eliminate the "steamer" as a popular favorite, as soon as gasoline cars could be simplified. While inexperienced in electrics, he realized they could not cope with the wretched roads of the American countryside; and, hence, could not capture the market which arose in his imagination. So he set to work to apply the gasoline engine to road transportation, as several others were doing in other parts of the country.

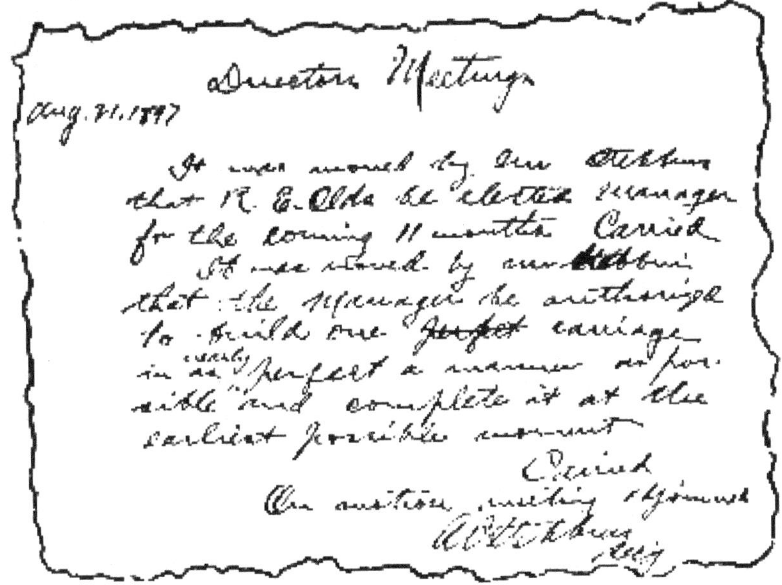

Directors Meeting

Aug. 21, 1897

It was moved by Mr Stebbins that R. E. Olds be elected manager for the coming 11 months Carried

It was moved by Mr Stebbins that the manager be authorized to build one ~~perfect~~ carriage in as nearly perfect a manner as possible and complete it at the earliest possible moment

Carried

On motion meeting adjourned

A C Stebbins Secy

Facsimile of minutes of directors' meeting, August 21, 1897, authorizing construction of first Oldsmobile

Olds worked hard to get a car ready for the Chicago race, but failed to complete it in time. However, he states that it was finished before the end of 1895, and adds these details: "This car had high wheels with one-and-one-half inch rubber tires. The engine formed the reach and was carried on the running gear."

The work of putting Michigan on the automobile map as the first quantity producer of gasoline cars began in 1895 with Mr. Olds as the chief figure, somewhat assisted by Frank G. Clark, whose father owned a Lansing carriage factory. Mr. Clark, interviewed in 1922, states that he and Mr. Olds worked together on the first Oldsmobile which appeared in 1897 and is now in the Smithsonian Institution at Washington. This car was the fruit of two years of experiments conducted chiefly after their regular hours of labor. Mr. Clark claims to have built the body, and testifies that the front axle was made in the carriage shop and that he and Mr. Olds worked together on the spring suspension and drive. The correctness of this narration at all points is questioned by Mr. Olds, but the appearance of Mr. Clark's name as the owner of 127 shares in the company which Mr. Olds soon formed to promote the new car, indicates that there was collaboration of a fruitful kind. However, Mr. Clark soon left the venture, and to Mr. Olds goes the credit of making a commercial success of the enterprise, as well as for the mechanical excellence of the early product, with the financial assistance of S. L. Smith.

Once the experiment gave assurance of success, Olds organized, with the assistance of bankers whose good-will he had earned by his all-around dependability, the Olds Motor Vehicle Company, Inc., with a capital of $50,000 divided into 5,000 $10 shares. Ten thousand dollars had been paid in when the papers, dated August 21, 1897, were filed on September 9, following. The purpose of the company was stated to be "to manufacture and sell motor vehicles." At the first meeting of the board of directors, as recorded in the minutes, Olds was empowered to "build one carriage in as nearly perfect a manner as possible."

The 1897 Oldsmobile on display in Washington is the oldest General Motors car in existence. It carried four persons, on two seats, both facing forward. The famous curved-dash had yet to make its appearance, the dash on the 1897 model being angular and clearly of buggy origin. This car is the only one of that model and year, as Olds Motor Vehicle never reached a production basis. However, the car ran, and ran remarkably well considering the handicaps its young inventors labored under, more than thirty-five years ago.

Among the stockholders of Olds Motor Vehicle was S. L. Smith, a copper magnate of Detroit. Mr. Smith deserves to be remembered as the first man of large means to peer into the future of Michigan automobile production, catch a glimpse of its possibilities, and finance a new venture in a large, vital way. When the Olds Motor Vehicle Company was later united with the Olds Gasoline Engine Works to form the Olds Motor Works, S. L. Smith's name led all the rest in capital contributed, and he remained a power in the company until his death. The Olds Motor Works came into being as a result of his backing. For some time Mr. Olds had realized that Lansing, as it stood in the late 'nineties, was not a practical location for a large manufacturing establishment. There were not enough skilled machinists and not enough houses to accommodate an influx of new inhabitants. The Lansing bankers already interested in Oldsmobile pulled wires in the East, and sites were considered as far away as Newark, N. J. But when it came down to cases, Eastern capital still fought shy of automobiles. On the way home, therefore, Mr. Olds stopped in Detroit to discuss the situation with Mr. Smith, already one of his stockholders. The old but vigorous millionaire, desirous of entering his sons, Frederic L. and Angus S. Smith, in business careers, advised locating in Detroit and backed his advice with enough cash to settle the issue.

The Olds Motor Works was promptly incorporated for $500,000 on May 8, 1899, "for the manufacture and sale of all kinds of machinery, engines, motors, carriages and all kinds of appliances therewith." The place of operation was designated as Wayne County, Michigan, with offices in Detroit. The capital stock was $500,000; 50,000 shares at $10 par. The record shows that S. L. Smith held 19,960 shares of the 20,000 originally issued and paid in, the other four shareholders, Olds and Sparrow of Lansing, James Seager of Hancock, Michigan, and F. L. Smith of Detroit, holding ten shares each. The new company acquired all the assets of Olds Motor Vehicle and the original Olds Gasoline Engine Works of Lansing, Olds Motor Vehicle being discontinued February 29, 1900.

In some sources the capitalization of 1899 is placed at $350,000 with $150,000 paid in, but the records of the Secretary of State must be accepted. The difference may be that between shares authorized and shares sold, a residue of $150,000 being then unissued. Later distributions of stock to the shareholders of the original company gave Mr. Olds and his associates' substantial holdings.

Mr. Smith's wealth and reputation drew in other influential Detroiters. The Company gained the assistance of Henry Russel, one of Detroit's most famous attorneys, who became a large stockholder. The company built on Jefferson Avenue East, near Belle Isle Bridge, where the Morgan & Wright plant of U. S. Rubber is now located, the first American factory especially

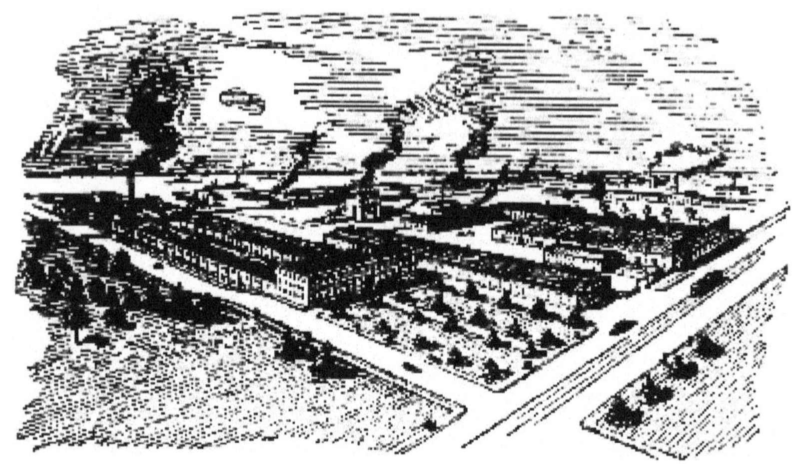

First factory erected in United States for automobile manufacture. Olds Motor Works plant, Detroit, Michigan, begun 1899

designed for automobile production. Of his early trials in Detroit, let Mr. Olds testify:

It was our plan at that time to put out a model that would sell for $1,250. I had fitted it up with some very up-to-the-minute improvements pneumatic clutch, cushion tires, and electric push-button starter. We thought we had quite a car, but we soon found that it was too complicated for the public. That first year we ran behind about $80,000.

The prospects of the industry were not very bright. Winton was making some cars down at Cleveland, Ohio, and Duryea, Haynes, and Apperson were all in the market. But the public persisted in the idea that it was not a practical proposition and would be a thing of the past within a year or two.

Finally, after a long sleepless night, I decided to discard all my former plans and build a little one-cylinder runabout, for I was convinced that if success came it must be through a more simple machine.

It was my idea to build a machine which would weigh about 500 pounds and would sell for around $500. The result was the curved dash "Oldsmobile," weighing 700 pounds and selling at

$650.[1] My whole idea in building it was to have the operation so simple that anyone could run it and the construction such that it could be repaired at any local shop. We rushed a few of them out as fast as possible, and they tested out so well I decided to put them on the market immediately.

We sold 400 the first year, which was considered a wonderful achievement for that period. Having felt our way carefully, I decided that the only plan to recover from the slump we had had the first year would be to come out with an announcement that the following year we would build 4,000 machines. I thought this would restore confidence in the industry and I staked all on the success of my plan.[2]

There is an element of luck in most successes, and what seems bad fortune at one time may become the basis of later triumph. Something like that came to pass in the fire which destroyed the Olds plant on March 9, 1901. One of the workmen at the new factory pulled his forge underneath a gas bag; the gas caught fire, and in an hour the plant was in ruins. But Fate, in the person of James J. Brady, a young timekeeper who was later to become one of Detroit's leading citizens, rushed in and rescued the only curved-dash runabout that had been built. It was the one tangible asset left. There was no good in going on with the other models; they vanished from the Olds list overnight. Most of the patterns for the new runabout had been burned, but new patterns were made from the rescued car. Just a month after the fire the Olds force had constructed a new runabout and drove it to the hospital where Mr. Olds lay ill. Before the end of the year, more than 400 of the famous runabouts had been built and sold.

The fire crisis in Oldsmobile history had another effect on General Motors history. To hasten the resumption of production Oldsmobile contracted with the Leland-Faulconer Company to make 2,000 motors for the runabout, thus intro-

---

[1] First priced at $600; soon raised to $650. Five electric cars were built and sold in 1899 and 1900.

[2] Sketch of R. E. Olds by O. D. Foster in Automotive Giants of America. B. C. Forbes Publishing Co., New York, 1926.

ducing the Lelands into the automobile business and turning their thoughts in a channel which led directly to the creation of the Cadillac.

The production figures of Oldsmobile one-cylinder curved-dash runabouts are, in round numbers: 1901, 425; 1902, 2,500; 1903, 4,000; 1904, 5,000.[1]

In addition, 6,500 one-cylinder straight-dash runabouts were built in 1905. No other automobile production record

Detroit to New York in seven and one half days
ROY D. CHAPIN in Oldsmobile, 1901

of the period approached this one, in its early achievement of quantity production. In 1902 when Olds was building and selling 3,299 cars, and when he could have sold 4,000 if suppliers had been able to fill his orders, less than 1,000 automobiles were registered in New York, the richest and most populous state in the Union. Among the leaders in the New York list were

---

[1] These figures are agreed upon by Messrs. F. L. Smith and Roy D. Chapin. For 1901 and 1902 they differ somewhat from figures on unit sales as furnished by Olds Motor Works, perhaps because marine and farm engines, made in both Lansing and Detroit, may have been included as "pieces" in unit sales.

Oldsmobile, Locomobile, Mobile, Winton, de Dion, Columbia, and Gasmobile. No Fords, no Buicks, no Cadillacs, and no Oaklands as yet.

Oldsmobile sales in New York City took a great surge forward after the second New York automobile show in the autumn of 1901, to which Roy D. Chapin, then a tester for Olds (he would presently be sales manager), drove a curved-dash runabout from Detroit. This was the first Detroit-New York trip made by a light car.

Well equipped with spare parts when he left Detroit, Chapin needed most of them before he reached his destination, as the wretched roads he traveled almost shook his light car apart. He was forced to leave the muddy highways, and drive along the towpath of the Erie Canal, contesting with mule teams for the right-of-way. His night stops were Leamington and St. Catharines in Ontario, Rochester, St. Johnsville, Hudson, and Peekskill, New York. He lay up for major repairs at Peekskill, and drove into New York seven and one half days after leaving Detroit. On his way down Fifth Avenue the runabout skidded into the curb, damaging one of the wire wheels, in spite of which he made his haven at the hotel where Mr. Olds was anxiously waiting. The doorman would not admit the young man in his greasy garments, and Chapin had to find his way around the building and sneak in unobserved through the servants' entrance to find his employer.

A good deal of social history is compressed in that incident the contrast between the formal East and the free-and-easy Middle West, the tremendous urge and surge of young America typified in a youth of twenty-one attempting something that had never been done and being trusted by his elders to "put it over."

As a commercial venture the drive was a decisive success, achieving wide publicity which enabled Mr. Olds to make a contract with Ray M. Owen to sell 1,000 cars in New York City. Detroit's sales drive on the rich New York market had begun.

Detroit occupied no prominence in the industry until the Olds Motor Works announced a production of 4,000 cars in 1902. In that year, Detroit's first automobile and sporting goods show was held. Olds displayed a car already popular while the Henry Ford Automobile Company displayed a Ford-Tom Cooper racer never successfully brought into production.

Mr. Ford, years afterward, stated that he had been extremely lucky in his competitors they had left him practically alone in the field under a thousand dollars.

"Within a few years," said Mr. Ford, "Olds, Hupp, Buick, and E.M.F. got out of my way, one by one, in something like that order. All of them went into larger cars after making a success of small ones. I recall looking at Bobby Hupp's roadster at the first show where it was exhibited and wondering whether we could ever build as good a small car for as little money."

The trend of early manufacturers toward large cars can be explained as due partly to the desire for more power and easier riding on the abominable roads of the period and partly to the desire of manufacturers to be represented in the market by something dignified, costly and well designed. At any rate, Oldsmobile production in 1903, the year in which the Ford Motor Company was organized, was 4,000 cars, by far the largest production schedule of any American manufacturer. A clear priority on quantity production belongs to Oldsmobile.

This outstanding achievement by no means reflected merely good fortune. Partly, of course, it was due to the fact that, underneath the hostility of many persons and sections, there existed real need for automobiles and an intense interest in them. Still, the business had to be pushed through sales and advertising pressure, and there were grave manufacturing difficulties to be overcome owing to the imperfections of materials and the haphazard processes of that day. Mr. Olds was a fortunate combination of commercial sense and general mechanical ability, but he was not a precisian when judged by latter-day standards. His big job was to get goods to market; the whole success of his venture depended on that, and he took what would now be considered a rather rough-and-ready view of engineering research. He told Roy Chapin once that it was time to correct a fault when the fault made itself evident on the road. There was no time in those hectic days to set up an engineering system which would refine the car in advance of need or consumer interest.

Even on this dot-and-go schedule, extraordinary strength and durability marked the Oldsmobile curved-dash runabout. We have seen that in 1901 Oldsmobile was the first light car to make the rough passage from Detroit to New York. In that same year Milford M. Weigle and F. L. Faurote introduced into the United States postal service a wire-wheeled Oldsmobile runabout which is said to be the first gasoline car used in any postal service in the world. It carried mail on contract under tests so successfully that a fleet of Oldsmobiles was soon being used for that purpose. Credit for initiating this idea of the automobile in postal delivery belongs to H. H. Windsor, editor of Popular Mechanics and the R. F. D. News, who invited F. L. Faurote, then advertising manager of the Oldsmobile works, to address the annual convention of rural letter carriers at Indianapolis in 1906.

Mr. Weigle's reminiscences are one of the best records of early Oldsmobile achievements on track and road. In 1902 he won the blue ribbon for piloting an Oldsmobile in the first hundred-mile non-stop endurance race staged in this country. Between 1902 and 1904 he won three gold medals and twelve silver cups, and hung up in succession several world's records for light cars on dirt tracks. He recalls being arrested three times for driving sixteen miles an hour on Broadway, New York City, when the speed limit was fifteen miles an hour. In the endurance race of 1902 at Chicago, any driver who made a speed greater than fifteen miles an hour was disqualified.

In 1903 the Oldsmobile Pirate established a world's straightaway record for making five miles in six and one half minutes. A little later, H. T. Thomas drove the same car to a new mile record the first American car and driver to cover a mile of space in less than a minute. In that year Oldsmobile won the Tour de France.

Perhaps the most successful of Oldsmobile's efforts to make America automobile-minded was the cross country race of 1905 from New York to the Lewis and Clark Exposition at Portland, Oregon, where the good roads convention in the United

States was also to be held. The drivers contested for the honor of performing the first transcontinental journey across America in a light car and for a prize of $1,000 offered by the company. The story of that adventurous journey is one of the liveliest in American motoring records.

The cars left New York City May 8, 1905, driven by Dwight B. Huss and T. R. McGargle. Each car carried a mechanic who assisted in the driving, Huss' assistant being Weigle, an Oldsmobile driver and inspector for many years. His log book shows Old Scout pulling into Portland on June 21, after forty-four days on the road for a total of 4,400 miles and an average of one hundred miles a day. Some 350,000 persons witnessed the triumphant arrival of the victorious car which had not only traversed rain-soaked stretches of gumbo and stormy mountain passes, but had also ploughed through trail-less wastes. When Old Steady, delayed by even worse conditions on another route, appeared a few days later, the staunchness of Oldsmobile was demonstrated beyond all doubt.

In the same car, equipped precisely as before, Mr. Huss repeated his New York to Portland journey twenty-six years later, in 1931, this time continuing on down the Pacific Coast. On Old Scout's second expedition across the continent he found hard-surfaced highways in place of mud and cattle trails, supply and service stations everywhere along the line, and a rousing welcome. Ten million persons, it is estimated, inspected the ancient Oldsmobile, which was displayed under the auspices of every major automobile club along the route. Old Scout continues in service. A feature of the opening of the building erected by General Motors for the Century of Progress Exposition in Chicago was the appearance of Mr. Huss once more at the tiller of this famous runabout, which he drove from Lansing to Chicago, carrying a letter from the Governor of Michigan to the Governor of Illinois.

From overseas comes the tale of an even more ancient Oldsmobile owned by the famous Krupp family of Germany and still in service. Efforts to secure this veteran for exhibition purposes in Europe have failed; Krupps keep it "on the job." Other distinguished patronage came as soon as it was seen that a quantity production and low price were not incompatible with quality. Among early buyers were the Queen of England, the Queen of Italy, Sir Thomas Lipton, Mark Twain, Chauncey M. Depew, Maude Adams, and other celebrities in all walks of life. But even more important was the certainty that the common people were being initiated into the idea that the automobile was here to stay. The country doctor drove an Oldsmobile on his rounds, giving more prompt service and extending his effective range. The more progressive country merchant went to call upon his scattered customers in an Oldsmobile, and kept a barrel of gasoline handy to sell to other motorists. Farmers began to lose their hostility to the new mode of transportation as they saw more cars, and rode in them occasionally. When a single company could make and sell more than five thousand cars in a year, as Olds Motor Works did in 1904, it was clear that America was on its way to becoming the motorized country in which a pedestrian came to be defined as a person on his way from one motor car to another.

Under the conditions of the period, the infant industry might very well have come to grief financially, since banks were cautious, and some even hostile, toward the new industry. Olds Motor Works survived by putting the first quantity trade in automobiles on a cash or C.O.D. basis. Of its firm stand on credits, John K. Barnes says:

> ... the industry profited greatly by it. [Olds] explained to his agents that it was also to their advantage to get their money when they delivered the cars. Then the purchasers, he pointed out, would be more careful how they used the cars ; they would not run them into the ditch when something went wrong and telephone the agent to go get the car. That is one of the reasons why the industry as a whole has come through past periods of business depression with little difficulty.

The Company had a care, too, for the consumer, not merely as a prospect but also as a user after the sale had been made, a point of view then new to business but one that has been followed consistently by the whole automobile industry, with an emphasis on "service" conditioning the whole relationship of the manufacturer and the market. One of the first of these service efforts was the famous Oldsmobile "Don'ts."

Finding it necessary to instruct the uninitiated, Oldsmobile issued these "Don'ts," the mere recital of which indicates the abysmal ignorance of the 1900 public on things motor-wise:

> Don't take anybody's word for it that your tanks have plenty of gasoline and water and your oil cup plenty of oil. They may be guessing.
> Don't do anything to your motor without a good reason or without knowing just what you are doing.
> Don't imagine that your motor runs well on equal parts of water and gasoline. It's a mistake.
> Don't make "improvements" without writing the factory. We know all about many of those improvements and can advise you.
> Don't think your motor is losing power when clutch bands need tightening or something is out of adjustment.
> Don't drive your "Oldsmobile" 100 miles the first day. You wouldn't drive a green horse 10 miles till you were acquainted with him. Do you know more about a gasoline motor than you do about a horse?
> Don't delude yourself into thinking we are building these motors like a barber's razor "just to sell." We couldn't have sold one in a thousand years, and much less 5,000 in one year, if it hadn't been demonstrated to be a practical success.
> Don't confess you are less intelligent than thousands of people who are driving Oldsmobiles. We make the only motor that "motors"

Early Oldsmobile advertising reflected the manufacturer's natural desire to convince a nation of horse drivers that his product could be used as cheaply and generally as the older means of transportation. Illustrations frequently showed automobiles passing horses on hills, and elaborate tables of figures were presented to prove that automobiles would not reduce to insolvency a buyer accustomed to the upkeep of horses and carriages. Gradually the automobile men forgot the horse. Oldsmobile advertising began confidently to sound the message of the automobile for its own sake.

Oldsmobile "firsts" include:

The first steps in modern assembly line development by improved system of routing materials in process.
The first automobile manufacturer's house organ Motor Talk.
The first automobile dealers' house organ The Oldsmobile News Letter.
The first national convention of dealers gathered by an automobile manufacturer, held at Lansing, 1907.
The first comprehensive instruction books to users.
The first sales manual to dealers.

Among the newspaper and publicity men often at the Olds plant, and helpful in getting the Oldsmobile firmly entrenched in the public mind were Alfred Reeves, now general manager of the National Automobile Chamber of Commerce, John P. Wetmore, then automobile editor of the New York Mail, Joe E. G. Ryan of the Chicago InterOcean, Edward Westlake of the Chicago Evening Post and C. G. Sinsabaugh, then editor of Motor Age, and later of Motor, Motor Life, American Motorist and Automotive Daily News. These and other automobile editors and advertising men organized "The Goops," an informal organization whose publication Goop-Talk was financed by the Smith brothers of Olds Motor Works. Oldsmobile always enjoyed a good press. Several famous advertising men took the Oldsmobile's message to the public, among them A. D. Lasker, later head of the Shipping Board, George Batten, Charles Brownell, and E. H. Humphrey.

In the first automobile copy to appear in the Ladies Home Journal a one-column Oldsmobile advertisement Oldsmobile is described as "The Best Thing on Wheels":

The ideal vehicle for shopping and calling equally suitable for a pleasant afternoon drive or an extended tour. It is built to run and does it.
Operated entirely from the seat by a single lever always under instant control. The mechanism is simple no complicated machinery no multiplicity of parts. A turn of the starting crank and the Oldsmobile "goes" with nothing to watch but the road.

Price Including
Mudguards $650.00

Each part of the mechanical marvel is made from thoroughly tested materials of the highest grade. Built in the largest Auto-

> mobile factory in the world by the most skilled motor specialists and guaranteed by a firm whose twenty-three years in Gasoline Motor and Automobile Construction stand as the very highest guarantee of mechanical perfection.

While the story is well known in the automobile trade, the general reader may wonder why Oldsmobile forsook its position as the outstanding leader in quantity production to enter into the manufacture of larger and finer automobiles. Several versions are available, but the true one seems to be this: R. E. Olds never possessed control of the company bearing his name. As he explains it, the younger Smiths, lacking in the experience gained from hard knocks, wearied of making cars for the masses. They desired to branch out in the direction of larger and more luxurious cars, forsaking the humble curved-dash runabout with its established market, for a more ambitious program. It is only fair to state, however, that the Messrs. Smith foresaw strong competition in that field from the rising Ford enterprise and also anticipated some of the trends toward the more elaborate engineering of the future which would soon render the simple Oldsmobile of the Detroit era a thing of the past. At any rate, a division of opinion arose as a result of which Mr. Olds retired. He had made "his million" with almost unparalleled speed once he got under way. Only forty-one years of age when the break came, the prospect of a little leisure appealed to him after twenty years of intense effort, and he retired gracefully in 1903, being succeeded as general manager by Frederic L. Smith, also an aggressive leader.

However, Olds's reputation had reached such heights that presently, merely for the use of his initials R E O he received a large stock interest in the newly organized company of that name.

The Olds Motor Works returned to the place of its birth Lansing. Following the fire of 1901, a Lansing plant had been set up to assist the Detroit operations. This became the chief seat of activity in 1905 and the nucleus of subsequent developments in which Oldsmobile has since eclipsed its earlier records. Production of the famous runabout continued but experiments in other directions indicated the new management's interest in larger cars. From 1904 on Olds Motor Works pushed export trade, doing business in Russia, England, France and Germany, and becoming the first American automobile company to do a quantity export business through regular dealers and direct sales representatives.

The trend away from the one-cylinder engine of runabout fame began promptly in 1905, with the launching of the "double-action" Oldsmobile with a two-cylinder engine, which

Famous Oldsmobile curved-dash runabout America's first quantity car

instituted a steady climb toward engineering perfection. In 1906 Olds Motor Works exhibited the first medium priced four-cylinder car offered to the public. It brought out its first six-cylinder model in 1907 and marketed it in 1908, but the "four" remained the mainstay of production.

In 1908, when the newly organized General Motors Company of New Jersey bought the Olds Motor Works, the production was 1,055 cars of which 1,000 were "fours." Clearly, Olds Motor Works during the years following the move to Lansing had not operated as profitably as at Detroit. Production fell from 5,000 units in 1904 to 1,055 in 1908. The company owed S. L. Smith

Chauncey M. Depew, at wheel of Olds runabout, 1904

more than a million dollars and was otherwise not in healthy condition. When it came time to sell, General Motors is said to have paid a million dollars for some road signs. But, of course, the names on the road signs made the value, as the buyers very well knew. Oldsmobile still had prestige with the public, almost as much in 1908 as in 1905 when Gus Edwards was moved to write his famous song the only automobile song which has come down to us in full flavor from those distant days, "ln My Merry Oldsmobile." A few bad years, in which difficulties of new designs and change of location had to be overcome, could not destroy the reputation Oldsmobile had won for itself. The new models were intrinsically sound in design and the Olds plant was in excellent condition. All that Oldsmobile needed, at the lowest turn in its fortunes, seemed to be the magic touch of a salesman. Oldsmobile soon speeded up to the General Motors tempo.

Olds Motor Works was the first unit purchased by General Motors after W. C. Durant formed his new holding company around Buick. Official negotiations began on October 10, 1908, though the leaders had been talking "deal" for some time.[1] Mr. Durant's first proposal to F. L. Smith, the Olds representative, set a price of $5 a share on Olds stock, payable four fifths in General Motors Preferred and one fifth in General Motors Common, other obligations of Olds held by stockholders to be paid off in General Motors Common at par. Olds Motor Works was to have two seats on the General Motors directorate, and General Motors was to name a majority of Olds Motor Works directors. This was not accepted, but a counter proposal by F. L. Smith on November I2th, which called for $100,000 more than the original offer, clinched the big deal. This settlement provided for the delivery of 152,530 shares of Olds Motor Works and the claims of S. L. Smith for $1,044,173.89, in exchange for $1,654,293.89 in Preferred stock of the General Motors Company of New Jersey and $1,152,530 in its Common stock. Three of the five Olds Motor Works directors were to be designated by General Motors which agreed to protect endorsers of Olds paper against loss and to provide working capital. General Motors acquired all Oldsmobile patents, chief of which were those on tires, carbureters, and engines, the latter specified under the names of Sintz, Richards, and Scavenger.

In the final settlement, more Olds Motor Works' shares "having been turned over than the number specified, the Olds Motor Works' stockholders received $1,827,694 in General Motors Preferred, $1,195,880 in General Motors Common, and $17,279 in cash, a total of more than $3,000,000.

The leadership which Olds Motor Works in its early days gave to the whole automobile industry may be measured not only in the production figures of the company itself, but also in the school which it provided for the budding talents of men who since have arrived at positions of influence and power. Names once on the Oldsmobile pay

roll recall the romance of the motor car's early history in Detroit and Michigan. A list of graduates from the Olds Motor Works "would read like a roll call of the captains of the auto-

[1] F. L. Smith: Motoring Down a Quarter of a Century, p. 36.

mobile industry." In addition to those already named, the early Oldsmobile circle included as employees or suppliers:

> Roy D. Chapin, president, Hudson Motor Car Co.
> Charles B. King, said to have been the first man to drive a gasoline car on the streets of Detroit.
> John D. Maxwell, who later pioneered the Maxwell car.
> Howard E. Coffin of the Hudson Motor Car Company and the "idea father" of the War Industries Board with his Council of National Preparedness.
> H. T. Thomas, later chief engineer of Reo.
> Carl Fisher, builder of the Indianapolis Speedway; developer of Prest-O-Lite Company.
> B. F. Everitt, body manufacturer, and William E. Metzger, master salesman, both later in the Everitt-Metzger-Flanders Company.
> George and Earl Holley, developers of Holley carbureter.
> Benjamin Briscoe, founder of the short-lived United States Motor Company.
> Charles B. Wilson, Olds factory manager, organizer and president of the Wilson Foundry Company of Pontiac, at one time the world's largest producer of automobile castings. Also his brother, David Wilson.
> Frederick O. Bezner, R. B. Jackson and James J. Brady who left Oldsmobile with Chapin and Coffin to found Chalmers-Detroit, later Hudson.
> Charles B. Rose, president, American La France and Foamite Industries, Inc., New York City.
> Charles D. Hastings, chairman of the Board of the Hupp Motor Company.
> John F. and Horace Dodge (Dodge Brothers).

A complete list of the Olds pioneers who now occupy prominent places in the automobile industry would include at least 150 names. Olds Motor Works was a training school for men whose later activities resulted in such companies as Reo, Hudson, Chalmers, Hupp, King, Columbia, Owen Magneto, Perfection Springs, and others. Furthermore, Oldsmobile orders for material, spread through the machine shops, body works and supply houses, set hundreds of wide-awake Detroiters to thinking how they could supply those wants, improve on their merchandise, and gather part of the golden stream of profits which Oldsmobile had started in their direction. Detroit rode to wealth and large population down a path in the direction which Oldsmobile had indicated.

This determining influence of the oldest General Motors unit on the geography of motordom and the industrial history of America is now clearly acknowledged. As John K. Barnes wrote in the Motor World of April, 1921, "It was Olds's success in Detroit that fixed the center of the automobile industry in that city." In less than three years Olds Motor Works paid 105 percent in cash dividends and its capital stock had risen to $2,000,000. It is equally true that Olds Motor Works was the first to reach quantity production by applying the progressive principle of assembly to the manufacture of a single model gasoline-engine-driven vehicle, and the first to popularize the automobile with the American people, taking it from the classification of rich man's toy to that of everyman's servant.

# Chapter V

## BUICK: THE FOUNDATION STONE OF GENERAL MOTORS

WHILE Oldsmobile was the acquisition by which General Motors first challenged the attention of the country, Buick was the nucleus around which W. C. Durant, Buick's chief and one of the dramatic figures in the history of the industry, built up the far-flung structure destined to become known the world over as the "G. M." How he acquired control of Buick and made it a leader is an absorbing story which begins with David D. Buick, man of many talents.

David Dunbar Buick, whose name already had adorned the front of more than two million motor cars when he died in 1929, completely realized the traditional picture of the American inventor. He was a man of brilliantly progressive ideas, native mechanical ability, and little business caution. Time and again, he sacrificed the certainty of present profits to experiment expensively with new ideas. One victory gained, he was always ready to rush on to another without consolidating the ground already carried, with the result that his finances were usually strained and his backers often in distress.

Men of the lovable and creative type, who sowed more benefits than they could reap for themselves, bulk large in the history of mechanical progress. They were scouts on the frontier of invention in the early days of the automobile industry; they penetrated little known territory, pointed out trails which others followed to their profit, but not infrequently they were unable to win wealth for themselves. Life to them was chiefly an opportunity to experiment. Many men of this sort made their contributions and were forgotten, but Buick remains fixed in the public mind.

Mr. Buick already had one substantial achievement to his credit when he entered the automobile field. A member of the firm of Buick & Sherwood, manufacturers of plumbers' supplies in Detroit, he had developed a method of fixing porcelain on metal, which is the key to the low-priced modern bathroom. A steady-going business man would have realized on this manufacturing advantage by sticking to bathtubs, but to David Buick a bathtub must have seemed a dead and inconsequential thing in contrast with the gasoline engines which had long engaged his eager and inquisitive mind and which he began to manufacture in 1900. His partner Charles Sherwood was also of an adventurous turn. When Buick Auto-Vim & Power Company was established in Detroit in 1901, it soon absorbed the resources of Buick & Sherwood in experimentation and sales efforts.

"Auto-Vim," in the name, had significance, as David Buick hoped to adapt an L-head gasoline marine motor of his design to a carriage. The company made these L-head engines for boat and farm use with some success, but their slender profits were spent as fast as earned. Need for new capital brought about the organization of the Buick Manufacturing Company in 1902, with Mr. Buick as president, and it was under this name that the first steps were taken in the development of the "valve-in-head" motor. The L-head motor was soon scrapped in favor of the new motor.

Several excellent engineering minds seem to have contributed to the early development of this famous motor, but documentary evidence indicates that the first steps were taken by Eugene C. Richard, an engineer born in France, trained in Philadelphia, and connected with the various Buick organizations for more than a quarter of a century. A contract is in existence between Buick Manufacturing Company and Eugene C. Richard, dated May 23, 1903, covering his employment as "designer and inventor and head of the drafting department."

Under the Richard patent No. 771095 issued to the Buick Manufacturing Company as his assignee, one of the allowed claims covered

In an explosion engine, the combination of the cylinder-head, of induction and deduction valves, having their stems extending through said head.... .

The "valve-in-head" engine is usually associated with the name of Walter L. Marr, who entered the Buick circle a little later than Mr. Richard, and who rose through many years of service to become the chief of the Buick Manufacturing

First Buick, Detroit, 1902

Company and one of America's leading automotive engineers. He it was who gave this superior motor its distinctive name. Mr. Marr was born in Lexington, Sanilac County, Michigan. Six years after he went to work as an apprentice, he completed a one-cylinder gasoline engine on the Otto cycle. He continued his experiments in his spare time through many years, producing six different types of motors. With this background he built in 1898 a vehicle successfully driven by a four-cylinder gasoline engine of his own design. It is described as having some very novel electric ignition fixtures, including a jump spark attachment. This first car of Marr's was built at

Cleveland, then the oil center of the country, and in it he traveled to various nearby manufacturing cities to visit other designers of gasoline engines, chiefly of the marine type. He was developing a car of his own when he met and became associated with David D. Buick.

In the early part of 1903 Buick put a single cylinder 5 x 6" horizontal engine, designed by Mr. Richard, into a chassis of his own design. A little later a two-cylinder opposed engine was substituted for the one-cylinder type. This was the car that became the famous Model F Buick, after certain changes were made in the crank case. The engine was 4½ x 5" and had its valves mounted in removable cages in the cylinder head.

These early trials and tests took place in and around the small factory on Holmes Avenue, Detroit, and apparently both Mr. Marr and Mr. Richard were present. Mr. Richard was engrossed in the power unit; Mr. Marr, in consultation with David D. Buick on chassis and body construction and adaptations, made many alterations preceding the trial. Walter Marr was an extremely ingenious man in solving last-minute problems. By the time he came to Buick he had a .grasp on the practical difficulties involved in making a staunch and dependable self-propelled vehicle. He had worked diligently over the uncertain fuels of that period, distilling his own alcohol and petrol spirits. He had designed and made many ingenious tools by hand, and from raw material had constructed all the engine and chassis parts for several cars.

Mr. Richard was Marr's technical superior under his contract, but it is unlikely that Richard could meet as well as Marr did the problems presented by the rough-and-ready assembly methods made necessary by the imperfect materials and workmanship of the period. At any rate, it was not long before the positions were reversed and Marr became chief engineer. In that position he clung so steadfastly to the "valve-in-head" principles that he became the chief figure in its future development.

Bringing the Buick car this far along had strained the resources of Buick & Sherwood, and they had borrowed considerable sums from the Briscoe Brothers, Frank and Benjamin, Jr., then manufacturing sheet metal. The Briscoes took an interest in the Buick Manufacturing Company and, assuming charge of its finances, changed the name to the Buick Motor Car Company, in which the Briscoe holding was $99,700 out of $100,000, as the Briscoes felt compelled to assume control of all Mr. Buick's various interests in order to protect themselves. Unable to carry the load any longer, the Briscoes determined to sell the Buick.

The Buick Company was on the market. Who would buy it? Remember that in 1903 the American automobile industry represented chiefly hopes and dreams. A great many persons had lost money in.it, and few had profited. Olds was the only quantity producer and money-maker in the gasoline field. Leland & Faulconer had prospered, but the Cadillac Motor Car Company was still unborn. Henry Ford was just getting the Ford Motor Company started. Bankers looked askance upon the industry, and there was no reason why the investing public should risk capital on new enterprises of such grave uncertainty.

In the end the sale of Buick was effected largely by accident, and a freakish chain of circumstances resulted in a startling change in the way of life for a whole section of Michigan. Visiting relatives in Flint, Frank Briscoe heard from Mr. Dwight T. Stone, a local real-estate man and son of one of Flint's early industrialists, of a prospect named James H. Whiting, who might be interested in the white elephant that Mr. Briscoe had on his hands.

Flint was then a city of less than 14,000 inhabitants, a pleasant county seat, located on a flat plain where two railroads, the Grand Trunk and the Pere Marquette, happened to cross. From the standpoint of physical geography, it held no advantages over the other county seats similarly located in central and southern Michigan. To this day Flint is a city whose rise to world-wide fame confounds the economic geographers, but to one who knows its history, the causes of that rise are evident. Flint was located at the ford, or, as the French explorers and trappers called it, the Grand Traverse of the Flint River, whose upper reaches extended for many miles through one of the best stretches of Michigan's superb pine forests. It was therefore a natural site for the location of lumber mills. For a generation its river banks resounded with the whirr of buzz saws and the stream was filled for miles with boom impounded logs. Fortunes were made, Eastern capital was attracted, and the citizens of Flint became accustomed to certain ideas of which they never afterward lost sight. They saw that big business had its advantages; they developed an extraordinary pride in their community and the success of their industries, and they stood ready to follow daring leadership.

Before lumber vanished from the scene through the destruction of the forests, there came to Flint a young Canadian, William A. Paterson, to establish in 1869 the city's first vehicle-manufacturing plant. At first Mr. Paterson worked at his forge as a carriage blacksmith, but one day he threw down his sledge and decided to be a business man instead. The carriage industry which he introduced there grew to large proportions, as one firm after another was founded and gradually expanded, until Flint became one of the great centers of the country for the manufacture of popular-priced vehicles. Assembly methods were worked out which have quite a modern ring; as competition increased in intensity, these methods were refined, until it is possible that the Flint of 1895 led the country in the efficiency of factory assembly. Though the frontier receded westward and raw materials near at hand were used up, Flint carriage companies were nevertheless able to maintain themselves down to the time when the automobile ended the Horse Age.

All these vehicle manufacturers, however, saw the handwriting on the wall: the automobile would drive out the horse. One of the first to recognize this was James H. Whiting, president of the Flint Wagon Works, whose plant, extensive for those days, covered part of the present Chevrolet site in West Flint. Mr. Whiting was a cautious man in most respects, but, foreseeing the eventual throttling of the carriage trade, he acted with what must now be reckoned a quite remarkable boldness. He began to look about for a car which might become the basis for an industry which would use part of his plant. Thus far Flint had had no luck whatever with automobiles in a commercial sense. Two of its more enterprising citizens, Judge Charles Wisner and Dr.

H. H. Bardwell, had built experimental cars for themselves, but neither of these gentlemen ever let business interfere with science, and their cumbersome vehicles merely amused a populace which thought in terms of wheels and hoped against hope for practical results. It looked to many as if young Alexander B. C. Hardy, who will appear in this tale later, had "hit it" with the dashing Flint roadster, all red paint and shining brass, which he had begun to make in a little factory down by the Grand Trunk tracks, a factory once devoted to the manufacture of the now almost extinct whip-socket. Mr. Hardy had been nerved to this great adventure by a visit to the Paris Exposition, where he saw how far French cars were ahead of American cars. Returning home he mustered a small capital and soon had a smart roadster ready for the market. In the end, however, Hardy was forced to liquidate, largely through the opposition he encountered from the owners of the Selden patent. While he was operating, a frequent visitor at his shop was James H. Whiting, and undeterred by the young man's lack of success, Whiting kept looking longingly for an automobile to manufacture.

All through the history of American automobile manufacturing in its early stages, will be seen shining examples of the courage of ignorance. Here was James H. Whiting already well along in years, with no engineering experience and no clear conception of the problems involved in making, selling, and marketing automobiles. He thought that cars could be sold by the same salesmen who went out to sell buggies, road carts, and farm wagons. As for manufacturing, he would buy what was necessary, put it together, paint and upholster the job which was practically the procedure in carriage manufacturing. Other concerns no better equipped than the Flint Wagon Works were building cars in just that way, so why shouldn't the Wagon Works? In due course, carnage firms discovered that building and selling motor cars was an entirely different business from making and selling carriages. The automobile business required far more capital and called for standards of mechanical precision beyond anything required in the carriage trade, but all that remained to be learned. By the time Frank Briscoe wanted to unload the Buick, James H. Whiting was in a frame of mind to consider negotiating for it. Brought together by Mr. Stone, Whiting and Briscoe quickly made a deal whereby the Buick concern sent its car over the roads to Flint. These roads were so

Original Buick factory at Flint, Michigan, 1903-04

bad that, in order to negotiate the 65 miles between Detroit and Flint, Buick and Marr, who drove the car, had to cover 115 miles, with every mile a test. Machinery, patterns, and dies were brought to Flint and housed in a small, one-story building adjoining the Wagon Works. The two concerns then formed the Buick Motor Company.

To bind the bargain quickly, $10,000 was borrowed from Flint banks on the endorsement of a number of prominent citizens. Though of small proportion, this deal was a striking example of community morale in a small town. Larger banks and endorsers elsewhere would have been more cautious, but Flint took the game with a rugged confidence.

Benjamin Briscoe, Jr., will be remembered as one of the colorful and energetic figures in the early days of the industry, and the daring promoter of the United States Motor Company. This company was put forward as an automobile merger planned to become the chief rival of General Motors. When United States Motor Company collapsed, Mr. Briscoe must have regretted his sale of Buick for a song, since the Buick in the meantime had become the keystone of General Motors.

The original capitalization of $75,000 in Buick was financed by the Flint Wagon Works' declaring a dividend of $75,000 which was paid into the treasury of the Buick Motor Company, which in return issued stock to the Flint Wagon Works stockholders and also to the Buick interest. David D. Buick and his son Thomas received 1,500 shares between them[1]. The other large stockholders were James H. Whiting with 1,505 shares, Charles M. Begole with 1,000 shares, George L. Walker, 725 shares, William S. Ballenger with 707 shares. Mr. Begole and Mr. Ballenger were active in the Flint Wagon Works and later in Chevrolet.

Buick now had a home and business management whose caution would be likely to restrain the optimism of David D. Buick himself. The skeleton of an organization was put together. A three-story brick building was begun, which at first housed all of the company's activities, but later was used only for motor and transmission manufacture. Old Buick No. i sometime later called No. 2 still has a sentimental attraction for now aging employees who began their careers there. Greatly enlarged, the building has become part of the Chevrolet motor plant. Sixteen cars were built in 1903, 37 in 1904, priced at $1,200. These first Buicks were equipped with a storm front curtain, with a large celluloid window in it, to protect the driver.

The $37,500 set aside for development had been exhausted, and loans had been made to carry on the work. Mr. Whiting felt that a younger man was needed to master this new business, with its insatiable appetite for capital and its crying need for quick decision. By this time he perceived clearly that the Buicks,

---

[1] Mr. Buick left the Buick Motor Company in 1906, two years before General Motors was founded.

father and son, were neither of them business men and that their chief associates were likewise more interested in mechanics than in profits. The need of the moment was for a man full of energy and vision who also possessed a keen sense of market possibilities and the courage to think in large figures.

First Buick car built at Flint, Michigan, 1903. Model A

At a meeting of carriage manufacturers in Chicago, in 1904, Mr. Whiting told F. A. Aldrich, representing the Durant-Dort Carriage Company, of the difficulties he faced getting the Buick Motor Company swinging marketwise. Mr. Aldrich advised him that the man he should interest, the one man who fitted the specifications and was immediately available, was William Crapo Durant. "Billy" Durant was already a leader among his associates and in the opinion of Flint. Born in Boston, December 8, 1861, he was the grandson of one of Michigan's war governors, Henry H. Crapo, who had brought to Flint part of the capital amassed by his thrifty ancestors in New Bedford, Massachusetts, where they had followed the sea as mariners and shipbuilders to good purpose for some generations. Originally of French stock, the Crapos of New Bedford and Flint alike were rich, prosperous, and able.

From boyhood "Billy" Durant's chief interest was business. He might have gone East to college, but instead went to work early in his grandfather's lumber business, one of the largest of the many large lumber mills in Flint. Then, to get more action, he branched out before the age of twenty-one into insurance with an agency of his own. That suited him, because insurance was something you could go out and sell. No waiting around for customers to come to you, as in. the store. An almost feverish activity possessed him. "Billy" Durant above everything needed action. While possessed of a notable faculty for remaining calm in the midst of alarms, he seemed to require dramatic tension in business. Yet he had also the power of concentrating intently on work.

All this both Mr. Whiting and Mr. Aldrich knew, for W. C. Durant at forty-two was already the most talked of man in Flint. As they discussed his availability for the automobile business, they recalled the dramatic entry Mr. Durant had made into the vehicle field some fifteen years before, when he had pioneered the road-cart business which provided Flint with its initiation into quantity production and salesmanship. The young insurance hustler had bought, while in Coldwater, Michigan, for $50 the patent rights for a road cart which carried a good selling point in its improved suspension. He took into partnership Josiah Dallas Dort, a young hardware clerk, and the new firm contracted with W. A. Paterson for 10,000 carts at $8 each.

This was an unheard-of quantity, calling upon the manufacturer to adjust his plant and workers to a new system of assembly for such a large operation. But "Billy" went out and made sales rapidly at $12.50. The success-, of this flyer in road carts induced other manufacturers to follow in that field and to bring to it and other fields large-scale repeat operations. Durant and Dort used their earnings to finance the Durant-Dort Carriage Company, which swiftly advanced to a position of acknowledged leadership in the trade with an annual production of 50,000 "Blue Ribbon" vehicles, high earnings, and a strong cash position, which, as we shall see, has its bearing on the story of the Buick and General Motors. His success in the carriage business made Mr. Durant a millionaire before his fortieth year, placing him in a position where an average man might have been satisfied with both his fortune and his prestige. But W. C. Durant was no average man; when the carriage business settled down into stodgy matter-of-factness, he looked for other fields to conquer.

Thus far his natural bent toward commercial adventure had found expression in the sharp competitive building and selling of styled carriages, a trade wide open to the risks of fashion. If the popular note could be struck with a buggy, if its lines attracted rural swains or a town's social leaders, the manufacturer drove a thriving trade. By the turn of the century the fun was out of the carriage business. Mr. Durant went to New York City and was studying Wall Street and the Stock Exchange at close range about the time that Mr. Whiting was beginning to think that the Buick Motor Company needed a new management.

Mr. Durant came back to Flint the next summer to see just what there was to this Buick car which the Flint Wagon Works and James Whiting had brought to Flint. With no technical experience of his own to guide him, Mr. Durant applied the only test he could make, but he did so with a thoroughness which to this day is recalled in Michigan. He drove that two-cylinder Buick back and forth over a wide range of territory devoid of good roads save for a few gravel turnpikes built by toll companies. He put it through swamps, mud and sand, and pitch-holes for almost two months, bringing it in for repairs and consultations and then taking it out again for another strenuous cross-country run. He had every sort of mischance chronic in the motoring of the period, often, of course, being stalled in out-of-the-way hamlets for lack of repair parts or fuel and oil. During these enforced waits, perhaps in a country blacksmith shop which some day would be a garage, this impetuous and eager mind wrestled with the future of transportation.

The central idea of the motor car must have appealed to his temperament, for it emphasized qualities and powers like some of his own: speed, novelty, flexibility, the ability to "get there." Its possibilities for salesmanship and showmanship would also appeal to one who had proved himself already a most successful distributor of vehicles. The motor car, he could see, fitted the progressive American spirit like a glove. In addition, here was a piece of merchandise that could not be hid; the motor car would advertise itself on the street and at the curb. Probably Mr. Durant concluded early in these tests that if Buick was not the medium by which he would enter the automobile trade, he would get into that business in another way before long. But after the car had met his severe tests, Mr. Durant looked no further: Buick would do.

By November 1, 1904, the deal between Buick Motor Company and W. C. Durant was complete; on that day the capital was increased from $75,000 to $300,000, represented by 3,000 shares of $100 each. Holders of the old stock agreed to accept Preferred stock paying 7 percent with a 25 percent bonus of Common stock. The contract covering this agreement was drawn by John J. Carton of Flint, for many years attorney for the Buick Motor Company. The Wagon Works stockholders agreed to accept Preferred stock for their holdings. It was agreed that later Buick capitalization should be increased to $500,000 and that the Wagon Works' interest should receive $175,000. This was done. On September u, 1905, Buick's stock was increased to $1,500,000 $900,000 Common and $600,000 Preferred. Mr. Carton relates that while the Buick business was sound and there was a legitimate need for this large capitalization, he had some difficulty finding enough assets to justify the increase. He says: In the application presented to the Secretary of State, I listed all the assets quite generously up to the legal requirements, but nevertheless we were still $60,000 short, and this was taken up by the following item:

> "Ownership of invention of combustion engine construction not patented for business reasons $60,000."

> This schedule went through, but later attracted little attention so that at the next meeting of the state legislature a law was passed prohibiting the listing in such cases of any items intangible and not subject to execution. However, it was quite important that the company have the full amount of the stock to issue at that time, as Common was usually given as a 25 percent bonus with sales of Preferred. The fact that I was very well acquainted at Lansing, the state capital, may have been beneficial in getting such a hazy item passed.

Mr. Durant himself sold most of the stock. It is related that at the outset he sold $500,000 worth in a single day to his Flint neighbors. There can be no doubt that Mr. Durant was a most persuasive salesman. An aura of success hung around him. For more than twenty years all his undertakings had profited; early in his selling campaign he had indisputable proof that Buick was making money. Production was steadily increasing, and he could sell every car he produced for cash, F.O.B. factory. So keen was the demand that his problem was not so much selling cars as finding capital with which to erect buildings, install machinery, and create a distributing organization so that more cars could be built and delivered. However loyally Flint might buy stock, it was too small a city to finance the expanding enterprise; Mr. Durant had to go out into the highways and byways of Michigan for capital. In this search he was tireless. He saw an immense fortune, tremendous power, and a lofty reputation as an industrial pioneer almost in his grasp if he could only find the necessary capital.

Of course, it goes without saying that, in representing the golden prospects of his venture to investors, faith and sincerity accounted for his extraordinary success in getting promotion money from individuals. He would have been saved a tremendous amount of time and energy if he had been able to secure the services of a strong investment banking house to dispose of his securities systematically, but this was out of the question in that time and locality. Local bankers helped him all they could, and his persuasive personality drew temporary aid from larger banks outside of Flint, but what he needed was a large fixed capital invested for a long pull, and this he could secure only through further personal effort.

Meantime, as the builder and leader of Buick, he found a host of problems other than financial demanding solution. Since it was early apparent that Buick's facilities in the western end of Flint would be inadequate, offices and assembly operations were transferred to a factory in Jackson which had been used by the Imperial Wheel Company. Imperial Wheel was part of the Durant-Dort family, as were the Flint Varnish Works and the Flint Axle Works. The three companies had already been located in the north end of Flint, where, with a view to future expansion of these and other companies, Mr. Durant had purchased the 220-acre Hamilton farm for $22,000. Thus he had in hand an excellent site for the expansion of Buick itself with adequate trackage on the Pere Marquette railroad, and good location as respects drainage, water supply, and general accessibility. He planned to sell part of this area as building lots, laying out for that purpose Oak Park Subdivision. But of course an immense amount of organizing work had to be done on the tract and its approaches, as well as in plant construction, before Buicks could be produced there. In the meantime the Jackson plant held the fort while Buick motors continued to be made in West Flint. Jackson, indeed, might have continued as the chief seat of Buick if capital could have been found there as easily as in Flint.

This geographic division of the business increased the labors of the leader. We can picture Mr. Durant at this time as a man desperately hurried, spurred by ambition and responsibility to feats of almost superhuman endurance, driving at breakneck speed over wretched roads between his two plants, holding conferences, making quick decisions, seeking out and encouraging new dealers, scouring the country for supplies and building plants, subduing raw land to industrial and residential uses and feverishly seeking new capital. This spare, small man seemed to draw upon irresistible sources of energy.

WILLIAM C. DURANT Founder of General Motors

He worked more hours than any of his employees, did with little sleep, yet came to his labors fresh and smiling every morning. There was a gaiety and resilience in him which overcame all obstacles. The press began to speak of him as the "Little Giant." His worshipful associates might call him "Billy," but among themselves they fell into the habit of calling him "the Man." "The Man says," was the common preface as they passed his orders along from one to the other. Sometimes "He says" would be a sufficient indication of authority. Both forms were proof of the loyal and willing acceptance of that authority. He was the first among equals rather than the autocrat, and no captain has ever been followed by more devoted troops. The camp followers, the local public, and the business men alike hung on his words.

One factor in developing this amazing and truly affectionate loyalty was his lack of concern for individual gain, the natural ease and buoyancy with which he played the prince in distributing bounties. There are innumerable evidences that he cared little for money for its own sake. His own tastes were simple, he had no time to spend money; already well off, he had serene confidence that he would always be successful and that nothing could stop him from amassing an immense fortune in the automobile game. I use the word "game" advisedly: if he was not the man who invented that adventurous expression to describe the early activities of what has since become a most precise and responsible business, he at least played that great game most completely as an adventure of the human spirit.

As an example of his lavish disregard of personal gain and his willingness to share profits with those whose backing had braced him in the past, Mr. Durant is said to have turned in to the Durant-Dort treasury at one time some $300,000 worth of Buick stock, voted to him personally in return for his work in promoting the company. In completing his layout for the approaches to the Buick industrial site, he paid $4,000 for land offered at $1,800 simply because he knew that this land as part of his grand objective was worth that much and more. The instances of his largesse could be multiplied indefinitely. He explained his generosity toward Durant-Dort stockholders by saying that he had been on the Durant-Dort pay roll during the period in which he was organizing Buick, though the fact is that he was drawing a merely nominal sum from his old company. Of course, the essential fact is that he enjoyed doing these things, and the power to do them was his compensation.

With a swift expansion program in hand and no banking connections equal to the situation, there were times when the good-will built up so generously brought important returns in timely assistance. Mr. F. A. Aldrich, secretary of the Durant-Dort Carriage Company for many years, shows from his records that Durant-Dort furnished Buick with capital in its early stages. Mr. Aldrich says:

> Owing to a decreased trade in horse-drawn vehicles Durant-Dort Carriage Company needed less capital; both our treasury and credit were in excellent shape, while Buick needed assistance. Hence in the spirit of "team-work" proverbial as applied to Durant-Dort and under full recommendation of Mr. Dort, president, we made in one way or another large investments in Buick stocks and also later in General Motors stock. At first these stocks were held in our treasury but later on legal advice they were placed in my name as trustee. We made several disbursements of this stock to Durant-Dort stockholders as dividends.
>
> The records show numerous instances of our helpfulness. We indorsed Buick paper in Chicago on one occasion, and the loan falling due at an unfortunate time, Durant-Dort had to sell Buick stock at distress prices to pay it off. On June 4, 1906, Durant-Dort bought $100,000 worth of Buick stock. There was a close financial relationship between the two companies then, and it continued after General Motors was formed. On Feb. 20, 1909, while General Motors was still in its infancy, I was authorized to buy and hold 10,000 shares of its Preferred and Common stock. Later, when Mr. Durant took on Chevrolet as an independent venture, the Durant-Dort treasury furnished him funds. In fact, within reason, our resources were always at his disposal.

The Flint Wagon Works also helped Buick get on its feet. Five of its directors loaned the Buick Motor Company at one-time $20,000 each to match an equal sum loaned to Buick at the same time by the Durant-Dort Carriage Company. Thus in one way or another, with occasional rescue loans and a vigorous search outside of Flint for capital, Buick expansion was financed.

Even when allowances are made for the newness of the automobile business and the suspicion in which it was held by the banking world in general in those early days, it does not appear that Mr. Durant ever quite deserved the reputation for financial genius which at one time clung to his name. Certainly, in the formative years, finance was his weak side. While he could make money in his operations, and raise a good deal of money by his personal force and the confidence which he inspired, he never seemed able to budget his operations accurately in advance and build up reserves. His vision was always running far ahead of his treasury, so that there was always the possibility that his affairs would approach the ragged edge of necessity if a turn came with an unfavorable market for his goods, or the well-springs of capital suddenly ran dry. His inventories and commitments were usually in excess of his present power to pay, but he had an immense faith that by the time he had to pay for them he would find the money somewhere. Either the market would provide it or stock would be sold; in a pinch he could go to friendly corporations or individuals. He kept the golden ball in the air by sheer dexterity and courage through six straining years of exceedingly rapid expansion. Looking backward upon the activities of a quarter of a century ago it can be seen that the notable human qualities behind this triumph also had their defects, which eventually caused Mr. Durant's retirement from the vast business which he originated. But it can also be appreciated that his qualities were precisely those needed to get a foundation laid with whatever tools and materials were ready at hand. Probably no other man could have built up Buick in four years to a point where, as an acknowledged leader in the industry, Buick became the rock on which General Motors was founded.

Courage is the key-word for this Buick surge to market leadership. Buick dared to produce in large quantities when most of its competitors were proceeding cautiously on restricted schedules. It pioneered in the development of attractive retail stores in large centers, and drew able, ambitious men into both wholesale and retail selling. There, perhaps, was Mr. Durant's greatest contribution to the technique of automobile administration.

For a man so vastly daring it was inevitable that as Buick production rose, further expansion should seem not only desirable but indeed necessary. Vital supplies had to be safeguarded both as to volume and prompt delivery. Competition was then less of a wrestle for markets and more of a race against time. The public would take Buicks as fast as they could be turned out; delay in delivery of even a minor part might cost a tremendous sum. Even to this day, no automobile manufacturer controls the production chain of all supplies from their primary forms to their incorporation in a completed automobile ready for the road; yet in this industry utter dependence on certain forms of goods was so essential that practically all the survivors in the stern battle for existence waged during the past thirty years are those who have been working toward self-determination, seeking positions where their operations could not be shut off by shrinkage of those essential supplies.

For instance, consider engines. In the early days of the industry many automobile manufacturers bought all or part of their power units. While these units may have been entirely satisfactory in price and quality, nevertheless, the automobile manufacturer soon realized that his production schedule was at the mercy of circumstances beyond his control. A stoppage in his supplier's plant, arising from any one of a number of causes, tied up his own plant. This risk being too heavy, the tendency has been for car manufacturers to take over engine manufacture. Some have gone a considerable distance toward controlling supplies from raw materials to the finished product, yet no manufacturer has been able as yet to process all the materials used in automobiles, because of the wide range of those materials and the special skill and large capital required to bring them into economic use. The drift has been toward self-sufficiency, yet complete self-sufficiency has not been attained and probably never will be. But in this evolution nearly all those manufacturers who depended altogether on assembling the products of other enterprises have either perished or have been absorbed. The survivors are those firms which accepted the responsibility of making for themselves goods which others would have been glad to make for them, but which for various substantial reasons it seemed vital to control throughout the entire process of production and assembly. Mr. Durant realized the value of broad organization before he entered the automobile field. The Durant-Dort Carriage Company had gone further

1904 Buick, priced at $1,250, top and lights, $125 extra

than any of its competitors in organizing subsidiary, or at least dependent, companies. It had fathered companies for the production of wheels, paint, varnish, and axles; through others it owned in whole or in part extensive timber holdings in distant states. It was natural that, faced with the market possibilities of the automobile and the difficulty of securing supplies of the right sort as required in his hot haste for action, Mr. Durant should leap to the conclusion that he needed broader organization than Buick, big enough to include not only other motor-car producers but also makers of essential parts. The need to control supplies was keenly felt in 1907, when Buick, which had concentrated successfully on two-cylinder cars, added four-cylinder models to the line. In 1908 diversification was carried even further, with two two-cylinder models, and four four-cylinder models. One of the latter the famous Model Ten started Buick on the heaviest production it had yet known, and its success was no doubt one of the elements encouraging W. C. Durant to envision a General Motors. Frederic L. Smith's reminiscences Motoring Down a Quarter of a Century indicate that his first talks were with Mr. Smith at Lansing, and that the very name, General Motors, was thus early discussed.

Flint meantime was booming as Buick drew labor from all directions. Responding to the pull of high wages, men hurried there from all quarters of the compass, from other industrial cities, from the farms of southern Michigan and the forest areas further north. Tool makers came from Providence and Hartford. The population of the city doubled in five years. House-building could not keep pace with the flood of arrivals. While Buick factory No. 10, then the largest industrial building in the world, was under construction, the neighborhood resembled a mining camp. Living quarters were at a premium; the same bed would be rented to a night-worker by day and a day-worker by night. Shacks, hastily thrown together to provide some sort of shelter, housed families who were having their first taste of prosperity. Farms were subdivided right and left, townspeople built houses as fast as they could, spurred by rising values as well as by public spirit. One could see all the evidences of rapid municipal growth, the difficulties of absorbing a large, new population swiftly into an old one. Persons of foreign blood congregated in colonies Polish, Hungarian, Serbian.

What one could not see as readily, unless he knew the Buick shops, was the terrific task which faced the Buick organization in molding this medley of raw and transient labor into an efficient working force, its members well disposed toward one another and toward management. Flint was an open-shop town, and that tradition, bolstered by high wages and the opportunities for advancement offered by a new industry, held firm against the few and withal rather weak efforts to unionize the plants. A dynamic and dramatic leadership helped to maintain that tradition until employee morale could be built up to a quite remarkable peak, until men began to see that Buick, springing from the soil of the Hamilton farm, would be an enduring institution in whose plants they could find steady and profitable employment during normal times and which in fact proved for years more resistant to business depression than the average manufacturing plant.

Buick in 1908 manufactured 8,487 cars, occupied the largest automobile plant in the world, and had a net worth of $3,417,142. It had never missed a dividend on its Preferred stock.

General Motors, immediately after its organization in September, 1908, took over Buick Motor Company for $1,500 cash, Common stock of $1,249,250, and Preferred stock of $2,499,500 a total of $3,750,250, a conservative valuation to which Buick had grown from $75,000 within the remarkably short space of four years.

Although its manufacturing processes would be considered haphazard and inefficient in the light of modern technology, they were abreast of the best practice of the day. Buick possessed a spirit in its personnel and a reputation with the public which made it a tower of strength from which its bold organizer, after surveying wider fields, could advance toward his great objective the formation of the General Motors Company.

# Chapter VI

## OAKLAND AND PONTIAC: OLD AND NEW

PONTIAC, another great seat of General Motors manufacturing, began its industrial development as Flint did, with the building of horse-drawn vehicles. The first Pontiac buggies and wagons were built by a blacksmith named King on the site of the present Fisher plant near Bagley Avenue. W. F. Stewart, who later went to Flint and rose to eminence there as a body-builder, bought King out and started his career in Pontiac, later selling his site to O. J. Beaudette who sold in turn to Fisher Brothers. The site has a continuous history of vehicle manufacture for more than half a century.

Another Pontiac pioneer was R. D. Scott, a Canadian from Guelph, Ontario, who established himself near the Grand Trunk tracks. W. A. Paterson, pioneer in the vehicle industry of Flint, learned his trade in Scott's shop in Guelph, followed him to Pontiac and then went "up country" to Flint. At first Scott's trade was altogether local, but in 1889 or thereabout, he began to branch out with road-carts.

Lee Dunlap, who went to work for Scott in 1889 and continued to be a factor in Pontiac industry well into the automobile days, explains the swing from small production to large in the Michigan carriage field, as follows:

> Until that time, carriages and wagons had been manufactured by hand, a few at a time and a few in a place, with the result that the costs were relatively high. In various parts of the country, it t

> was discovered that through division of labor and quantity production, costs could be greatly reduced, with the result that within four or five years there was a considerable boom in the trade and an almost complete stopping of custom building throughout the country districts. An incident will serve to illustrate the trend to larger marketing. I sold to Sears-Roebuck, shortly after they organized in Minneapolis, the first buggy they sold by direct mail, and within a few years, at the height of the carriage industry, every mail brought large orders from them.
>
> Under the encouragement of prosperity, the entire industry proceeded to overbuild. By the early years of the twentieth century, it was quite apparent that the buggy business had seen its best days, especially as far as Michigan was concerned. When automobiles were still few and far between there were other parts of the country nearer the raw materials, which seemed to have an advantage over the Michigan factories. Practically every carriage manufacturer began to look around for some new development. The nearest one at hand seemed to be the automobile.
>
> Of course you and I know that there is no striking similarity between the carriage and the automobile except that both of them travel on wheels. We had worked out a system for progressive assembly and a quite efficient division of labor, but none of us knew anything about machine operations, except in a very limited way. We seldom used blue prints and close measurements were unnecessary. Nevertheless, the automobile industry located in Pontiac, entirely because Pontiac had been a city which manufactured horse-drawn vehicles. In Jackson, Flint, and many other cities the same tendency was in evidence. That the similarity in the two lines was more apparent than real is proved by the fact that the carriage men are not now the factors in the automobile trade that they were in the beginning, their places having been taken by men of more engineering experience.

The largest of the Pontiac carriage factories was the Pontiac Buggy Company, which Edward M. Murphy, S. E. Beach, and Francis Emmendorf had incorporated in November, 1893. Although incorporated for only $25,000 paid in, it built a factory then reckoned large on Oakland Avenue, where the Pontiac, Oxford & Northern tracks crossed that thoroughfare, on land now occupied by the Pontiac Motor division.

All his old associates ascribe to "Ed" Murphy extraordinary powers of organization and business drive. Born in Wayne,

Michigan, he climbed the ladders of success largely by his own efforts and came early into business authority. In 1898, after the Pontiac Buggy Company had enjoyed its share of the boom, he brought Lee Dunlap into his orbit by establishing the Dunlap Vehicle Company, to manufacture a somewhat lighter grade of buggies than Pontiac had built. Mr. Dunlap came over from the C. V. Taylor organization. Pontiac Buggy also formed the Crescent Carriage Company in 1903. In 1904 Mr. Beach sold his interest to Mr. Murphy and bought the latter's interest in the Crescent and Dunlap plants, but later went back into the Pontiac organization, remaining until the change from carnage to automobile production. When this change came Mr. Murphy was sole owner of Pontiac Buggy Company, but his associates had interests in the allied carriage plants.

Not only was the carriage trade falling but also production costs were rising, owing to the automobile manufacturers in Flint and Detroit drawing Pontiac's skilled workmen away. From its beginning the automobile trade paid practically double the wage rates customary in carriage production. A carriage trimmer might get two dollars a day, an automobile trimmer four dollars for work roughly similar. Naturally, the automobile business drew the best workers, and the new wage standards dealt the declining carriage trade a heavy blow.

Facing decreased production and increased cost for carriages, Mr. Murphy began looking around for an automobile to manufacture, or, to be more exact, to assemble from purchased parts. He had heard of a two-cylinder car designed for Cadillac in which the latter did not seem to be particularly interested. Tests proving satisfactory, on August 28, 1907, Mr. Murphy organized the Oakland Motor Car Company for $200,000; 20,000 shares of Common stock, par $10. New money was furnished by James Dempsey of Manistee, Michigan, a wealthy retired lumberman. The Murphy carriage plants were acquired by Oakland at various times. On September 25, 1908, the capital stock was increased to $300,000 by the addition of $100,000 in Preferred stock. The two-cylinder design not proving highly successful, the company brought out in 1908 Model K, a four-cylinder car powerful for its time and underselling all competitors. It became a hill-climbing champion, winning the Giants Despair Climb at Wilkes-Barre and other contests at Baltimore, Maryland; Paris, Illinois; and Jefferson Hill, Long Island, New York. The capital outlay being larger than anticipated, the close of 1908 saw the new venture perilously close to disaster. Mr. Dunlap speaks of Oakland as "broke" when W. C. Durant bought it early in 1909 for his new General Motors Company, but perhaps this is merely relative to commitments. Another source says that the cash position of Oakland was good, capital being doubled in a year. The production figures are 278 cars for 1908, 1035 for 1909[1]

Mr. Murphy explained to his group, some of whom objected to the sale, that he was selling obsolete buildings. From Mr. Durant's side, one of the advantages was Oakland's organization. Like everyone else in the Michigan carriage trade, he had a great liking for "Ed" Murphy and probably expected that Mr. Murphy, who was still a young man of great vigor, would become a leader in the new "automobile game." This prospect was defeated by the latter's early death. Other members of the Oakland group were men who have since become important figures.

On January 20, 1909, General Motors directors authorized the acquisition of a half interest in the Oakland Motor Car Company. By February 23rd, Mr. Durant reported the acquisition of 15,000 Oakland shares, also that he expected to acquire up to 21,000 shares at $11 a share. Mr. Murphy took stock and notes for his interest, but others had to be paid off in cash and by June 5th the treasurer reported that he had paid out $200,856 in cash for 18,783 shares. Hardly was the purchase completed than Mr. Murphy passed away and Lee Dunlap became general manager. Production advanced to 4,000 cars for 1910, which required a heavy plant expansion program, not without its bearing on the financial difficulty which General Motors experienced in the latter year.

Mr. Dunlap's narrative shows the speed at which anyone who followed Mr. Durant in those days was forced to travel:

---

[1] figures furnished by National Automobile Chamber of Commerce.

When Mr. Durant visited one of his plants it was like the visitation of a cyclone. He would lead his staff in, take off his coat, begin issuing orders, dictating letters, and calling the ends of the continent on the telephone-, talking in his rapid easy way to New York, Chicago, San Francisco. That sort of thing was less common than it is now: it put most of us in awe of him. Only the most phenomenal memory could keep his deals straight ; he worked so fast that the records were always behind.

On this visit of which I am thinking, early in IQIO, I expected he would stay several days as we were to discuss the whole matter of plant expansion. But after a few hours, Mr. Durant said, "Well, we're off to Flint." In despair I led him on a quick inspection of the plant. Instantly he agreed that we would have to build, and asked me to bring the expansion plan with me to Flint the next day. There wasn't any plan, and none could be drawn on such short notice, but his will being law and our need great, something had to be done.

So I called in a couple of our draftsmen to help me and that night we made a toy factory layout existing buildings in one color, desired buildings in another. We drew a map of the whole property, showing streets and railway sidings, and then glued the existing buildings to it in their exact locations. Feeling like a small boy with a new toy, I took this lay-out to Flint and rather fearfully placed it before the chief. I needn't have been alarmed at our amateur lay-out. He was pleased pink. We had a grand time fitting our new buildings into the picture as it was spread on his desk. We placed those new buildings first here, then there, debating the situation. When we agreed as to where they should go, he said, "Glue them down and call W. E. Wood."

Mr. Wood came in after a few minutes and received an order for their construction. In the whole history of America, up to that time, buildings had never arisen as swiftly as those did. Contractor Wood had men, materials, and machines moving toward Pontiac within twenty-four hours, and we were installing machinery in part of the structures within three weeks. But, of course, we could not be equally swift in paying for them. That was something else. But for the time being none of us worried too much over that; we figured the "Little Fellow" would find the money somewhere. Which he did, in the end, though we know there was plenty of trouble before the bills were receipted.

These early years in the automobile business were marked by tremendous personal activity and a very grave shortage of capital. Anyone going direct from the carriage manufacture to

> automobile manufacture could have little conception of the large use of capital required in the new field ... I was with Oakland all through the 1910 "pinch" when the plant was frequently visited by members of the Creditors* Committee.

Mr. Dunlap was succeeded as general manager of Oakland by George E. Daniels, one of the earliest associates of W. C. Durant in the automobile business and the first president of the General Motors Company of New Jersey, holding office for a short time directly after incorporation. After resigning the presidency, which he occupied merely as an. interim officer, Mr. Daniels remained with General Motors having charge of Cartercar. The bankers' reorganization placed him at the head of Oakland where he remained until 1914.

Production rose in 1913 to a new peak of 8,618 because of two advanced models a fast, light "four" with an electric self-starter, and a "six" Oakland's first priced at $2,450, a moderate figure.

As part of the change wrought by W. C. Durant's return to power in 1916, Mr. Fred W. Warner, who had succeeded Mr. Daniels as general manager, became a vice-president of General Motors. Under him price reductions and improvements in design brought Oakland into a swift run of prosperity as reflected by these unit sales figures:

| 1916 | 27,000 cars. | The v-8 introduced |
|---|---|---|
| 1917 | 35,000 cars. | The "light six" sold at $795 |
| 1918 | 30,000 "light sixes." | First offering of closed cars |
| 1919 | 52,000 "light sixes" | sold on a rising market |

Automobile prices reached a peak in 1920, owing to scarcity of materials caused by war. Pioneering in the light car field with closed bodies, was so much of a marketing experiment for Oakland that dealers were required to take one closed car with each carload shipped from the factory. Now the closed car is standard and the open car the exception. The post-war boom brought prosperity to Oakland, as it did to other manufacturers. In General Motors at that time lack of central coordination, allowing great latitude in inventories and production programs, created difficulties which as they multiplied brought the second Durant administration of General Motors to an end and a change of management to Oakland.

After Mr. Warner resigned in 1920, George H. Hannum served as president and general manager for six years, to be succeeded by Alfred R. Glancy who continued in the saddle until 1931. These were years of remarkable expansion in the Pontiac area, the General Motors plants spreading northward along the railway and an entire new factory layout being built beyond Harris Lake in what had been open country a few years before. During the 1920-1930 decade Pontiac grew faster than any other industrial city in its census classification. The acute shortage of houses, from 1919 on, led the Corporation into an extensive housing development through the Modern Housing Corporation.

The Oakland line, after a highly successful record with "fours," entered the u six" field in 1915, the "eight" field in 1916, and remained a leader in quantity production down to 1931, when it was discontinued. The Pontiac car, introduced as a light, low-priced "six" in 1926, received such a warm market welcome in the next few years that under the management of I. J. Reuter, who took hold in 1931, the huge Oakland plant was concentrated on Pontiac production. Pontiac appeared in 1932 with an eight-cylinder motor. Its manufacturing operations were under the direction of W. S. Knudsen, president of Chevrolet, from May, 1932, to October, 1933. Mr. Knudsen at that time became executive vice-president of General Motors in charge of car and body manufacturing in the United States and Canada. H. J. Klinger, vice-president and general sales manager of Chevrolet, was then advanced to general manager of Pontiac.

Among the innovations which Oakland-Pontiac takes pride in originating or early adopting are oil and fuel filters, air cleaner, crankcase ventilation, automatic spark control, interchangeable bronze-backed main bearings, harmonic balancer, oil-tight universal joints, honed cylinders, full pressure lubrication, and rubber spring shackle bushings. Oakland was the first division to bring Duco-finished cars to quantity production; the year, 1924.

The Pontiac Motor division, in succession to Oakland, now occupies 231 acres within the corporate limits. The experimental unit fronts on Oakland Avenue. The main division, to the north just inside the city limits, is one of the most advanced industrial lay-outs in the world, and entirely new since 1927. This huge grouping is in startling contrast to the modest factory building in which Oakland started to manufacture motor cars in 1908, when only 50,000 square feet of space housed its initial activities.

A breezy description of how some of the burning questions connected with new car models are threshed out and what a new model in this case the Pontiac for 1932 involves financially is given in Fortune for December, 1931. Perhaps the account is sufficiently accurate to give one a general impression of what is involved in a change of models.

The scene is a luncheon club in the Fisher Building, Detroit, where half a hundred General Motors executives are reported as present. The reader is supposed to be accompanying Mr. Richard H. Grant:

> Mr. Grant's sales problem will begin with the oldest thing in automobile salesmanship: namely, the automobile. Should the 1932 Pontiac have free-wheeling? Ride control? Should the foot throttle be changed at the cost of 35 cents per car ($35,000 added to production costs)? The Pontiac engineers have been maintaining forty experimental cars on the road to prove these things, some of them embodying features that will not appear until 1933 or 1934. In the end, during the coming ten months, Mr. Reuter will have spent $1,000,000 for research on the 1932 car, to which should be added Pontiac's share of the research done by General Motors Research Laboratories, by Fisher Body, by Delco-Remy (ignition) by Harrison Radiator, by New Departure (ball bearings), by AC (spark plugs, gauges, etc.) by Kelsey-Hayes Wheel, and by headlight, shock-absorber, tool, and die manufacturers throughout the land: a total research bill of $2,000,000 spent by the industry for this car alone. After the tireless engineers have estimated the cost of every one of the 15,000 Pontiac parts, and after they leave submitted blue prints to George Christopher, in charge of Pontiac manufacture, Mr.

> Reuter will find it necessary to spend $1,400,000 in retooling for this model, and the retooling of Fisher Body will cost that unit $1,000,000 more. New machinery will cost Pontiac $300,000 and to revamp the floor plan will add $150,000.

All through General Motors history, groups of vigorous men have been making similar decisions of great import which affect the Corporation, its products, and the world. Change the names to those of other cars and persons, place them in other scenes, and you have a fair idea of the group planning which has been going on in General Motors for twenty-five years. All over the broad geographic range of the Corporation, both in the United States and abroad, plants have been growing, as the Pontiac plant grew, because men who knew what they wanted could agree on programs and carry them through.

# Chapter VII

## CADILLAC: THE TRIUMPH OF PRECISION

THE two most vital trends in American machine industry Middle West daring and New England craftsmanship met in the founding of Cadillac, an enterprise whose success has had a profound effect on the whole automobile trade and through it upon American industrial history. The first of these influences has been discussed in the chapters on Oldsmobile and Buick; the second entered the scene with Cadillac, whose roots reach back into the New England scene of traditionally competent workmanship and native mechanical ingenuity.

New England's early eminence in American industry produced an imposing array of inventors and an army of able mechanics. Indeed, until almost the turn of the present century, New England dominated the American machine industry, and for a time it seemed likely to capture first place in the rising automobile trade. The state of Massachusetts recognized the importance of Blanchard's steam carriage, built in 1825, by legislative enactment giving that vehicle the right to use the roads. The first gasoline "horseless buggy" of the modern era was also built in Springfield by the Duryeas. The Pope interests in Hartford came early to the support of the automobile, and both as manufacturers of vehicles and part owners of the Selden patent, wielded great influence through the infant years of the industry.

With this early start, the wonder is that the automobile business slipped away from New England to the Middle West. There were several reasons for this shift, and one is this: New England inventors and mechanics went West themselves. New England was always a seed-bed of talent in the industrial arts; as the young men of New England marched West they took its best traditions with them.

HENRY M. LELAND

The career of Henry Martyn Leland, founder of the Cadillac Motor Car Company, is an outstanding example of the influence of New England machine shop practice upon the Middle West. Mr. Leland was born at Danville, Vermont, February 16, 1843. He was early apprenticed in the Crompton-Knowles Works at Worcester, Massachusetts. As a skilled mechanic he was taken into the Federal Arsenal, at Springfield, Massachusetts, and later worked in the Colt revolver factory. The New England arms factories were the first to apply fully the practice of assembling interchangeable parts, the great Eli Whitney establishing this principle. By the time of the Civil War the machining and assembly of arms parts had been highly refined, with the result that Leland early became accustomed to a high degree of precision, laying the foundation for an ideal of accurate machine work which later distinguished the output of his Detroit factories.

After a stay of twenty years with the famous Brown & Sharpe Manufacturing Company at Providence, Rhode Island, Mr. Leland went "on the road" in the Middle West where the company desired to introduce their machine tools to the growing mechanical industries of that section. It was in this way that he became acquainted with Detroit, its industrial leaders and possibilities.

Moving to Detroit in 1890, he established the Leland, Faulconer & Norton Company which made machine tools, grinders, gear cutters, etc., and also did custom work along those lines. In 1895 the firm became the Leland & Faulconer Manufacturing Company, with Charles A. Strelinger as secretary. Mr. Strelinger was a leading hardware merchant whose immense trade in bicycles led him to take a keen interest in every type of rapid transport. The year 1896 is notable in Leland annals for the establishment of a grey iron foundry which scored a remarkable success by introducing more closely machined castings than could be had at that time elsewhere, so that its products commanded premium prices, sometimes as much as thrice the usual price.

In its plant on Trombly Avenue the company added gears for chainless bicycles to its other lines, working out a process for grinding these case-hardened gears to closer standards. Another activity embraced the manufacture of gasoline motors of five, ten, and fifteen horsepower for marine uses, a line of production which inevitably brought the company into the newly developing automobile business. His experience with marine motors began in the East, and all through his early days in Detroit Mr. Leland was an enthusiastic prophet of the almost unlimited future of internal combustion engines.

Although it had made machinery and parts for other automobile companies earlier, Leland & Faulconer did not enter quantity production of automotive material until it began manufacturing transmission gears for the one-cylinder motor car designed by R. E. Olds the famous curved-dash runabout. After the Oldsmobile factory had been destroyed by fire, need to restore production quickly led Mr. Olds to turn over to Leland & Faulconer part of his motor manufacture. This was seized upon as an opportunity to show the world to what close standards a gasoline engine could be built. An associate of Henry M. Leland says in this connection:

> At the first automobile show in Detroit, the Olds display contained two cars, one powered by Olds, the other by Leland. I recall that Henry Ford pointed out to us as a curiosity the fact that the Leland motor was operating under brakes in order to bring it to the same pace as the Olds motor. Our motor developed 3.7 horsepower as against 3 for the Olds-built motor. This superiority was due entirely to closer machining. H. M. Leland went seriously to work on the Olds motor and by introducing larger valves and improved timing system we were able to build this one cylinder up from 3.7 horsepower to 10.25. On taking the improved article to the Olds Company, we were dismayed by the refusal to use it. Mr. Olds was getting all the business he could handle and I suppose such a radical change in power plant would have necessitated alterations in many directions. The point is significant however as a key to a future of accuracy methods in manufacturing.

While Olds was rushing rapidly into quantity production and Leland & Faulconer were gaining their reputation for precision, a third group of Detroit men, with ample capital at their disposal, formed the Detroit Automobile Company in 1899, in an endeavor to pioneer a merchantable passenger car. These men were William H. Murphy, Lem W. Bowen, Clarence A. Black, and A. F. White, all citizens of the first rank, who had as their chief guide and mainstay in the mechanical department no less a person than Henry Ford. Mr. Ford's ideas then ran more to speed than to comfort, and after he had produced two racing cars which made quite exceptional records but were not at all what his backers visualized as meeting the market demand, the Detroit Automobile Company group came to Leland & Faulconer and agreed to use the latter's improved motor which the Oldsmobile organization had not seen fit to adopt. With their assistance the Cadillac Automobile Company was organized on

August 22, 1902, with a capital of $300,000. Thus Cadillac became established at Cass & Amsterdam avenues, Detroit, a site it continued to occupy for many years.

The selection of that name was a logical one for men versed in Detroit history. Antoine Sieur de la Mothe de Cadillac was the leader of the French expedition which made its first settlement at Detroit in 1702, just two hundred years before the

Cadillac "one-lunger" 1902 From Duncan s World on Wheels

founding of the new automobile company. In his high courage, enterprise, and ability the founders saw the very qualities which they hoped to bring to their fledgling corporation. His coat-of-arms became the emblem of the new company.

Work on the first Cadillac automobile had been begun in September, 1902 a one-cylinder car known as Model A, of which two units were completed in March, 1903. In the following twelve months Cadillac built and shipped 1,895 finished automobiles. From the completion of that first car the future of Cadillac has never been in doubt. This famous "one-lunger" was then the last word in superior workmanship. The engine was located under the front seat; a hand crank at the side served as a starter. The driver sat to the right; a real steering wheel offered an advanced feature, nearly all the other cars of the day steering by tiller control. The car had no running board, but patent-leather fenders imparted an air of distinction. Four passengers might ride, two in front and two in the rear. Heavy brass kerosene lamps and a hand-operated bulb horn cost extra.

A situation then arose which was not uncommon in those days. While Leland motors were giving excellent satisfaction, the Cadillac Automobile Company was unable to do equally well as regards chassis and body manufacture. It brought its troubles to the Leland organization. As the latter could not afford to see art outlet for their motors closed, they accepted, with a good deal of misgiving, the proposition to undertake the general management of Cadillac. The consolidation of Cadillac Automobile Company and Leland & Faulconer Manufacturing Company into the Cadillac Motor Car Company was completed on December 27, 1904. Henry M. Leland became general manager.

In 1904 the company introduced the Model B one-cylinder Cadillac. In all, 16,126 one-cylinder Cadillacs were manufactured on the two models. The one-cylinder cars formed the bulk of early output demand continuing after the "30" of four cylinders was introduced in 1905. A spectacular proof of the long life of these one-cylinder cars was given in 1922 when on the twentieth anniversary of the company, Lucien R. Burns, who had been with Cadillac since its organization, drove one of the first cars it had ever built from Detroit to New York where he had long before unloaded the first Cadillac shipped into the city. More than 67,000 of the "305" were manufactured. Both the one-cylinder and the "30" had the notable feature of interchangeability of parts.

The proudest achievement of early Cadillac history, the winning of the Dewar trophy, awarded by the Royal Automobile Club of London, England, for the greatest advance by any motorcar during the year, flowed directly from this achievement of parts interchangeability. Until the London demonstration, the world believed that English, French, and German workmanship stood superior to American and that the rise of the automobile industry in this country received its impetus from other factors than those involved design and manufacture such factors as the

Cadillac cars drawn up in London, 1907, for tests which won the 1908 award of the Dewar Trophy

vast distances of America, its large population and the broad distribution of wealth. Cadillac, practising the highest standards of close workmanship, now revealed that automobiles could be made of interchangeable parts just as effectively as firearms could be.

A graphic picture of that triumph and explanation of what it meant to the reputation of American cars abroad is contained in a letter in the Automobile Trade Journal of December, 1924, from F. S. Bennett of F. S. Bennett, Ltd., 24 Orchard Street, London, W. I., the pioneer agent of Cadillac in the British Isles. A half-page Cadillac advertisement in that trade organ in 1903 led Mr. Bennett into correspondence with the company, which resulted in the setting up of an agency. He writes:

> I would not like to say just how many million dollars have changed hands as the direct result of this advertisement, and it goes much further than this, as it can be rightly claimed to have been the birth of the industry of American cars in Europe.

The early history of American cars over here might be said to be the early history of the Cadillac car. In the very early days a few American cars came over here but did not make good. The Cadillac was the first to establish itself and in doing so established the American-made automobile in this market. For several years I plowed a lonely furrow, being the only man who did not falter in his belief in the American car.

It was not until what is now known as "the famous Cadillac standardization test" that the prejudice, still surviving, of dumped American bicycles and a certain number of poorly made cars, was successfully overcome.

This test in 1906 consisted of three Cadillac cars being assembled in an open shed at Brooklands track from a medley of parts representing the dismembered components of three Cadillacs taken from the dockside by the Royal Automobile Club officials as they arrived from America and taken to pieces by them and jumbled in the shed mentioned.

Then the Cadillac mechanics assembled the cars from the heap with no other tools than wrench, hammer, screw driver and plyers. The three cars were immediately put into the hands of the Royal Automobile Club official observers and went through a 500-mile test on the Brooklands track, finishing with a perfect score.

That was the first demonstration here in the art of motor car standardization and it created a profound impression. For this, the Sir Thomas Dewar trophy awarded for the most meritorious performance of the year was awarded by the Royal Automobile Club to Cadillac. The Cadillac is the only automobile company that has won this trophy twice.

The effect of this test on the public mind makes it stand out as historic and unique, and even this morning, as I write this, I received a press clipping with reference to this test carried out eighteen years ago. I can truthfully say that during the eighteen years, no week has passed in which a clipping from some newspaper in some part of the world has not arrived referring to this test.

It had the effect of giving the Cadillac car in particular, and the American-made car in general, a place in the sun in this country. On this side of the water it answered completely the adverse criticisms against the American-made car and opened wide the gate for many American manufacturers to come into this market.

The result proved a triumph for Cadillac methods, also a revelation of the truth that quantity production and quality production by no means need be antagonistic. The lesson was plain enough: high standards of accuracy could be secured in a large factory geared to high production as well as in a small shop where each part received individual attention.

Cadillac won the Dewar trophy a second time in 1913 as a result of its pioneering in electrical starting, lighting, and ignition, being the first car so equipped. Mr. Leland's keenness to relieve the need for cranking spurred C. F. Kettering of Dayton Engineering Laboratories to his notable experiments and eventful triumph in this revolutionary creation.

The Dewar Trophy, twice awarded to Cadillac, 1908 and 1913

By 1908, when General Motors was formed, Cadillac had secured a position of unquestioned eminence in the automobile world. Its one-cylinder car had been a tremendous success, and the four-cylinder "30" had reached an annual production of between 7,000 and 8,000 with the price at $1,400. The company was well managed, highly prosperous, and financially conservative. Although capitalized for $1,500,000, this represented largely the plowing back of profits. It has been said that only $327,000 in cash went into the enterprise from all sources. Already Cadillac had achieved an outstanding reputation in the fine car field, and it was this reputation which led Mr. W. C. Durant to attempt adding Cadillac to his newly formed General Motors Company.

Mr. W. C. Leland, who had charge of the negotiations for Cadillac, said that no written option was ever given in the course of the transaction which covered the better part of a year:

Our first price was $3,500,000 good for ten days. Mr. Durant did not return for six months and then said he would accept that figure. However, the situation had changed in our favor in the meantime because Cadillac had been making exceptionally good earnings, the country had rallied from the slump of 1907 and we could see only fair weather ahead. Consequently we asked them $4,125,000, this to hold for another ten days. Mr. Durant was of course tremendously busy during all this time and I suppose he had difficulty in finding the money for what was then a colossal deal as we had asked cash. At any rate the second ten-day period expired before Mr. Durant again returned, whereupon he was given a quotation of $4,500,000 which he accepted. The transfer was made on that basis on July 29, 1909, in cash, the transaction being the largest negotiated in Detroit up to that time.

It should be added in explanation that Cadillac was really purchased by the Buick for General Motors to which transfer followed shortly after. The net worth of Cadillac at the time of acquisition was $2,868,709, the difference between that figure and the purchase price representing the good-will value which Cadillac had built up in its seven years of life. The sales price represented slightly more than $300 a share because, in addition to the cash involved, $275,000 was paid in Preferred stocks. It is interesting to record that Benjamin Briscoe at one time had an option on 60 percent of the company's stock at $150 per share. But the option expired, owing to his inability to find cash, shortly before Mr. Durant entered the scene with his offer of $160 a share. If the latter had acted promptly on his first option he would have saved a million dollars.[1]

However, the price as finally fixed was not at all unreasonable, viewed in the light of Cadillac's earnings at the time of these negotiations, for a net of $1,969,382 was reported for the year ending August 31, 1909. The company's earnings continued at or above this rate for the next ten months during which Cadillac turned into the General Motors treasury $2,000,000. Both Henry M. and W. C. Leland remained with Cadillac until 1917.

From its beginning Cadillac has followed this slogan, "Craftsmanship a Creed, Accuracy a Law." Various later developments of Cadillac in its evolution toward its present position will be described elsewhere, but as the story of General Motors unfolds, the reader will do well to recall that Cadillac standards have influenced all General Motors cars and products, and indeed the whole automobile industry.

[1] The full purchase price was $5,669,250. See p.223

# Chapter VIII

## THE BIRTH OF GENERAL MOTORS

ALMOST from the moment when he reorganized Buick and increased its output, Mr. Durant seems to have begun looking toward wide horizons. Hardly had he moved Buick assembly to Jackson than he organized the Janney Motor Company there. Realizing that axles were almost as important as motors, he induced the Weston-Mott Company to move from Utica, New York, to Flint, allotting it a strategic site in his new industrial area next to Buick. With Weston-Mott came Mr. C. S. Mott, a future director and vice-president of the Corporation. Mr. Durant also drew to Flint the Frenchman, Albert Champion, and encouraged him to locate there the production of the famous AC spark plug. Forced at the outset to lean on local independent makers of springs and bodies, Mr. Durant looked forward to the time when those activities would be part of a larger picture. General Motors was beginning to take shape in his mind.

Mr. Durant was not the only man thinking along those lines. Others perceived the difficulties and dangers of independent operations, and visioned from afar the profits to be garnered by a great automobile merger. On the staff of A.L.A.M. (Association of Licensed Automobile Manufacturers) was Hermann F. Cuntz, a mechanical engineer who acted as contact man between the Selden patent licensees and the central office in New York. He was personally acquainted with all the manufacturers operating under the Selden patent. He knew what they had in their treasuries as well as in their plants, whether their organizations were strong or weak, and what their prospects were for future business. Possessed by the idea that a merger of certain products was both feasible and advisable, he interviewed Anthony N. Brady, one of the powers of American finance, who had built a vast fortune through consolidating traction lines and public utilities.

C. S. MOTT Founder, Weston-Mott Company

Mr. Brady being favorable to the idea, Mr. Cuntz negotiated with a number of manufacturers who proved willing to sell control at favorable prices. But Mr. Brady held off, prophesying the panic of 1907 and saying that after the automobile trade had experienced a taste of the hard times in store for them the desired companies could be brought into the combine for less money. The financier was right about the panic, which arrived according to his schedule and found him prepared for it, but he was wrong in estimating its effect upon the automobile trade.

American motordom weathered the storm of 1907 better than any other line of business. Buick, and the other companies with which Mr. Durant had affiliated himself, came through without a scratch. In that bad year Buick production actually rose by 50 percent, and Flint was one of the few cities in the country unaffected by panic. Its banks went on paying out currency as if nothing had happened, although currency payments were suspended in many quarters.

The effect of this immunity on Mr. Durant was twofold. It immensely enhanced both his confidence and his prestige, and it placed various other manufacturers, whose businesses had proved less resistant to strain, in a frame of mind where they would be willing to listen to him when he should be ready to talk merger to them. But whoever talked merger to them effectively would have to do so in large figures. The break in values which Mr. Brady had forecast for the automobile world simply had not come to pass. Instead, Anthony N. Brady's opportunity to crown his business career with a stupendous industrial merger had passed while he hesitated. He still kept one eye on automobiles, however, becoming active in United States Motor Company and in 1910 a director of General Motors, but his chance to play the leading role in the industry had departed.

It is interesting to speculate a little on what might have happened if Mr. Brady, with his large resources and shrewd sense, had taken the lead in the industry in 1906, when he was considering the plan presented to him. Probably his entry would have been effective in bringing the automobile business into the great financial marts under proper sponsorship far earlier than it appeared there. Automobile securities might have appeared on the New York Stock Exchange in 1908 instead of in 1911, when the voting trust certificates of the General Motors Company were the first automobile securities listed by the Stock Exchange. The whole automobile industry might have been given a definite and enduring trend toward the East, with the early New England companies growing instead of dwindling as they actually did, and the Middle Western companies not making the headway which, historically, they have made. If Mr. Brady had organized General Motors instead of Mr. Durant, the Corporation might have lost much of the dramatic interest now attached to it.

With Buick standing like a rock through 1907 and swinging into 1908 at a pace which promised a strong financial position at the end of the year, Mr. Durant was ready to proceed with his merger. He planned the erection of a corporation which would be merely a holding company, of which there were many examples among the "trusts" of the period. We shall see how General Motors changed from a holding company to an operating company under the pressure of economic forces too strong to be withstood, as the automobile made its way into large production and broad relations with the consuming public. In general, American economic history shows that while holding companies meet with little difficulty in the public utility field, where monopoly and state regulation are involved, they do less well in the manufacturing field, where large labor forces must be maintained and style goods sold in open competition.

Spurned in Chicago by a shortsighted luminary of the Bar, Mr. Durant took himself to a young lawyer of New York, Curtis R. Hatheway, of Ward, Hayden & Satterlee, who drew up articles of incorporation for the General Motors Company of New Jersey. The company was empowered to use its capital of $12,500,000 (authorized September 28, 1908) in the purchase of the securities of automobile and accessory companies.

The certificate of incorporation of the General Motors Company was signed by the incorporators, George E. Daniels, Benjamin Marcuse, and Arthur W. Britton on September 15, 1908, and filed in the office of the clerk of Hudson County, New Jersey, the same day. The following day, it was filed in the office of the Secretary of State of New Jersey at Trenton. As the New Jersey statute declares that the filing in the latter office determined the date of the body corporate, the latter date, September 16th, has been accepted as marking the birth of General Motors.

The incorporators met on September 22nd, adopting bylaws under which they became the first directors and officers, as follows:

George E. Daniels, president
Benjamin Marcuse, secretary and treasurer

Directors:
Arthur W. Britton
George E. Daniels
Benjamin Marcuse

On October 20, 1908, the by-laws were changed to provide for a board of seven directors and the original officers were succeeded by the following who held office for considerable periods:

W. M. Eaton, president from October 20, 1908, to November 23, 1910
W. C. Durant, vice-president from October 20, 1908, to November 16, 1915
C. R. Hatheway, secretary and treasurer from October 20, 1908, to January 26, 1911

It will be noted that Mr. Durant appears as vice-president, though he was actually the founder and active head of the organization. By placing Mr. Eaton, an influential business man of Jackson, in the highest office, Mr. Durant left himself free for more active duties. The three incorporators were all out of the directorate by the end of the year, Mr. Britton being succeeded by Henry Henderson, and Messrs. Daniels and Marcuse by Messrs. C. R. Hatheway and Frederic L. Smith on December 16th, after which meeting the board consisted of:

| | |
|---|---|
| W. C. Durant | elected October 20, 1908 |
| W. M. Eaton | " " " " |
| C. R. Hatheway | " " " " |
| Henry Henderson | " " " " |
| W. J. Mead | " " " " |
| Frederic L. Smith | elected December 16, 1908 |
| Henry Russel | " " " " |

Theodore F. MacManus in Men, Money, and Motors gives a vivid recital of the origin and naming of General Motors, which exemplifies the offhand way in which the dashing founder arrived at his decisions:

> The automobile business was a hazardous business. Durant appreciated this. His azure dreams of power were often disturbed by nightmare flashes. Fly-by-night concerns with no objective save a skimming of the market and immediate profits for their promoters were everywhere. Durant realized there had to be stabilization. Early in 1908 he proposed to Ford, Couzens, Briscoe,

> and Olds a consolidation of Ford, Maxwell-Briscoe, Reo, and Buick.
>
> Ford and Couzens played with the idea, matched their wits against the wits of the others and when Durant appeared most hopefully they tossed in this stipulation:
>
> "We will go in only on condition that we receive three million dollars, in cash."
>
> Not to be outdone by the Ford and Couzens ultimatum, R. E. Olds got to his feet and pronounced sentence on the consolidation:
>
> "If you do that for Ford you've got to do likewise by Reo. We will expect three millions, in cash, also."
>
> Durant waved his hands. The meeting ended. The project was abandoned.
>
> Durant, however, was not easily discouraged. Calling aside Benjamin Briscoe, he said:
>
> "Let's go it alone. We two."
>
> Briscoe was willing and the two men went to see George W. Perkins, of J. P. Morgan & Co. The bankers agreed to underwrite $500,000 of the new $1,500,000 capital required. A charter was tentatively drawn up and the consolidation was to be called the International Motors Company.

These negotiations fell through, but since a representative of the banker had suggested the name "International Motors" Mr. Durant crossed out the word "International" and wrote in "General."

Mr. MacManus writes:

> "How does that look?" he asked, holding it so Briscoe could see.
>
> "Fine. You mean you'll call your organization the 'General Motors Company'?"
>
> "Yes."
>
> Briscoe left.[1] Durant sat there, studying the sheet of paper on which he had written almost carelessly had written a name that has come to be familiar around the world.

---

[1] Briscoe's United States Motor Company was dissolved through receivership in 1912. Its backbone was the Maxwell-Briscoe, which Benjamin Briscoe organized in 1904 with $42,000 in cash. MacManus says that Mr. Durant

Mr. Durant considers that here and there in Men, Money, and Motors he is not reported with entire accuracy by Mr. MacManus, who is somewhat given to dramatic rhetoric while Mr. Durant was ever precisely polite in his choice of words in conversation. As to the name, General Motors, Mr. F. L. Smith says in his published reminiscences that he contributed it some time before in a Lansing conference looking toward the merger of Buick and Oldsmobile.

Amid what seems almost a conspiracy of silence General Motors was born. The New York Times of September 16th, conned by Mr. MacManus, revealed that:

> President Theodore Roosevelt was in a jovial mood. "Bully!" he cried when reporters told him that Charles Evans Hughes had been nominated for Governor of New York state on the Republican ticket ... Mike Donovan, an athletic trainer, was on his way to the Roosevelt home at Oyster Bay and when approached by newspaper men he vowed: "I am going to write a book on Teddy he deserves it!"
>
> Justice Davis decided against the city in its case against an autoist who had been arrested in Central Park the preceding December 16: "The charge against this man is driving with skid chains and the law does not prohibit their use," declared the Court.
>
> The White Star Company announces the construction of two new ocean liners, the keel to one ship is now being laid and it will be called the Olympic; the keel to the second ship will be laid later and it will be called the Titanic.
>
> New Jersey had a flyer who was critical of the Wright aeroplane as compared with his own; Wilbur Wright remained in the air over LeMans, France, for thirty-nine minutes and broke distance records by flying twenty-six miles.
>
> Jack Johnson and Tommy Burns were matched for the heavyweight championship of the world and agreed to fight in Sydney, Australia.

---

offered the equivalent of $5,000,000 in General Motors stock for Maxwell-Briscoe, saying that this bloc would have been worth $200,000,000 later and in the meantime would have paid $35,000,000 in dividends

But there was no mention of the incorporation of General Motors, and no wonder, since the original incorporation was for only $2,000, a nominal sum soon increased.

Except in Flint, the fledgling corporation seems to have attracted little attention at the start. Horseless Age, then the leading journal of the automotive trade, gave no extensive notice until December 30th, when it was stirred by the Olds negotiations to publish the following:

> THE GENERAL MOTORS COMPANY LAUNCHED
> The formation of the General Motors Company as a New Jersey corporation, and offers of an exchange of stock made to stockholders of the Olds Motor Works, of Lansing, Michigan, have started anew rumors of the consolidation of automobile manufacturing interests. Considerable secrecy is maintained in regard to the new company, which is said to be capitalized at $12,500,000, divided into $7,000,000 Preferred stock and $5,500,000 Common stock, each share having a par value of $1. At present the new company is said to embrace the Buick Motor Company, of Flint, Michigan, and the Olds Motor Works, of Lansing, Michigan. Reports from Lansing, Michigan, state that the stockholders of the Olds Motor Works have been approached with a proposition to exchange their shares for shares of the General Motors Company, the basis for the exchange being $400 in Preferred stock and $100 in Common stock of the General Motors Company for each $1,000 of the Olds Motor Works stock. There is said to be considerable opposition on the part of some Olds stockholders, although it is said that holders of three-fourths of the stock have assented. Attorney C. R. Hatheway, of the law firm of Ward, Hayden and Satterlee, is said to be engineering the deal, but refused to make any statement at present. The General Motors Company has an office in the Terminal Building, Forty-First Street and Park Avenue, New York City, but the manager was said to be "extremely busy" and could not be seen. The plan is said to be to continue the different works as at present under their proper names, assigning to each the manufacture of certain types of vehicles. The General Motors Company will act as a holding concern, and appoint the directors of the subsidiary companies.

On September 29th General Motors bought from W. C. Durant 18,870 shares of Buick Common and 1,130 shares of

Preferred at $150 a share, payable two thirds in General Motors Preferred and one third in Common. The new holding company purchased the W. F. Stewart Company body plant, directly across from Buick for $240,000 in stock, and leased it to Buick. Then began the Olds negotiations, detailed elsewhere, which were the first notices given the automobile world outside of Flint that a colossal merger had been begun.

When Oakland entered General Motors within the month, the automobile world did not need to be reminded again that here was something worth watching. Buick, Oldsmobile, and Oakland three of the largest plants and three of the "best names" in the industry had been brought together in less than three months. Further acquisitions soon followed. A few notes amplifying the minutes of the young corporation indicate how simply the three original companies were "digested" and more companies added:

> February 23, 1909, General Motors declared its first semi-annual dividend of $3.50 on its Preferred stock a dividend never missed from that day to this. This dividend was payable April I on stock of record, March 20.
>
> On June 5, the directors heard the good news that all Buick Common stock had been converted into General Motors stock on the agreed basis of $150 a share, payable two thirds in Preferred and one third in Common, and also all but fifteen shares of the 5,000 outstanding Buick Preferred, a remarkable example of stockholder confidence in a new enterprise. On September 10 the directors voted to increase the capital stock to $60,000,000 200,000 shares of Preferred and 400,000 shares of Common.
>
> The Olds union was completed by the purchase of the Seager Engine Works in September and the last of Oakland's stock passed to General Motors about the same time, though the price of the last purchase rose considerably, as the directors authorized the final purchase at $30 whereas the first blocks had been secured at $10 and $11 a share.

In what must be reckoned a terrific outburst of corporate energy, the following units were added to the General Motors family, either through purchase of all stock or a substantial interest, in less than two years:

Champion Ignition Co., Flint (later AC Spark Plug Co.)
Weston-Mott Company, Flint
Reliance Motor Truck Co., Owosso
Rainier Motor Co., Saginaw
Michigan Motor Castings Co., Flint
Welch Motor Car Co., Pontiac
Welch-Detroit Co., Detroit
Cadillac Motor Car Co., Detroit
Jackson-Church-Wilcox Co., Jackson
Michigan Auto Parts Co., Detroit
Rapid Motor Vehicle Co., Pontiac
Cartercar Co., Pontiac
Ewing Automobile Co., Geneva, Ohio
Elmore Manufacturing Co., Clyde, Ohio
Dow Rim Co., New York City
Northway Motor & Manufacturing Co., Detroit
National Motor Cab Co.

Also stock interests were taken in Maxwell-Briscoe, the United Motors Company and Lansden Electric.

By the end of 1909 General Motors had acquired or substantially controlled more than twenty automobile and accessory companies. Two other prospective purchases narrowly missed fire Thomas and Ford. October 26th, A. H. Goss offered to sell to General Motors the entire capital stock of the E. R. Thomas Company of Buffalo, and on the same day the directors authorized the purchase of the entire capital stock of the Ford Motor Company of Detroit for $8,000,000 $2,000,000 down and the balance in one and two years, "if arrangements can be made to finance."

Fateful "if!" Apparently arrangements could not be made to finance. Mr. Ford held out for cash, the young holding company did not have the cash and bankers could not be found who would risk $8,000,000 on an industrial property which has since become one of the most valuable in the world.

> Henry Ford and James Couzens were at the Belmont Hotel, in New York City [writes T. F. MacManus in Men, Money, and Motors]. They were together in their room and Ford, suffering

> from lumbago, was lying on the floor because there was no comfort for him in bed. The telephone bell jangled and Couzens, picking up the receiver, heard a voice:
> "This is Billy Durant. I'd like to see you."
> "What about?" "I can tell you when I see you."
> Returning upstairs within a half hour Couzens ... said:
> "Billy Durant wants to buy the Ford Motor Company."
> "How much will he pay?"
> "Eight million dollars."
> "All right. But gold on the table!" snapped Ford.
> "How do you mean that?"
> "I mean cash."
> Durant came back the following morning, and Couzens delivered Ford's message. The two men shook hands on the proposition and Durant left to raise the necessary money. With him he had a resume of the Ford business and its prospects. The capital stock of the Ford company was $2,000,000 and it had a surplus of $1,180,000, exclusive of goodwill and patents. Its earnings in 1908 were $2,684,000 on a business of $9,000,000 and it was planning to do a business of $15,000,000 in 1909 including an output of 21,000 automobiles. Durant had reached an agreement with Couzens to pay $2,000,000 in cash and the balance in one and two years.
> On October 26, 1909, Durant received the sanction of his board of directors for the acquisition of the Ford business, went back to his bankers and was informed:
> "We have changed our minds. The Ford business is not worth that much money."

Yet, as between the purchase of Ford and the purchase of the E. R. Thomas Company, both of which failed to materialize, probably the latter would have attracted the more public attention in 1909. Mr. Ford had not then become the public figure he is today, and the Thomas Flyer was a phrase on the lips of everyone, as the nation had followed its victorious race from New York to Paris, three quarters of the way around the world, via Siberia, a contest waged on three continents and over wretched roads, the sternest test met by an American car up to that time.

General Motors swallowed so many companies in its first two years that acute indigestion followed as a matter of course. Since the whole industry was young in engineering experience, Mr. Durant bought certain companies of unproved productive power simply because they were believed to have basic patents or features which might be important later. Cartercar, for instance, was purchased while the selective transmission and the friction drive were still in rivalry. The possibility that the latter might develop made the Carter patents seem worthwhile. Elmore, too, was thought to have basic patents on a two-cycle engine which might prove valuable. Other "buys" were made to assure regular supplies of material and parts.

Some of the purchases were soon seen to be untenable. Welch of Pontiac, for instance, entered the General Motors family on June 5, 1909, and was on its way out within five months, decision to sell being taken on October 26th. Welch-Detroit remained longer, though it never could be brought to a profitable position. The Welch brothers, both noted engineers, had developed a large, heavy car of advanced design, which was expected to take place as the price-leader of General Motors. But Fred Welch, the driving force in the enterprise, was drowned in Lake St. Clair, and without him the operation languished to the point where it had to be cut adrift.

Others could not be brought quickly to the point of carrying themselves and required capital outlays burdensome to the holding company. Ewing, Elmore, Cartercar, Rainier, and Dow Rim Company are examples of projects that cost General Motors more than they returned. No doubt, several of these properties were "unloaded" on General Motors in the rush of 1909, but the sellers probably did not wait for the enormous profits which would have been theirs if they had held for a long period the securities they then received. For instance, Dow Rim Company was purchased on November 27, 1909, for $28,000 in General Motors Preferred and $20,000 in notes, Mr. Alexander Dow electing to take notes instead of an amount of Common stock which later had a value, according to his own estimate, of several million dollars. Dow Rim was written down to $i on the company's books in 1911.

Mr. Goss, whose effort to sell the E. R. Thomas Company to General Motors came to naught, sold the Elmore Manufacturing Company of Clyde, Ohio, to the Company for $600,000 in General Motors Preferred at par. Mr. Goss was later active in Buick and General Motors affairs. Another addition destined soon to fall by the wayside was the Ewing Automobile Company of Geneva, Ohio, purchase of which was authorized on October 26, 1909, and written down to $1 on June 14, 1911.

Less disastrous but nevertheless troublesome was the Rainier purchase. The big Rainier car not having found a market, the directors on March 29, 1909, incorporated the Marquette Motor Company in Michigan for $300,000 to take over all the Rainier assets, the idea being to use the Saginaw plant for the production of a light Marquette car. Some of the difficulties encountered there will be related in another chapter.

Net sales of General Motors, by all its units, for the first year of operations, ending October i, 1909, were $29,029,875, and net income available for dividends, $9,114,498, a truly impressive record which was the result of the combined labors of 14,250 persons. No wonder General Motors was hailed at the end of its first year as the lustiest industrial infant ever born in America, and a stock dividend of 150 percent on the Common stock, declared November 15th, seemed to be entirely justified.

The directors at their memorable meeting of October 26, 1909, found that the Company had four reliable producers. Buick and the newly purchased Cadillac could certainly be classified as dependable; Oakland was also doing well and Olds had excellent prospects. The situation seemed rosy in the extreme, but within a few months changed rapidly for the worse.

In spite of substantial earnings and encouraging stock sales, a cool survey would have revealed the dangers of the situation. Because of the rapidity with which the various units had been gathered into the fold, there could be no certainty of uniform bookkeeping as yet and probably no exact knowledge of inventory commitments. Consequently, the liabilities were very likely larger than appeared and the surplus smaller. Furthermore, of course, plant and material assets could not be made liquid easily. Nevertheless, there was the comforting thought that the company now had stock to sell, the authorized capitalization having been increased on September 10, 1919, to $60,000,000; 200,000 shares of Preferred stock and 400,000 shares of Common, all $100 par. The directors' faith in Mr. Durant's ability to sell stock was enough to keep the company on its way buying and building.

In January, 1910, the directors authorized purchase, up to $400,000, of shares of the Randolph Motor Car Company, a commercial vehicle manufacturer, and added $100,000 to its capitalization.

At the same meeting a fateful decision was reached to invest in Heany electric stocks by trading in 8,290 shares of General Motors Preferred and 74,775 shares of General Motors Common. John Albert Heany claimed to be the inventor of the modern tungsten filament electric light, which, in fact, worked a revolution in illumination. Unfortunately for General Motors, his patent claims were not upheld by the Patent Office. Even if all had been plain sailing, buying into the Heany companies was buying into a lawsuit. On May 5, the Company plunged deeper into the Heany venture.

One of the historic moves of the Company in 1910 was the incorporation in Canada for $1,200,000 (12,000 shares) of McLaughlin Motor Car Company, Ltd. At Oshawa, cooperation between the McLaughlin carriage interests and Buick had brought Buick profitably into the great Canadian market. Of the 12,000 shares in the new McLaughlin company, 5,000 were taken by Buick. Out of this company General Motors of Canada, Ltd., would in time emerge.

The faster a man or an institution travels the more serious is a stumble. To maintain the tempo of that surprising period both a high ratio of earnings to sales and capital inflow were necessary. Michigan had supported the enterprise loyally with stock subscriptions, but Michigan was not then a rich state in liquid resources; the cream of capital had already been skimmed. Mr. Durant had to go further afield for money. While Buick, Oakland and Cadillac were making substantial earnings and Olds was being revived, various of the lesser companies were losing. General Motors was increasing its volume but not its earnings, yet expansion had been predicated on more profits. By May, 1910, Flint knew something had happened when new construction suddenly stopped at Buick and lay-offs became chronic. Nevertheless, the greatly daring chairman of the executive committee reported on June 21st to the directors that he had certain negotiations with Willys-Overland Company under way, looking to acquisition. The directors must have gasped when they heard that, for General Motors now needed to be saved from the penalties of too rapid expansion rather than be expanded still farther.

Mr. Durant rallied to the feat of saving his great enterprise with his usual intrepidity. He dropped everything else to search for money. Until this pinch, he had perhaps not taken bankers quite as seriously as business leaders must take the guardians of public funds. It would be pleasant now to have behind him the inexhaustible resources of J. P. Morgan & Company, a connection which had been possible at an earlier stage.

The situation had changed with such rapidity that adequate relief was hard to find. In talks to bankers, the extraordinary earnings of Cadillac were cited as contributing $2,000,000 to General Motors coffers within the year after its purchase. Always the reply came back that Cadillac could have the money if it stood alone, but as it could not be unscrambled from General Motors and since this was intended to be a loan to General Motors it was, therefore, not forthcoming. Even Cadillac itself was in danger. An emergency joint meeting of the boards of the First National and Old National banks of Detroit loaned Cadillac $500,000 a few hours in advance of a pay roll which otherwise could not have been met. This whole episode reveals an astonishing degree of banking blindness both as to the safety of a specific loan and as to the possibilities of the industry. Modern practice accepts the earnings of subsidiaries as the element properly to be considered in establishing the credit of a holding company. Bankers now understand, also, that loans to going industries are part of their own social responsibility, since such loans preserve employment and community values. The fact that Mr. Durant had prophesied that 300,000 automobiles would be produced annually in America had been a shock to many conservative bankers, who began to prophesy doom for the Republic if that incredibility should come to pass. The automobile "craze" was likened to the bicycle "craze" of a few years before, the passing of which had brought financial disaster to many companies. It was in the face of such hostility that the young General Motors Company went out hunting for credit.

Buick pay rolls were also troublesome. Harry K. Noyes, its Boston distributor, saved the day for Buick by expressing to Flint suitcases full of specie and currency. Mr. Noyes also collected company money, banked it in his own name and remitted direct instead of through banks, so that the money would be available for pay rolls: otherwise it might have been applied on an overdraft.

In desperation Mr. Durant went West after the East failed him. The West must help him, he thought, because the West needed motor cars; it was in the West that he had built up Buick trade so phenomenally. But Kansas City, St. Louis and Chicago bankers declined to aid. Returning discouraged from this journey, Durant and Goss picked up A. B. C. Hardy in Chicago. The latter relates the following as evidence of his hero's sunny courage under pressure:

The train stopped in Elkhart, Indiana, in a pouring rainstorm. Far down the dark and dismal street shone one electric sign BANK. Durant shook Goss, who was dozing dejectedly in a corner. "Wake up, Goss," said the leader. "There's one bank we missed."

The Chicago negotiations almost saved the day. A loan of $7,500,000 was arranged with the Continental and Commercial Savings and Trust Company. Then it was raised to $9,500,000. Then it was rescinded. The longer the situation waited the worse it grew, partly because decreasing confidence brought demands, partly because investigations revealed growing liabilities, a condition easily explainable in view of the independent operations of the various units and the lack of uniform accounting. One item in the minutes makes it quite clear, for instance, that the directors, on September 19th, did not know how much money Oakland owed. By the time $7,500,000 had been borrowed, $9,500,000 was needed, and by the time $9,500,000 was available, $12,000,000 was needed.

Retrenchments were hurriedly begun; the number of employees declined by 4,250 to a total of 10,000. The irony of the situation lay in the fact that net sales were nearly double those of the preceding year, reaching a total of $49,430,179. But the business had been done at a closer margin, net income being only $10,225,367 or slightly more than $1,000,000 above 1909. The rate of profit had declined while commitments had increased; between the two, General Motors was caught as in a vise.

JAMES J. STORROW

By September nearly all those interested had given up hope, but Mr. Durant saw one avenue of escape. His previous efforts to raise money had been based on a continuance of his management. In the East he found that he could save the company if he stepped out. J. & W. Seligman & Company of New York and Lee, Higginson & Company of Boston, were interested on that basis, not otherwise. The money centers of the East were conservative; Mr. James J. Storrow of Lee, Higginson, who arranged the rescue, represented a school of financial thought which could not approve the dashing methods and hairbreadth adventures which characterized General Motors in its infancy. With Mr. Albert Strauss of Seligman's he decided, further, that the situation had gone so far that not only a change of management but also a trusteeship was required. The terms were hard, but in no other way could a receivership be avoided, so pressing was the need for funds. Mr. Durant swallowed his pride and stepped down.

The arguments against receivership were of a sort to appeal to both the bankers and Mr. Durant. Times were good but not too good. A crash of such proportions might have unsettled the country, and surely would have been disastrous in areas where General Motors operated, especially in Pontiac and Flint which had been growing rapidly on the basis of Buick and Oakland expansion, and where even the shadow of a shut-down already had created distress and political unrest. Mr. Durant was quite aware of these local considerations which weighed heavily on the citizens of Michigan who had backed his enterprise. The trusteeship seemed to be the only way out.

On September 9th, when the directors rescinded the Chicago loan, they authorized the sale of the Michigan Auto Parts Company and Welch-Detroit, probably on the hint that a disposition to trim ballast would be acceptable: ten days later they were willing to sell Marquette. Then, on September 26th, after three weeks of anxious negotiations, the board authorized the loan of $15,000,000 from J. & W. Seligman & Company of New York and Lee, Higginson & Company of Boston.

The situation proved to be better than the prospect, and only $15,000,000 of the $20,000,000 authorized 6 percent notes went out. The company executed a blanket mortgage of all its Michigan property through the General Motors Company of Michigan, a corporation set up to hold title. As part consideration for the loan, $4,169,200 in General Motors Preferred stock and $2,000,000 worth of Common, both at par, were delivered to the bankers. The company received $12,750,000 in cash in return for its notes and stock given under the loan contract, dated November n, 1910. Preparing these documents and the exchanges necessary under them required several weeks, but banker control can be considered effective in the interim, although the new board of directors did not take office until November 15th.

On that day the following directors retired from office:

A. M. Bentley of Owosso
E. R. Campbell of Flint and New York, son-in-law of W. C. Durant
Wm. M. Eaton, also president
Harry G. Hamilton of Pontiac
C. R. Hatheway of New York, also secretary and treasurer
Henry Henderson of Scotch Plains, New Jersey
Schuyler B. Knox of New York
W. C. Leland of Detroit
R. S. McLaughlin of Oshawa, Ontario
W. J. Mead of Lansing
John T. Smith of New York City

Thus ended, after less than two years of phenomenal growth, the first phase of General Motors history. The young giant, like so many young giants, had overreached and come to grief. Many thought General Motors would never rise again. Saddled with what seemed at that time an unpayable debt, and all its property pledged as security, there remained only the intangibles of good-will and a strong demand for cars, to give the Common stock any value whatever. To turn these factors to substantial account was the task of a new management inexperienced in automobile manufacturing and merchandising, though expert in finance and well founded in business.

The glamor of bold dreams and brave deeds marked the first phase: steady aim and calm judgment were to mark the second. General Motors paid a stiff price for the lifesaving funds made available by the bankers. If the latter had retained until 1929 the $2,000,000 worth of Common stock they received in 1910, its value would then have been more than ninety times the face value of the loan. The 1910 General Motors loan revealed on a large scale the high cost of financing automobile enterprises.

Viewed in retrospect, the 1910 loan seems to have been unreasonably costly. Where $8,000,000 might have pulled the Company through, the latter had to pay more than 7 percent on $12,750,000, the bonds having been taken at 85, with the utmost security given to the lenders under mortgage and trust deed by a company with substantial earnings and excellent prospects. In addition, General Motors had to repay $2,250,000 which it never received, and paid a large bonus in stock. The transaction is hardly explainable except as placed against a background of extreme banking distrust of the automobile industry in general and also of driving personality behind the General Motors expansion program.

## Chapter IX THE BANKERS TAKE THE WHEEL

THE new management, taking hold in the autumn of 1910, faced a complex task. It must sell or liquidate subsidiaries which seemed either hopeless or unnecessary to the corporate operation as a whole, set up efficient management in weak plants worth retaining, command the loyal cooperation of the labor and management staffs, and integrate control over an organization built up on the basis of decentralization. When General Motors gave a blanket mortgage and a trust deed, its position as a holding company was at once invalidated. Since it, and not its subsidiaries, owed the $15,000,000, practical consideration enforced the exercise of more authority from headquarters, which were temporarily located on Woodward Avenue, Detroit, across from the Pontchartrain Hotel, a site now occupied by the National Bank of Detroit. About the middle of 1911, the Company leased the upper floors of the new six-story Boyer building at the southwest corner of Brush and Congress streets, directly south from the Wayne County Building.

The new board consisted of:

Anthony N. Brady of New York
Emory W. Clark of Detroit
W. C. Durant of Flint and New York
A. H. Green, Jr. of Detroit
J. H. McClement of New York
M. J. Murphy of Detroit
Thomas Neal of Detroit
James J. Storrow of Boston
Albert Strauss of New York
Nicholas L. Tilney of New York
James N. Wallace of New York

These gentlemen elected James J. Storrow as interim president while making up their minds as to his successor. Mr. Storrow, fortunately, was a man of great force and vigor, and, in addition, he was somewhat automobile-minded, continuing his interest in motors until his death many years later. Under him the work of reorganization got away to a swinging start. Almost immediately it was decided to dispose of Seager Engine, Welch-Detroit, Michigan Auto Parts, the National Motor Cab Company, and Ewing. The board acted with equal promptness to clarify the financial structure by purchasing stocks in various other General Motors enterprises which had accumulated in Buick's treasury during the hectic days when Mr. Durant was buying properties with any funds available.

Having weighed the situation, Messrs. Storrow and Strauss, whose union of interests dominated the board of directors, selected Mr. Thomas Neal of Detroit as president. Mr. Neal was president of the Acme Lead and Color Works, one of Detroit's successful industries, and had a high degree of business sagacity. Standish Backus, a young lawyer, became secretary and James T. Shaw, treasurer. It was a distinctly Detroit management. Mr. Durant was retained on the board and given the important post of chairman of the finance committee which he held for a year.

Banker control from the Atlantic seaboard had its discouraging features for the practical men who were bearing the company's burdens in plant and field. Banker Strauss desired to move machinery from the Welch-Detroit plant to the Marquette plant in Saginaw by water, having discovered that both plants were near the waterfronts of their respective cities. It took a statistical report to demonstrate to him that moving cargo from shop to shore would cost more than the direct rail haul from one factory sidetrack to another. For two years expansion had been the watchword, with efficiency of operation a minor consideration. Inventories were woefully out of balance; improper storage of supplies had caused great waste. In one plant thousands of tires were found exposed to heat and sunlight; in another valuable machinery lay rusting out of doors. Tons of unsaleable merchandise had been built to faulty specifications or of faulty materials. However, new managers, grappling with these discouragements, soon brought improved order and spirit into plant operations. These readjustments required, of course, considerable time, and an immense amount of patient, detail work on the part of plant managers and their staffs.

THOMAS NEAL

During the two flush years of General Motors life, manufacturing, sales, and accounting had not kept pace with growth of volume. The new management accordingly began to coordinate policies and to put into use standardized accounting and reporting. Henceforth, at least, General Motors would know how its cash position stood from day to day. It set up the General Motors Export Company, a Michigan corporation with $10,000 capital, to handle all General Motors products in the foreign field, this being perhaps the first operation in which General Motors presented a completely united front. A General Motors magazine, the Insider, made its appearance, the first of many corporation publications designed to spread the General Motors message among its own people and advance the solidarity of the enterprise. An experimental and testing laboratory was set up. New managers were installed in many plants. Economy became the watchword, and under that banner began thorough reform.

There was plenty to be done. In the mopping-up process the following central office projects were wound up:

General Motors Securities Company.
General Motors interest in United Motors Company sold to United States Motor Company, the rival holding company organized by Benjamin Briscoe. A little later the Company's interest in United States Motor Company was written down to $i.
Efforts to sell the Company's interest in Seager Engine failing, that asset was also written down to $1.
Ditto with the Maxwell-Briscoe interest.

These decisions took some time to mature, but in 1914, by the process of dissolving the subsidiary corporations, General Motors finally rid itself of Randolph, Peninsular, and Welch-Detroit. It sold the Reliance plant at Owosso to the American Malleables Company, and the Michigan Motor Castings Company was absorbed by Buick.

Elmore's capital stock was reduced from $60,000 to $6,000. As the spare parts business of both Elmore and Peninsular had been sold for $44,000, the Elmore write-off to $6,000 placed a bargain price on the Clyde, Ohio, plant. In five years the value of the Elmore investment had shrunk from $600,000 to $6,000 in the ratio of $100 to $1, at a time when the automobile business was generally prosperous and the Elmore car was considered thoroughly creditable. Actually, the plant sold for $50,000 when a buyer for it appeared some years later.

It will be noticed that in this period of retrenchment General Motors concentrated on Michigan producers, a trend applicable to the entire industry, as it advanced in age and capitalization. Although in later years the Corporation widened its activities geographically with excellent results, the great automobile producers in its orbit still have their chief plants and operating headquarters in Michigan.

Hopes being still held out for Rainier, General Motors increased its holding in the Saginaw company by $110,000 of stock, paying about half in cash and half in surrender of claims.

Full ownership of Rainier was completed on May 29th. Another move with enduring results was the incorporation of General Motors Truck Company on July 22, 1911, as a sales company to handle the product of Rapid Motor Vehicle Company of Pontiac, Michigan and Reliance Motor Truck Company.

The most hopeless task of all remained in the liquidation of the Heany assets. Mr. Durant, seeing the white light of truth where once he had seen only the rosy dawn of mythical profits, reported adversely on Heany at the meeting of May 15, 1911, and the board resolved to grant Heany no further advances. The Heany plant at York, Pennsylvania, was of small value compared with the prospective value of Heany's patents, now seen to be in jeopardy when carefully weighed in view of the General Electric suit.

The Heany fiasco cost General Motors enormously, estimates running from $5,000,000 to $12,000,000. Its denouement marked the withdrawal of Mr. Durant from the chairmanship of the Finance Committee on November 11, 1911, when he was succeeded by Mr. Storrow. Mr. Durant, although remaining on the board, was now entirely divorced from the management and devoted his energies to founding Chevrolet, on which he again rode into power four years later. He remained, however, a director.

General Motors voting trust certificates, the share for share equivalent of the deposited Common shares, were listed on the New York Stock Exchange on July 31, 1911, the first automobile securities listed by that body. That listing meant that stockholders and the public generally henceforth would have clear statements of the company's financial position at regular intervals, and it therefore became part of the management's plan to rebuild public confidence in General Motors' structure. At this time the net worth of the company was approximately $35,000,000, although the authorized capital was $60,000,000.

The banker management, even at that early date, caught the significance of the relationship of corporate financial responsibility to automobile sales. The automobile buyer, more definitely than the consumer in other lines of merchandise in general use, seeks reassurance of the manufacturer's stability, because the service he receives and the resale value of his car alike depend unmistakably upon the continuance of the manufacturer in business. General Motors' position in this respect suffered a body blow from the uncertainties reigning during the height of the busy selling season in the midsummer of 1910, and every effort to reestablish confidence was a step toward better sales. Banking opinion had been satisfied at once by the character of the new board, and this confidence worked its way down again to the sales staffs and the buying public. More efficient plant management was steadily improving the quality of the cars offered, and as the feeling grew that General Motors had escaped its Waterloo, business improved. The tide turned faster than the conservative men at the center of things fully realized. Although they were paying off the trust notes regularly and adding to surplus, General Motors stock was offered for sale to several of the officers in 1911 with no takers. But W. C. Durant watched the situation out of the corner of his eye, considered the stock ridiculously undervalued, and as he could find the means, began to buy back toward control.

One factor in the upturn was the growth of the export business. The capital of General Motors Export was increased in January, 1912, from $10,000 to $100,000. Another transaction of moment was the purchase of the remaining outstanding shares of Weston-Mott from C. S. Mott. This transaction which involved also 3,000 shares of Brown-Lipe-Chapin stock and several notes, totalled more than $2,000,000. Brown-Lipe-Chapin, manufacturing gear differentials at Syracuse, New York, later became a division of the Corporation.

General Motors of Canada, a sales unit, was born with a capital of $10,000 and General Motors completed its ownership of Cartercar. Great things were expected of Cartercar, a friction-drive car. It occupied the former site of the North and Hamilton carriage factory in Pontiac, and had as president George E. Daniels, who will be remembered as the first nominal president of General Motors Company and later head of Oakland. But Cartercar, in spite of its early promise, soon faded from view. The friction drive simply could not be popularized, and the chance for it had passed with the rapid improvements in gear shifting.

Engineering improvements came along swiftly, and General Motors' priority in some of these hastened the recovery. Buick introduced its first closed car a limousine in 1911. Other improvements were the electric horn, slip detachable rings, and the worm-gear drive for trucks. But the greatest mechanical triumph of the period was Cadillac's introduction of the Delco electric starter, described in the Insider in 1911, to become standard equipment in 1912. This epoch-making achievement opened a new market for automobiles by making motoring cleaner, safer and easier, and, in particular, met the driving needs of women, resulting in a striking market advance.[1]

The process of shearing off disappointing properties and consolidating others continued. Buick's Jackson plant was sold to Imperial Wheel, and Michigan Auto Parts went to Northway. Welch-Detroit, of which the Company had long been weary, went to the Peninsular Motor Company. Into the Peninsular shell had now been poured three companies Rainier, Marquette, and Welch-Detroit. It was a grand property physically but had little luck until it went finally to Chevrolet.

The 1911 figures reveal the young giant twisting in the coils of bitter circumstance but making headway. Net sales were off nearly $7,000,000 and net income cut to a third of 1910 profits only $3,316,251 was available for dividends and reinvestment. For a true comparison the latter figure needs enlarging by one fifth as in 1911 the close of the company's fiscal year was pushed ahead to July 31st. Even so, the shrinkage was large. Nevertheless, the preferred dividend was paid promptly, and as the Common

stock had never paid a cash dividend the company's credit held firm. Pay rolls began to creep up, the number of employees advancing from an average of 10,000 to 11,474.

Through this period of reorganization Buick had made extraordinary progress under Charles W. Nash and his works manager, Walter P. Chrysler. Both were extraordinary men,

---

[1] 'See Chap. XIX: "Research: the March of the Open Mind."

self-made, strong-willed yet versatile, possessing qualities of leadership, financial acumen, and shrewd market wisdom.

CHARLES W. NASH

Charles W. Nash had risen from manual labor to be general superintendent of the Durant-Dort Carriage Company. In Flint they say he began his rise by mowing a lawn so well that Mr. Durant noticed his thoroughness and gave him a steady job at the factory, polishing lamps at seventy-five cents a day. If Durant had taken Mr. Nash into Buick with him in 1904, the whole history of General Motors might have been otherwise, but Nash remained with Durant-Dort. However, Durant recommended Nash to James J. Storrow for the Buick post in 1910. That Mr. Storrow accepted this suggestion proves his openness of mind, for Mr. Nash lacked automobile experience, and even those who knew and respected his ability wondered how he would fare in so novel an assignment. But if "Charlie" Nash did not know motor cars, he knew men and he knew factory organization. In a matter of months after taking hold, Buick was functioning efficiently, rolling up a record so outstanding that in November,

1912, he was the unanimous choice for the presidency of General Motors.

Mr. Nash was fortunate in having Walter P. Chrysler with him during this period. Mr. Chrysler's rise is one which richly deserves all the notice it has received, but the story is so well known that it need not be repeated here. A product of Western railroading, he was brought by Nash to Buick from the American Locomotive Works at Pittsburgh. A genius at plant organization, he soon lifted that burden from Mr. Nash's shoulders, and when Mr. Nash became president of General Motors, Mr. Chrysler succeeded to the chief responsibility of Buick, there to remain until 1920.

In the nearly four years of the Nash leadership, General Motors made a remarkable financial "comeback." Good earnings enabled the Company to anticipate notes, add to surplus and enlarge plants. For 1912 earnings held steady, but in the next year more than doubled on a 30 percent increase in sales, sure evidence that the reorganized General Motors was finding itself as an efficient manufacturer.

But an even more striking index of growing efficiency is to be seen in the ratio of net sales to number of employees. In 1913 each employee of General Motors produced on the average $4,236 in net sales value; in 1914, when there was a considerable drop in employment but only a small drop in sales, the average was $6,037.

From 1913 to 1916 inclusive General Motors earned, available for dividends, nearly $58,000,000, doubling its profits each year from 1914 to 1916, as follows:

| | |
|---|---|
| $ 7,249,734 | in I914 |
| $14,457,803 | in 1915 |
| $28,789,500 | in 1916 |

While this astonishing earning power reflects the general prosperity of the country in the "munitions boom," following the onset of the World War in Europe, it also reflects the outstanding fact that General Motors plants and sales systems had been thoroughly reorganized and were ready to go ahead with the general improvement of business.

WALTER P. CHRYSLER

During these three busy years net sales rose from $85-373,303 in 1914 to $156,900,296 in 1916 and the amount reinvested annually in the business from $6,201,055 in 1914 to $17,010,437 in 1917, while the average number of employees rose from 14,141 to 25,666.

At the beginning of this run of luck and good management combined, the price of the Common stock as represented by voting trust certificates still hung around $25 a share. Common stock began to appreciate in the spring of 1914, doubled in value before the New York Stock Exchange closed at the opening of the World War, and gained persistently thereafter, eventually reaching a high of $852 a share. When the present Delaware Corporation was formed in 1916 it exchanged five shares of the Common stock for one share of the Common stock of the New Jersey Company. Owing to stock dividends and divisions, each share of the old Company Common equals roughly 100 shares of Corporation Common and was worth in November, 1933,

$3,000. At the peak price of 1929 one share of the old Company Common was worth more than $9,000.

There was no headlong expansion, however, on the basis of improvement. The only substantial capital increase in 1913 was that of General Motors Truck Company capitalization from $10,000 to $250,000. Elmore, Peninsular and Welch operations were wound up. With their passing the General Motors passenger car units had been whittled down to Buick, Cadillac, Oakland, and Oldsmobile, and of all the truck units absorbed by General Motors in its period of great expansion only one remained: General Motors Truck Company.

General Motors had developed a clear outline by the time the banker reorganization was complete. It would grow tremendously, of course, and add many units, but they would be chiefly accessory and parts companies supplying the great automobile manufacturing establishments which dealt directly with the public. Never again would General Motors give substantial evidence of a desire to acquire a leadership in the industry based merely upon buying competing companies. Fear of monopoly aroused by swift early expansion had been allayed and never reappeared. Henceforth General Motors would strive for a larger share of the automobile business of the country, not by buying up competitors, but by improving the output of its four great producers, to which Chevrolet would soon be added as a result of an extraordinary situation which no policy could foresee. This in the passenger car field; as other opportunities developed, purchases of corporations in those fields were made and, of course, the whole supply and service situation has been rounded out.

The point is that the general plan of 1913, which ultraconservative men worked out under the pressure of necessity, still stands, though still subject to revision as circumstances dictate. General Motors' evolution from a holding company to an operating company, while not completed for several years, was begun under banker control, and the greater part of its expansion since has been through the growth of its leading component divisions rather than by purchase of rival companies. The country consequently no longer thinks of General

Motors as a horizontal trust capable of infinite expansion until it dominates the essential industry of vehicular transport, but rather as an organization with certain trust features of advantage but still subject to the restraints and incentives of competition.

As it stands, General Motors is neither a vertical trust nor a horizontal trust, but in adapting itself to large scale production in a land where competition is estimated the life of trade, it has taken on certain aspects of both horizontal and vertical organization, an adjustment dictated by the necessities of cooperation on the one hand and by a determination to avoid even the appearance of monopolistic intent on the other.

For the beginnings of this sane evolution the bankers who ruled General Motors from 1910 to 1915 are responsible. How that control passed, amid scenes and circumstances of intense drama, will be told in the two chapters to follow.

# Chapter X

## CHEVROLET: THE CINDERELLA OF MOTOR-CAR HISTORY

THE rise of Chevrolet is one of the epics of American industry, with almost fairy-tale aspects. Like Cinderella, Chevrolet rose from the ashes, went to court, was cheered by the populace, married and lived happily ever after.

We have seen that Buick began its Flint career at the western end of Flint in the Flint Wagon Works Plant; also that while new buildings rose in northern Flint to house the amazing growth of that company, only one Buick structure stood in western Flint. Buick motors were still being made in this old or No. 2 Plant, while across the tracks Flint Wagon Works engaged in turning out a few of the Whiting cars. All of these buildings have since been demolished to make way for larger plants of Chevrolet.

In 1909, Louis Chevrolet, bearing a name destined to become a household word in America, was experimenting on a small car in Detroit and also with a six-cylinder one. He was being financed by W. C. Durant who had employed him as a racing driver several years before. Chevrolet was a picturesque figure in his racing days, as he brought Buicks down the stretch with his huge Gallic mustache blowing in the wind. When Louis and his brother Arthur came to Flint, Mr. Durant tried them out on a small dirt track maintained by Buick in the rear of its factory. Louis won the contest, and was chagrined when Mr. Durant immediately selected Arthur as his personal chauffeur on the ground that Arthur had lost by taking no chances while Louis

Chevrolet had won by taking all chances. Both Chevrolets became members of the famous Buick racing team which contained Bob Burman, Louis Strang and other celebrities. But Louis Chevrolet had talents beyond mere speed; he had ideas about design. He considered that America, with improving roads, would receive favorably a light car which, like the French favorites of the day, combined beauty with modest price. This idea appealed to Mr. Durant who accordingly backed the Chevrolet experiments in Detroit.

The first Chevrolet, built in April, 1913, has traveled 237,462 miles

In 1910, Mr. Durant lost control of General Motors. In the retrenchments which followed, Buick No. 2 Plant was abandoned and motor operations moved to the north end of Flint. Flint Wagon Works had discontinued manufacturing the Whiting car, and as the carriage business showed a steady decline under the competition of the rising automobile trade, Mr. Whiting was anxious to liquidate his company. The opportunity came. Proving that one setback was not enough to keep him out of the automobile industry, Mr. Durant bought the Wagon Works property, plant and contents, and organized the Little Motor Car Company to occupy the property. The company was named for William H. Little who had been general manager of Buick under Mr. Durant. Mr. Little, Charles M. Begole, and William S. Ballenger signed the articles of incorporation as of October 30, 1911. Capital stock authorized was $1,200,000 $700,000 Common and $500,000 Preferred. Stock subscribed was declared to be $615,600 Common, and $207,600 Preferred. Only $4,827.42 is represented as paid in cash, the balance being paid in real estate and materials, inventory, etc., the value of the Wagon Works plant being set up at $350,000. The Preferred stock was paid in exchange for Wagon Works machinery, equipment, and accounts and bills receivable. Little, therefore, started from scratch, with an old plant full of horse-drawn vehicles and mortgaged to Mr. Whiting for $200,000. When it is recalled that Little became the foundation of Chevrolet, and Chevrolet grew to larger volume than any other motor manufacturer in the world, industrial history has nothing to match the march from poverty to power.

Little Motor Car Company at first built only one model, a small four-cylinder runabout, selling around $650, which placed it in competition with Ford. It was attractively designed, but weakly powered.

The Wagon Works plant was in confusion. The problem facing the management was selling a large inventory of carriage parts and work in process at the same time that it rehabilitated the property for automobile production. In the meantime Louis Chevrolet continued in Detroit his experimental work on four- and six-cylinder cars, the "six" being known as Model C which was finally accepted as ready for production. A plant was rented on West Grand Boulevard, and there the first Chevrolet cars were produced for the market.

On November 3, 1911, the Chevrolet Motor Company of Michigan was organized with Louis Chevrolet, William H. Little and Edwin R. Campbell as incorporators. Up to that time, Chevrolet had built only four or five experimental cars, and the new company's production in Detroit was never large. A sensation was created in Detroit real estate circles when vacant property was purchased directly across from the Ford factory and a large billboard was erected saying that a complete modern plant would be built on that location. This was never done, however, as it was soon seen that quantity production could be more economically effected in Flint where considerable strides had been made.

The most important of these was the incorporation of the Mason Motor Company on July 31, 1911, with Arthur C. Mason, Charles Byrne and Charles E. Wetherald as incorporators. Followed by a large number of former Buick employees, with

Chevrolet "Royal Mail" roadster, 1914

amazing speed they began motor production in the old Buick No. 2 Plant rented for the purpose. There, as contracted for, the motors for both the Little and Chevrolet cars were built by the Mason Motor Company. Mr. A. B. C. Hardy succeeded William H. Little as general manager of the Little Motor Car Company at Flint on January 21, 1912. Mr. Hardy found the plant still full of everything for farm use from wagons to whip sockets.

The Little mortgage was very tightly drawn. It had to be reduced in certain proportions as any building or machinery item was changed or moved, with the Whiting interests consulted at each step. Mr. Hardy says that only $36,500 in new cash was put into the Little Company of which $10,000 represented the claim which the Weston-Mott Company had against the Wagon Works for axles.

> Mr. Mott's willingness to take our stock was a real help and gave us all courage. As part of the deal, we were bound to pay off all debts of the Wagon Works, so the problem I faced was that of turning a set of old fashioned carriage buildings into an automo-

> bile plant on only $26,500, getting the rest of the money out of the sales of horse-drawn vehicles and their parts. For the first year we were in several kinds of business at the same time.
>
> Our plan was simple to the point of innocence. We took the small motor, had it revamped and improved by Mason and put it into a small roadster for sale for $650 which was somewhere near Ford's latest price. The Little had good lines but was underpowered. Nevertheless, we sold 3,500 Littles in my first year and also made up and sold several thousand buggies, many thousand sets of wheels and a raft of miscellaneous carriage parts and accessories. All this in addition to putting the plant in fair shape for the production of automobiles. We made all of our bodies and most of our sheet metal parts.

Mr. Little spent most of his time in Detroit where Chevrolet was getting into small production. Chevrolet sales offices were opened in Chicago, Philadelphia and Boston, these three being the first of many retail stores later operated by Chevrolet. Its large six-cylinder car was priced at $2,500 and upwards. The capital for the Detroit operation came largely from material and parts suppliers. The plant in which Chevrolet operated on West Grand Boulevard in Detroit had been taken over from Thomas Corcoran, a lamp manufacturer of Cincinnati. For those days it was an exceptionally good plant, even luxurious from the standpoint of offices, and there Mr. Durant made his headquarters for a considerable time.

The production record hung up at the Little plant in Flint, however, far exceeded what had been done in Detroit. The Flint enterprise actually was making money, while the Detroit operation was losing it. Consequently, Mr. Durant planned a much heavier schedule for Flint during 1913 based on a production of 25,000 four-cylinder Littles and the introduction of a six-cylinder Little ; but the 25,000 figure was reduced to 5,000, and orders were placed for materials on that basis. A small assembly plant had been leased at the corner of 57th Street and nth Avenue, New York City, to which parts were sent to be assembled. The matter of financing these shipments was disputed between the Flint and Detroit organizations which were separate corporations with different sets of stockholders. Messrs. C. M. Begole and W. S. Ballenger of the Wagon Works remained active in the Little Company, which succeeded in paying off its floating debts, reducing its mortgage very considerably, and building up a small surplus not enough, however, to finance extensive shipments to New York and Detroit.

The logic of the situation indicated the advisability of concentrating the manufacturing at Flint. Consequently, the Chevrolet Motor Company of Michigan discontinued operation in Detroit and moved from there in August, 1913. For a time Chevrolet occupied the old Imperial Wheel Company plant on property now owned by Buick. General management of both the Little and Chevrolet companies was given to A. B. C. Hardy. All former sales offices were retained. Mr. Durant further consolidated his Flint position by extending his interest in the allied Mason Motor Company.

The year 1914, which was to inaugurate the World War and witness the revival of general business activities in the United States, found both the Little and Chevrolet cars in production in Flint. Chevrolet introduced its famous Baby Grand touring car and Royal Mail roadster. The organization had become highly efficient through the necessity of making its own way on very little capital. The used car problem had not yet developed and instalment selling was still in the future. Chevrolet could sell for cash every car it produced. So great was the demand that if a shipment was not taken off the railroad track promptly by the consignee, someone else in the community could be depended upon to lift it without delay. As the money rolled in, Mr. Durant transferred his offices from Detroit to New York to push sales from that center and also to work out a plan which he had formed for recovering control of General Motors. The latter had snapped back into prosperity under the Nash management. Profits were large, interim payments on its funded debt were made promptly, and large reserves accumulated. Yet, no one of the Eastern bankers seemed to appreciate the full potential value of General Motors as thoroughly as did Mr. Durant. Chevrolet expansion went on at top speed. During 1915 the Little Motor Car Company discontinued production. Following an exchange of real estate, all Chevrolet operations were concentrated in the western end of Flint. The Baby Grand touring and Royal Mail roadster, volume cars from the beginning, ran into large production.

Model 490, Chevrolet the basis of the present Chevrolet

The selling force took decisive steps forward. In 1914 a wholesaling organization had been opened in Oakland, California. In the following year wholesale offices, the first of the Chevrolet zones, were opened in Kansas City and Atlanta. Agreements were made with Russell Gardner, Sr., for the assembling and sale of the Model 490 in St. Louis territory, and with R. S. and G. W. McLaughlin for similar production and sale in Canada. Chevrolet had no financial investment in either company, but had a profit-sharing contract. Chevrolet entered the Eastern production field with the "490" by buying the Maxwell Motor Company plant at Tarrytown, New York, the New York City plant being continued for production of Baby Grand and Royal Mail models. Most of the assembly operations were set up with the assistance of local capital, with the result that the still rather small Chevrolet Company bloomed into a nationwide operation in a remarkably short space of time. Chevrolet Motor Company of Delaware, through which W. C. Durant regained control of General Motors by an exchange of stock, was incorporated in Delaware on September 23, 1915, with a capital of $20,000,000, raised to $80,000,000 the following

December. With very broad powers, including the right to hold, own, and deal in securities, it acquired all the stock of Chevrolet Motor Company of New York, Chevrolet Motor Company of Michigan, Chevrolet Motor Company of Bay City, Chevrolet Motor Company of Toledo, Ohio, Mason Motor Company of Flint, with contract interests in Chevrolet Motor Company of Canada, Ltd., Oshawa, Ontario, and Chevrolet Motor Company of St. Louis.

The officers were R. H. Higgins, chairman of the board; L. G. Kaufman, chairman of the Finance Committee; W. C. Durant, president; A. B. C. Hardy, vice-president; E. R. Campbell, second vice-president; W. C. Sills, treasurer; J. T. Smith, secretary. Among the directors were H. M. Barksdale of Wilmington, Delaware, a du Pont representative, and L. G. Kaufman, president of the Chatham Phenix Bank, New York City. Mr. Kaufman entered Chevrolet financial councils through his ready acceptance of Chevrolet loans at the instance of Nathan Hofheimer who had been a large stockholder in the Heany enterprises, and who followed Mr. Durant into Chevrolet.

The balance sheet of December 31, 1915, showed net working capital of $7,368,572 and capital stock outstanding of $19,752,300, which had been increased to $23,663,800 on February 29, 1916, as a result of the stock transfer previously outlined.

Before demonstrating his control of General Motors through Chevrolet, Mr. Durant offered a proposition to have Chevrolet taken into the General Motors family as a unit. That declined, a show of strength was necessary. Cinderella Chevrolet stepped forth in her new-won splendor to the applause of the multitude.

Chevrolet Motor Company of Michigan took over the capital stock of the Mason Motor Company. The year 1916 saw Chevrolet still expanding. A new assembly plant was built at Flint and another started at Fort Worth, Texas. The Mason Motor Company increased its motor manufacturing space at Flint. A new axle plant was also added. Two notable purchases had been made of properties which greatly enlarged Chevrolet's manufacturing resources. The National Cycle Company and the National Motor Truck Company at Bay City, Michigan, were purchased for small parts production and Warner Gear Company at Toledo, Ohio, for transmission and gear production. Retail stores were opened in many cities. Chevrolet assembly began at Oakland, California. A profit of $278,000 was derived from sale to the Ford Motor Company of Highland Park property bought five years before. Additions to the manufacturing, assembling, and sales facilities came fully in use in 1917 and made that year notable in spite of the difficulties which manufacturing in general encountered through the participation of the United States in the World War and the setting up of quotas on materials.

Chevrolet leased an eight-story office building at 224 West 57th Street (1764 Broadway), New York City, and moved the executive offices to that location on December I, 1917, later exercising an option to purchase. In preparation for its absorption into the General Motors Company, Chevrolet in 1918 began to buy the outstanding stocks held locally in its enterprises in various sections of the country, and their several production and sales agreements were cancelled out in this way at figures highly profitable to the local investor. In 1918 General Motors purchased all the assets and assumed all the liabilities of Chevrolet Motor Company (Delaware).[1]

In all American industrial history there has never been anything to equal the rise of Chevrolet from an experiment in 1910-11 to the position it held when merged with the General Motors Corporation in 1918. It had made $6,000,000 in six years and had amazed the automobile world by securing control of General Motors Company of New Jersey. Moreover, it had worked out a broadly based system of production and nationwide assembly upon which it could build its quantity leadership of the future. Starting with almost no cash and an antiquated plant, Chevrolet's swift and dazzling ascent to profit and power caught the attention of the country. How Cinderella Chevrolet came to the court of public opinion unheralded and unsung, to ride away in her coach-and-six to the cheers of the populace, will be told in the following chapter.

---

[1] for later history of Chevrolet, see Chap. XVI.

CHEVROLET

# Chapter XI THE CORPORATION ESTABLISHED

IN ITS lusty youth the automobile industry displayed strong resistance to the down swings of the business cycle. When banks were failing in 1907, and many sections had only scrip for a circulating medium, the automobile cities of Michigan hardly knew that distress was stalking the land. Pessimists were always talking of the "saturation point" but conveniently it kept receding.

Business in general began to slow down in 1913, but General Motors' reorganization had given the Company such momentum that its earnings kept rising. Likewise, Chevrolet passed swiftly through infancy into assured earning power. The situation had all the elements of drama; indeed, one could be sure of drama wherever William C. Durant concentrated his attention. On the one hand stood the organizer of General Motors, bereft of power in his old enterprise, but already in the field with a new, expanding venture, built on a "shoe string" in the sense of capital structure but paying its way, growing through earnings, the bright faith of its captain and his uncanny ability to find capital when it was needed. On the other hand stood General Motors, directed by more conservative men, with the power of ultimate decision resting in bankers more interested in getting the Company out of debt than in taking unnecessary risks.

Sometimes the conservative view is really the short view, the optimistic, the long view. It proved to be so in this case. With their eyes on the ledgers rather than on the future, the bankers never quite comprehended the potential value of General Motors. Having mopped up the debris of the initial expansion they failed to grasp the importance of what remained in General Motors after the house-cleaning had been done. Few realized that

General Motors' earning capacity far outran its debt commitments, and among those few was the man who had founded the company. W. C. Durant began buying General Motors.

In his market campaign for control from 1913 to 1915, Mr. Durant proceeded from the firm base of a large personal interest in the stock. He had sold none of his original holdings. Members of his family held large blocks. Other blocks were held by old friends and business associates who had clung to their holdings through the period of falling prices for the stock. Mr. Durant's liberality in buying companies for stock had endeared him to beneficiaries upon whose loyalty he could rely.

Altogether there was, both in property and sentiment, the basis for considerable buying, credit being comparatively easy. General Motors certificates trebled in price by the time the New York Stock Exchange closed at the outbreak of the World War and, when the Exchange opened four and a half months later, General Motors was one of the few stocks to have appreciated in the interval. Thereafter the stock rose in value rapidly as the munitions boom gathered headway. The World War played into Durant's hands by bringing new money and new men into the stock market, and by stimulating confidence in business. While it was generally understood that Durant was buying, the Street was skeptical how far he could go, and the bankers in control never seemed to have read the danger signals. Their resources, of course, far outweighed those of the opposition, if they had ever been mobilized for resistance.

General Motors in 1915 was a rich prize. The statement of July 31st, of that year, shows net working capital of $31,141,238, cash of $15,527,124, and net profits for the year of nearly $15,000,000. Only $2,328,000 of the $15,000,000 gold notes issued in 1910 remained unpaid, the mortgage debt having been reduced by approximately $1,000,000 in the fiscal year ending July 31, 1911; $1,500,000 in 1912; $1,500,000 in 1913; $3,000,000 in 1914; and $5,500,000 in 1915. By comparison Chevrolet was still a pigmy yet plans were afoot to have little Chevrolet gain control of the General Motors giant.

This, of course, had to be done carefully. The campaign was cleverly timed, and had support from sources highly respectable. Mr. Durant received substantial banking aid from Louis G. Kaufman of the Chatham Phenix Bank, New York. The open market was buying on a tremendous scale, as is evident from the high and low prices for the General Motors voting trust certificates representing Common stock. The range of prices was:

| Year | High | Low |
|---|---|---|
| I911 | 51¾ | 35 |
| 1912 | 42⅞ | 30 |
| 1913 | 40 | 25 |

Then the spread began, with a 1914 low of 37⅜ and a high of 99. Buying for control took the stock in 1915 from a low of 82 to a high of 558. But unlike so many other market movements, victory was followed by only a slight recession and the scoring of a new high figure low for 1916, 405; high, 850. No matter how rapidly Mr. Durant and his friends "bulled" the market General Motors stock remained worth what they paid for it. But, in all probability, no market campaign which was nothing more than speculative would have resulted in the recapture of General Motors control by its founder.

As part of the plan Chevrolet Motor Company of Delaware was organized for $20,000,000, but the papers were not filed until September 23, 1915. Chevrolet of Delaware was the horse Durant rode to battle. The word went out, quietly, that Chevrolet would be traded for General Motors at the ratio of five to onc,[1] and those who had followed Durant by investing early in Buick and General Motors began to send in or bring in their stock for transfer. Mr. Hardy relates how A. M. Bentley of Owosso brought in a large brief-case of General Motors certificates to the Chevrolet headquarters in New York and was willing to turn them in without even a receipt. Reliance had gone the way of many other promotions, but Mr. Bentley still had the General Motors stock with which it had been purchased. The close

[1] The offer held until January 25, 1916, when a change was made by which four shares of Chevrolet Common stock instead of five were exchanged for one share of General Motors Company of New Jersey Common stock

friendly relations Mr. Durant had established with the old stockholders, to many of whom he had sold stock personally, stood him in good stead now. With the value of their holdings increasing daily under his market generalship, with both Chevrolet and themselves rising richer each morning, they forgot the long wait for dividends and the temporary embarrassment of the company now happily cured.

In the final check-up, to be absolutely certain, Mr. Durant was himself one of four men who passed the certificates from hand to hand, each one calling out the names, numbers and holdings. So fortified, the certificates being brought in in baskets, Mr. Durant entered the stockholders meeting of September 16, 1915 on the seventh anniversary of incorporation absolute master of the situation. A dividend of $50 a share on the Common stock was declared by the directors payable October I5th, one of the most substantial ever paid by a large American corporation. This dividend had been earned under the old management, and the intention had been to declare a generous distribution, but out of deference to the coming of the new management action on it had not been taken.

The advantages of this disbursement for the victors were obvious, when the circumstances of the stock accumulation are recalled. General Motors Common had never paid a cash dividend, though Preferred dividends had been fully maintained. To the original stockholders Common had gone as a bonus, but many of them had bought more Common in the course of the campaign for control and others, through the years of adversity, had sacrificed Preferred to cling to Common with a stubborn faith in the Company's earning power. Also, the expense of the campaign had been considerable, especially for brokerage on Street transactions. Heavy borrowings had gone into the stock, and bankers behind Durant breathed easier. As the Chevrolet treasury would receive a huge sum, and part of it would stay there, Chevrolet changed instantly from an adventure to a made property.

Before maturing this coup, Mr. Durant is said to have proposed that Chevrolet be taken as a unit into the General Motors family, but this was declined. Whereupon the trap was sprung.

Described as the coolest man in the room during this momentous meeting, Mr. Durant continued to wear the velvet glove over the iron hand for some time. On November 16th the following directors were elected, representing the du Pont interests: F. L. Belin, Pierre S. du Pont, J. Amory Haskell, and John J. Raskob. Among other new directors were Arthur G. Bishop, president of the Genesee County Savings Bank at Flint, and Louis G. Kaufman. The new board contained strong financial representation.

Messrs. Strauss and Storrow, perhaps seeing the handwriting on the wall, had retired from the board in the preceding June. They were now joined in retirement by Joseph Boyer of Detroit, president of Burroughs Adding Machine Company, Robert Herrick, Edwin D. Metcalf, Nicholas L. Tilney, and Jacob Wertheim. Several survivors of the old bankers' directorate lingered on, trying to resolve in their minds the future of this dramatic union of Chevrolet and General Motors, which to many had all the earmarks of a corporate mésalliance, since a small, new company had gained control of a larger and older one, an industrial and financial giant of the first magnitude.

Pierre S. du Pont was elected chairman of the board, a position he held for over thirteen years, from November 16, 1915, to February 7, 1929. Mr. L. G. Kaufman took Mr. Storrow's place as chairman of the Finance Committee. The Central Trust Company of New York was directed to cremate "all of the 6 percent first lien five-year sinking fund notes as are hereafter received," the full amount necessary to meet the last of these having been deposited in advance of their maturity on October 1, 1915.

All wondered what Mr. Durant had in mind when he did not take the presidency immediately. Probably his hesitatation in that regard flowed from two sources: his admiration for Mr. Nash both as a man and an executive, and a desire, noticeable from the birth of General Motors, to remain in the corporate background and work through others. During the interval in which control of General Motors remained vested in Chevrolet, this matter of the presidency remained in abeyance, but eventually it had to be faced, and on June 1, 1916, Mr. Nash resigned and Mr. Durant succeeded him.

JOHN J. RASKOB

It is said by men close to the situation, that Mr. Durant saw Mr. Nash go with regret and that he himself had to be urged to accept the presidency. There can be no doubt of the first. The two men had worked together for many years in the Durant-Dort Carriage Company. The Nash record in the presidency was a really superb achievement, so noteworthy as to recommend him. to instant backing in an independent venture. But in the process of making that record Mr. Nash had grown to a point where the prospect of subordinating himself, even in the most silken of controls, could hardly be attractive. Also as an intensely practical man who had labored tremendously to correct industrial ills due to unrestrained optimism, he dreaded seeing optimism in the saddle again. Perhaps, too, he was a little chagrined at the ease with which Mr. Durant had overcome the management in the recapture. Though urged to stay, Mr. Nash went his way to new success.

Other members of the old board waited, in the expectation that the Durant-du Pont alliance would not hold. There was at least this ground for that expectation: the du Pont directors, schooled in a big business of ancient origin and close relations with government, held steadfastly to certain high conceptions of corporate responsibility to stockholders and public. They were descendants of an elder American industrialism, and while intensely alert in technology and finance, they had never taken their projects to the public as promotions. The birth pangs of General Motors, as presided over by Mr. Durant from 1908 to 1910, would have seemed almost incomprehensible to them if they had been watching closely that dynamic scattering of stock and accumulation of properties. Now that they were associated with the most daring promoter of the age, the question arose how long that association could last. Rather studied attempts were made to divide the two camps, but these came to nothing. The du Pont interests in General Motors grew until it became roughly equal to that of Mr. Durant and resulted in a sharing of control. In 1918, through direct or indirect ownership, the du Pont holdings were approximately 28 percent of the outstanding Common stock of the Corporation.

The reasons behind the du Pont entry into General Motors are part of a most remarkable chapter in American industrialism. For generations this gifted family, of aristocratic French descent, had been manufacturing powder and other explosives at Wilmington, Delaware, their plant becoming the chief reliance of the government in every American war from that of 1812 onward. Although developing explosives for peace-time pursuits as well, each war found the du Pont plants expanding at a rate which rendered their full-time use in peace something of a problem. Consequently, from the base of explosives, the du Ponts gradually widened their activities in the direction of general industrial chemistry, building up research staff organizations which were ever peering into the future for new things to manufacture.

In the peaceful years before the World War, there were several forces at work to limit the powder business. The frontier had vanished under the steady march of population, with an accompanying decline in hunting. While the use of explosives was growing in agriculture and engineering, du Pont had to divide that trade with several strong competitors. The feeling grew that the future of its business depended upon cultivating close relations with industries using chemical products, and preferably with the industry likely to grow both in proportions and in its increasing use of chemical products. The automobile filled these specifications. It is probable that the du Fonts would have entered the automobile business in some significant way even if there had been no World War.

The war, of course, tremendously accelerated this trend. Almost from the first clash of arms in Europe, orders for powder and other explosives poured into the E. I. du Pont de Nemours Company. As the war continued, the volume of these orders kept increasing, and Allied needs were so urgent that du Pont expanded its facilities on a large scale in order to meet them. Large profits resulted as a matter of course, but always this question arose as the highly conservative du Pont organization considered the future: What is to be done with all these plants, all these workmen when peace is declared? The du Ponts knew, better than most, that wars have a way of ending suddenly, and also that at the close there follows a sharp strain in readjusting a great industry to the requirements of peace. Moreover, the du Pont philosophy of employment takes account of the difficulties suffered by staff and labor through a forced change of scene and occupation. Therefore, they decided to make an investment in the automobile business, taking, as already related, the successful flyer with Mr. Durant on Chevrolet during his drive for General Motors control, with immediate recognition in the directorate. Subsequent analysis revealed a firm basis for further investment of surplus earnings. The du Ponts also had products to supply; still growing, the automobile trade was distinctly a peace-time business, and the phenomenal increase in closed cars indicated an expanding market for paint, varnishes, artificial leathers and other du Pont products, either then available or in prospect.

The alliance had equal advantage for Mr. Durant. His brilliant double-barreled success, first with General Motors and then with Chevrolet, had commended him to everyone save the country's most influential bankers. With them he still needed conservative sponsoring, not from any suspicion of his intentions, but because he was thought of as an optimist easily carried away from a solid footing by the undertow of dreams. The conservative banker-control of General Motors having passed in the heat of battle, the financial world was comforted at seeing the steady and dependable du Pont interests well to the fore when the change came. Their presence insured that stability would have a strong voice at the council table.

In particular the du Pont connection gave reassurance to the financial community that General Motors would continue, even though its control had been secured by Chevrolet. The relation of the two companies for the future puzzled Wall Street greatly for a time. Jonah having swallowed the whale in perhaps the most startling reversal of form ever witnessed on the Stock Exchange, it was now apparent that getting the General Motors whale into the open again was a matter of importance to investors.

Effecting this transformation took some time, as it involved nothing less than the dissolution of the General Motors Company of New Jersey. The last year of the old company's existence saw its Common stock on a regular quarterly dividend basis, with the first dividend declared January 5, 1916.

The same meeting began to clear the decks with the sale of the old Elmore property at Clyde, Ohio, for $50,000, as heretofore mentioned, and the dissolution of General Motors of Michigan, organized in 1910 to hold the company's real estate for the benefit of the trust mortgage. The Imperial Wheel Company's plant, next to Buick, was bought for $80,000 plus certain Detroit real estate, a transaction which completed Buick's holding on Hamilton Avenue, the "front street" of its vast plant at Flint.

After the resignation of President Nash had been accepted with regrets, Mr. Durant became president on June 1, 1916. He would have preferred remaining in the background as he had done before and letting someone else have the place of honor. There were special reasons why Mr. Durant did not wish to burden himself with the presidency. He disliked being tied down to one duty, and scarcely had he carried the General Motors battlements than he began assembling various accessory and parts companies into United Motors Corporation incorporated May 11, 1916, although negotiations for some of the properties had been begun the preceding year.

Other fields tempted him in this exuberant period when fortune and prestige rode high on the wings of success. Farm transport, tractors, and all sorts of mechanized farm implements these took hold of his imagination. Also gasoline-driven highway construction machinery something might be done there. From his first entry into the automobile world, Mr. Durant naturally took a keen interest in pushing good roads. It was he who started state Senator H. S. Earle, "Good Roads" Earle, on his career of highway propaganda which made his name almost a household word through the Middle West and left enduring monuments in magnificent highways. When Mr. Durant's attention riveted itself on a general idea, he seemed to leap around all its various facets and discover something practical to do about each of them. Perhaps his very unwillingness to assume detail executive responsibility came from a correct reading of his own nature; he may have seen himself then, as others have since, as one whose outstanding talent was the gift of seeing opportunities and starting projects which, once begun, could safely be left to others. One can perceive in him a creative spirit which would fret itself against the chains of prudence and tradition, building on a vast scale gigantic projects which he had difficulty in managing after they were built. Probably he would have preferred going on making mergers to being president, but the weight of advice and his just pride in his accomplishment overcame his dislike for routine executive responsibility.

At the close of its last fiscal year, July 31, 1916, the General Motors Company of New Jersey owned the entire capital stock of:

Buick Motor Company, Flint, Michigan
Cadillac Motor Car Company, Detroit, Michigan
General Motors Company of Michigan
General Motors Export Company
General Motors Truck Company, Pontiac, Michigan

Jackson-Church-Wilcox Company
Northway Motor & Manufacturing Company, Detroit, Michigan
Oakland Motor Car Company, Pontiac, Michigan
Olds Motor Works, Lansing, Michigan
Weston-Mott Company, Flint, Michigan

In addition it owned 62.5 percent of the capital stock of Champion Ignition Company of Flint and 49.85 percent of the McLaughlin Motor Car Co., Ltd., of Oshawa, Ontario.

The estimated production capacity of its automobile plants was given as:

| | |
|---|---|
| Buick | 100,000 |
| Cadillac | 20,000 |
| General Motors Truck | 6,000 |
| Oakland | 30,000 |
| Oldsmobile | 15,000 |

The significant items, as a result of the year's operations, were as follows:

| | |
|---|---|
| Cash and Cash Investment. . | $22,762,574.86 |
| Net Profits | $28,812,287.96 |
| Net Working Capital | $43,664,671.40 |
| Dividends Paid | $11,779,122.99 |
| Paid on Debt Reduction | $2,328,000.00 |

On October 13, 1916, General Motors Corporation was incorporated in Delaware to acquire all the stock of General Motors Company of New Jersey, a basis of exchange being established at one and one third shares of new Preferred for one share of the old Preferred and five shares of the new Common for one share of the old Common stock. The new corporation's capital structure on October 31, 1917, consisted of:

Preferred stock: 6 percent cumulative $100 par, $20,000,000 authorized, $19,674,800 issued.
Common stock: $100 par, $82,600,000 authorized, $82,558,800 issued, of which $4,685,500 was in the Corporation's treasury.
The outstanding stocks, plus their proportion of surplus, in affiliates and subsidiaries now owned by the Corporation, totaled $1,380,430.73.
There was no funded debt.

John J. Raskob became a member of the Finance Committee on November 21st, and a little later its chairman, in which position he exercised a potent influence for the next ten years. A man of keen vision, it is said that he took the lead in interesting the du Ponts in General Motors and many of the far-sighted plans in the next expansion era of General Motors may be traced to his initiative. Once more General Motors began to lop off properties, as in 1910, but there was a vast difference in the point of view of the organizers in the two periods. In 1910 all had been skepticism regarding the future of both the Company and the industry, and at that time the motive was chiefly economy with a view to debt payments. Now the Corporation, free and clear of funded debt, was in the hands of hopeful and sanguine men, ready to expand their business to match the leaping trade of a country grown so prosperous that not even the epochal political campaign of 1916 could reduce automobile sales materially. They were clearing the decks in order to take on more cargo, razing old structures to make way for new and larger ones.

The Cartercar plant at Pontiac was sold for $35,000, a fraction of the sum paid for the enterprise when it seemed important for General Motors to own basic patents in the friction drive field, because motor-car design might turn in that direction. Other dissolutions were Elmore, Oakland Motor Car, Ltd., of Canada, the General Motors Company of Michigan, the three of the Heany companies Heany Electric, Heany Lamp, and Tipless Lamp.

By the dissolution of the Buick, Oldsmobile, Cadillac, Oakland, Jackson-Church-Wilcox, General Motors Truck, Northway, and Weston-Mott corporations, authorized December 14th, and consummated at the turn of the year, the General Motors Corporation stood forth as an operating company with the above divisions, thus completing the evolution from the holding company of 1908 to the operating company of modern days. Each of the divisions named continued to operate independently in the sales field through its own sales company incorporated in the nominal sum of $10,000. Their general managers were:[1]

| | |
|---|---|
| Buick | Walter P. Chrysler |
| Cadillac | W. C. Leland |
| General Motors Truck. . | W. L. Day |
| Jackson-Church-Wilcox . | G. H. Hannum |
| Northway | A. L. Cash |
| Oakland | F. W. Warner |
| Oldsmobile | Edward VerLinden |
| Weston-Mott | C. S. Mott |

Although the old General Motors Company of New Jersey no longer functioned, its legal existence continued for some months. On August 1, 1917, its outstanding Preferred stock was retired at $101.75, and the Company itself was dissolved on August 3 of that year. Its last holding, 56,855 shares of General Motors Corporation Common stock, was acquired by the Corporation in connection with a plan worked out to interest valuable employees in the financial success of the enterprise.

It is in order to review the record of the New Jersey Company as it passes from the corporate scene. Through it was effected the first automobile merger. It came on the scene when the industry was still young and confused and when the automobile itself was still on trial before the public. When it departed, the automobile was an essential part of life, a reliance

---

[1]Messrs. Chrysler, Leland, Day, Warner, and Mott were vice-presidents of General Motors; Messrs. Hannum and VerLinden later became vice-presidents. For tenures see Appendix II

in the routine of daily living and also a dynamic tool, as events in Europe were showing. The industry's permanence and importance were recognized alike by the man in the street, the banker, the investor and the statesman. General Motors of New Jersey was in no small degree responsible for this change of attitude. It had weathered as severe financial storms as ever beset a young industrial merger, and had staged a startlingly profitable "comeback." Its activities had increased the population of numerous towns and cities, creating fortunes in the appreciation of land values. It had been free of serious strikes; wages rose as General Motors grew. The Company had never been accused of unfair commercial practices to throttle competition; through years marked by intense public criticism of corporations and numerous government actions aimed at corporations, General Motors escaped both attentions without question. It had been the first automobile company to tap the great banking reservoirs of public credit, the first whose securities were admitted to the Stock Exchange. Its Preferred dividend had been maintained through bad days and good and its Common stock had finally reached a dividend position after scoring sweeping advances as a result of high earnings. No American corporation of the period had passed through a more dramatic experience or emerged with a cleaner record.

## Chapter XII THE WAR YEARS

AMERICA'S declaration of war in April, 1917, forced a sharp readjustment of objectives on the automobile world. No one quite knew, at the start, what the government wanted from the industry in the way of goods or what quantities of raw material would be available for the production of cars for the civilian market. Large supplies were in process and storage, but after these had been assembled and consumed, what could the industry expect to run on? Uncertainty created anxiety. It was clear that the automobile would be used in war efforts, but how and when? The answers to those questions were momentous, particularly for the inhabitants of cities which had grown rapidly around the plants and were dependent upon daily work for daily bread. Ever since 1910 the total number of General Motors employees had been rising year by year, from 10,000 in 1910 to 25,666 in 1916. The year 1917 brought the first break in that advance, the number of employees falling slightly.

With as much speed as was possible in a country so unprepared for war both psychologically and industrially, the government set up a program for war manufacturing which kept many automobile plants busy, but of necessity there was a dislocation of activity. While certain plants ran night and day with increased staffs, others could operate only on part time, using a limited supply of materials allocated by government agencies conserving the various goods which were needed or might be needed for war purposes. An automobile production program is an immensely complicated operation, requiring the gathering of thousands of parts and accurate timing of supply arrivals. Absence of any one of a thousand parts, or shortage of some raw material, may throw out of alignment an entire schedule of operations involving thousands of men and millions of capital. By extraordinary efforts many of these difficulties of supply were overcome, and cars continued to reach the market with some regularity. With war activities and increased automobile production in the latter part, employment for 1918 rose swiftly to a new peak of 49,118, nearly double the names on the 1917 pay roll.

Such cars as came to market were readily sold. Sales rallied surprisingly after the first shock. Although millions of young men, the best potential buyers in the light-car field, were mobilized, and savings were being drawn on for war loans, nevertheless automobiles continued to sell. The purchasing power of the country was high as a result of war inflation and the consequent rise in wages. While automobile prices rose somewhat in answer to increased costs, they did not rise in proportion to food and clothing. The people had money to buy cars, and the automobile fitted into the high-speed picture of the war years, when time was the most important element in a life-and-death struggle and economy was a forgotten word. With millions of men withdrawn from employment, those at home increased their activities by using motor cars more freely. The incentive to save time would carry on long after the war was over and be one of the factors in the next great advance of the motor car in popularity. But, for the remainder of 1917 and the greater part of 1918, automobile production in general went forward under handicaps.

All General Motors' facilities were, of course, placed at the government's disposal the instant hostilities were begun. Of the twenty-three operating units eighteen were engaged on government contracts. The gross value of war products actually completed by the Corporation approximated $35,000,000. War production included: ambulances and trucks, 5,000; officers' cars (Cadillac), 2,350; artillery tractor engines (Cadillac 8-cylinder engine), 1,157; Liberty motors for aircraft work, 2,528 actually completed and delivered, with orders for over 10,000 on the books when the Armistice was signed. The Jackson-Church-Wilcox division, operating an entire plant on trench mortar shells, reached a production of 20,000 per day. Oldsmobile built field kitchen trailers. Buick constructed special factories for the production of the famous Liberty airplane motors. Associated plants made shell caps and other munition parts.

Buick war tractor 1918

It was a proud moment for General Motors when the Cadillac stock car, as previously noted, after gruelling tests in the mountainous wastes of the Big Bend country in Texas, was selected as standard for war use in July, 1917. With no mechanical changes, but painted khaki-brown, Cadillacs were standard transport for general officers and their staffs in the American Expeditionary Force in France. Cadillacs were also used by officer personnel at army bases and cantonments in this country. Liberty motors were also made by Cadillac.

In addition to the completed war orders, large quantities of war material were in process at the close of hostilities. Several plants had been constructed rapidly at government suggestion to accommodate war orders. Such was the inception of the Central Products division on Holbrook Avenue, Detroit, near the Northway division. There a drop forge plant was rushed to completion in 1918, with a capacity of fifty tons a day. This plant, soon expanded, became the nucleus of the present Chevrolet group in Detroit. The Central Foundry plant at Saginaw was also pushed rapidly forward, 126 acres being purchased for the site.

Mr. Durant tells a striking war story illustrating the uncertainties of the war years and indicating some of the sacrifices

General Motors made to relieve trying situations. The Corporation was .requested by the British government to prepare for the manufacture of a new airplane engine in large quantity. Accordingly property was purchased, construction begun and machinery ordered. After waiting months for the arrival of the promised sample engine from abroad, Mr. Durant was informed that nothing could be done. The first engine tested abroad had functioned perfectly, but others had failed and the plan was declared abandoned. "Submit your bills," said the Allies' representatives. "No," said General Motors through its president, "We will take the loss and try to find another use for the property. Under the circumstances our stockholders are not likely to object." Eventually the commitment was worked off without serious loss to the Corporation.

Meantime, the directors continued putting their house in order. The union of General Motors and Chevrolet was at last consummated on May 2, 1918, when General Motors acquired Chevrolet for 282,684 shares of Common stock, taking over all assets except 450,000 shares of General Motors Common in the Chevrolet treasury. The original Chevrolet shareholders realized a very substantial profit.

Further additions to the roster of General Motors being contemplated, as well as plant extensions, the certificate of incorporation was amended twice, raising the authorized stock of the Corporation to $200,000,000 on March 20, 1918, and to $300,000,000 on August 27th. June 26, preceding, the president and chairman of the Finance Committee were authorized to buy United Motors for $44,065,000, paying approximately three quarters of that sum in Preferred stock and one quarter in Common stock. United Motors had been organized by Mr. Durant, and under the presidency of Alfred P. Sloan, Jr., was soundly administered. Thus Hyatt Roller Bearings, Dayton Engineering Laboratories, New Departure, Harrison Radiator, Remy Electric, and Jaxon Steel Products entered the Corporation, and the assets of Perlman Rim came into the Jaxon company.

That the directors on July 26, 1918, were optimistic about the early and successful close of the war may be gathered from the fact that they voted on that day to invest $1,000,000 in a 40 percent interest in the Doehler Die Casting Company of Brooklyn and authorized J. A. Haskell to buy the stock of the Scripps-Booth Corporation of Detroit. The Scripps-Booth car, a runabout of advanced design, seemed a desirable addition, but other General Motors units soon overtook and surpassed Scripps-Booth, in excellencies of design, resulting in its abandonment.

Barn in Dayton, Ohio; first home of Dayton Engineering Laboratories Company., 1909

Midsummer, 1918, saw the World War enter its final phase, marked by the last despairing German drives which were finally checked at the Marne. While none knew how long the war would last, the Corporation maintained as an objective the quickest possible return to large production as soon as quota restrictions should be removed.

The World War gave General Motors a decisive turn of interest toward aviation. Production of Liberty engines and other essential aviation supplies in its plants emphasized the kinship between airplane and automobile production. Other manufacturers, not yet in General Motors but with whom the Corporation had close business relations, were also participating in the gigantic aviation enterprise of the government. Fisher Body, with a vast quantity production program under way, was building DeHavilands in its great Number 18 plant at Detroit; likewise the Dayton industrial grouping led by Mr. Kettering. The Dayton-Wright Airplane Company, the spearhead of the

Dayton aviation effort, turned out completed airplanes for a time faster than any other company has done before or since. On September 25, 1919, General Motors bought all outstanding shares of Dayton Metal Products Company and certain assets of the Dayton-Wright Airplane Company, as well as other Dayton interests.

The post-war situation, however, proved discouraging to aviation, as the chief customer, the government, severely curtailed its buying. The market was saturated with planes and materials. It seemed that years must elapse before aviation would take its place as a self-sustaining industry; in the meantime, development would proceed largely through the assistance of government buying for military purposes and government subsidies on mail contracts. Accordingly General Motors, while still acutely aware that aviation was a kindred industry, felt constrained to follow the policy of watchful waiting. It made such disposition of its aviation properties as seemed wise, and settled down to wait until 1929, when some of the "unknowns" had been eliminated by time from the aviation problem.

This conservative approach reflected the changed attitude which the Corporation adopted in other respects a little later. The public hopefully awaited a strong lead in aviation; and from many sources came pressing opportunities for General Motors to assume that leadership. But the cool heads in control understood that aviation was in for a long pull, during which investors seeking quick profits probably would be disappointed. While the standing of General Motors would have brought strong public support in the financing of a vigorous aviation program, it was believed that stockholders, early following General Motors lead, would take losses because they would not be disposed to wait for long-range plans to mature. Consequently the Corporation refused all offers to enter airplane production under those circumstances, but its interest in aviation continued and later revived, under circumstances and with results to be related in Chapter XXIII: "General Motors in Aviation."

General Motors decided to complete the United Motors consolidation by using a certain number of 6 percent Debentures in place of Preferred stock, and on November 7, 1918, appropriate action was taken. On December loth, the certificate of incorporation was amended to increase the authorized capital to $370,000,000 consisting of 200,000 shares of Preferred, 1,500,000 shares of Debenture, and 2,000,000 shares of Common, all par $100, to provide capital for the expansion swing which had been determined upon with the renewal of peace. Du Pont American Industries subscribed for $24,000,000 of the $100 par value Common stock of General Motors at $120 per share as of December 19th, which was approved on the last day of 1918 with the reservation that all Common stockholders could subscribe to the new stock at $118 a share up to 20 percent of their holdings. At these two meetings three other important moves were made:

1. General Motors bought the minority interest in three affiliated Canadian companies McLaughlin Carriage Company, Ltd., Chevrolet Motor Company of Canada, and the balance of the McLaughlin Motor Car Company, Ltd. for 49,000 shares of its Common stock, thereby creating the foundation for General Motors of Canada, Ltd., whose Oshawa and other plants are among the industrial prides of the Great Dominion.
2. The purchase of Lancaster Steel Products at Lancaster, Pa., was approved, on the basis of 500 Preferred shares and 3,555 Common shares and 15,660 General Motors Debentures held by Lancaster for 16,175 shares of General Motors Common and 5,000 shares of Debenture stock.
3. The purchase of United Motors, instituted earlier in the year, was approved on December 31, at a price of $45,000,000, for which purpose 99,564 shares of Common stock and 298,692 shares of Debenture stock were authorized. General Motors already owned 106,000 shares of United Motors, which stock was cancelled.

Moody's *Manual of Industrials* for 1919 gives the following outline of the various properties in the United Motors group and their functions:

| | | Incorporated | Square feet floor space | Products |
|---|---|---|---|---|
| Dayton Engineering Laboratories Co.. | Dayton, O. | July 22, 1909 | 500,000 | Starting and lighting equipment |
| New Departure Mfg. Co........... | Bristol, East Bristol and Hartford, Conn. | July 5, 1889 | 900,000 | Ball bearings, coaster brakes, and steel balls |
| Jaxon Steel Products Co............ | Jackson, Mich. | June 10, 1918 | 250,000 | Rims |
| Hyatt Roller Bearing Co........... | Harrison, N. J. | Nov. 7, 1892 | 750,000 | Roller bearings |
| Remy Electric Co................. | Anderson, Ind. | Oct. 5, 1901 | 300,000 | Starting outfits and ignition sets |
| Harrison Radiator Corp........... | Lockport, N. Y. | Nov. 23, 1916 | 300,000 | Radiators |

Bearings Service, Inc., United Motors Service and one half of the Common stock of the Klaxon Company were also acquired with the United Motors group.

Bearings Service Company had been incorporated June 24, 1916 and United Motors Service, Inc., on October 14, 1916, as distributing and service subsidiaries to make United Motors' product easily available to automobile users.

The history of Klaxon will be found in Appendix IV.

In 1918 General Motors, from both war orders and advancing prices on motor-cars, recorded the high net sales to date $269,796,829 but its margin of profit had been low and only $1,667,753 was left for reinvestment after paying dividends which had been established at the rate of $3 a quarter on the $100 par Common stock in the preceding year. Bonus awards for conspicuous services in 1918 were the equivalent of 490,238 shares of the present $10 par value Common stock.[1]

In the year of America's sustained war effort, 1918, the Corporation, due to government reduction of the business, barely earned its established dividend, but its contribution to employees remained high. Although resentment against profiteering was in the air, the record left General Motors free of criticism in that regard. As an essentially peace-time business, General Motors from top to bottom was glad to get back to work on its big job of making motor cars for a world no longer torn by war.

---

[1] See Chap. XXVIII on "Cooperative Plans," for a full account of the Bonus and Investment Funds.

# Chapter XIII THE EXPANDING CORPORATION

GENERAL MOTORS swung into the post-war boom of 1919 with a spirit which matched the boldness of the country's mood. Peace found the nation jubilant, united, and confident in its strength. New high wage levels had been reached, and it was hardly possible for a nation so circumstanced to believe that they would recede. Prices were high, also, but it was thought that the need of Europe for supplies would keep them up for some time to come. The automobile industry, in general, concluded that it must make up for time lost during the war interruption in production. Consequently, the year 1919 records an amazing expansion program for the Corporation in the course of which the authorized capital stock was increased from $370,000,000 to $1,020,000,000, represented by 5,000,000 shares of Common stock of $100 par value, 5,000,000 shares of 6 percent Debenture stock and 200,000 shares of 6 percent Preferred stock. General Motors became a billion-dollar corporation on paper, but less than a third of its Common stock was issued.

General Motors of Canada, Ltd., had been organized November 8, 1918, and before the end of the year its authorized capital was increased to $10,000,000. This strong Canadian company, located at Oshawa, Ontario, succeeded several Canadian companies theretofore representing various units of General Motors in the Dominion, as will be more fully explained in Chapter XVII.

General Motors Acceptance Corporation was incorporated under the banking law of New York on January 29, 1919, to finance instalment sales of General Motors products "through the proper application of the credit function." Twenty thousand shares of GMAC stock were purchased by the parent corporation at $125, enabling the new subsidiary to begin operations with a capital of $2,000,000 and a surplus of $500,000. The rise of GMAC is fully narrated in Chapter XXVII: "Financing and Insuring the Buyer."

At the meeting of February 27th, steps were taken to complete General Motors ownership of New Departure Manufacturing Company of Bristol, Connecticut, and Harrison Radiator Corporation of Lockport, New York, two companies which had come into the General Motors circle through the purchase of United Motors.

Two new accumulations were made, both of which bulk large in Corporation history, though for different reasons. In 1918-19 the Corporation bought into the tractor business, among its purchases being the Janesville Machine Company at Janesville, Wisconsin, and 122 acres of land on which a new plant was to be erected ready for operation in July, 1919. The Corporation's farm implement business is reported as 3,000 tractors and 56,400 farm implements in Poor's Manual for 1919, which also reports the Corporation holdings in Samson Sieve-Grip Tractor Company of Stockton, California, as $400,000 and in the Janesville Machine Company, Janesville, Wisconsin, as $1,000,000, both representing complete ownership of all Common stock issued by both companies. Later the Corporation's holdings in the Janesville Machine Company rose to $2,250,000, where it stood when that company was dissolved and its operations consolidated with the Samson Tractor division. The latter soon took over all the assets of the Samson Sieve-Grip Tractor Company. This mushroom expansion in the direction of motorized agriculture brought heavy losses.

In February also was begun the most significant enterprise the Corporation had as yet undertaken in the field of social research. A new spirit of brotherhood was abroad in the land, and General Motors was one of the first to respond to it. Owing to the uncertainties attendant upon the change from war activities to peace-time pursuits, the Corporation considered it necessary to have the basic needs and living standards of its employees studied, to the end that wage rates would be fair and living conditions acceptable to thousands of families likely to move into cities where General Motors was rapidly expanding its operations. Accordingly, the Executive Committee, consisting of Messrs. Durant, Haskell, and Chrysler, was directed to investigate industrial conditions affecting the plants of the Corporation. These gentlemen promptly appointed a research committee, which met in Detroit on February I7th and proceeded at once with an extensive study of the labor situation, living conditions and other industrial problems of General Motors plants. An imposing body of data was collected on bonus plans, group insurance, employees' committees, pension plans, etc. The findings of this Research Committee launched the Corporation upon a large-scale housing program in Detroit, Flint, Pontiac, and Lansing, Michigan; Bristol, Connecticut; and elsewhere, and also led it into other activities aimed at promoting the wellbeing and contentment of employees.

This elaborate report, one of the most comprehensive social studies ever made by an American corporation, is a testimonial to the energetic altruism of the chairman of the Research Committee, Mr. J. Amory Haskell, in whose kindly and judicial mind the well-being of employees ever was a first consideration. Mr. Haskell, after a long and impressive career in the explosives industry in competition with the du Fonts, had brought his company into the du Pont organization, and entered General Motors as one of the du Pont directors. A conservative man, yet with broad humanitarian interests, he was a force in General Motors affairs until his death in 1923.

This disposition to consider employee interests had another significant development in the founding of General Motors Institute of Technology at Flint, described in Chapter XXVII, on "Cooperative Plans." In March, 1919, General Motors bought additional stock in the Frigidaire Corporation for $56,366.50. Frigidaire, originally Guardian Frigerator Corporation, was bought by Mr. Durant in 1916 and is another example of his power to read the future. The rise of Frigidaire to first place in the quantity production of electric refrigerators will be told elsewhere.

Occurred then, also, the Corporation's entry into Muncie, Indiana, where it bought the Interstate Motor Company for $248,000 and paid $40,000 for additional land. For a brief period, the Sheridan car was made there. The property is now the seat of production for Delco-Remy batteries.

On April 24, 1919, the Corporation began extensive real estate and construction projects destined to have an important effect upon its history, since they diverted many millions of dollars from the treasury and were still uncompleted when the prosperity of the post-war boom waned. The most ambitious of these was the construction of the mammoth office building at Detroit, planned as the largest structure of its kind ever built. Covering the entire block bounded by Grand Boulevard, Cass and Second avenues and Milwaukee Street, it is now the central office of the Corporation. The Durant Building Corporation having been formed, General Motors authorized subscription to $3,000,000 of its stock on April 24th. Before it was completed, this building cost approximately $20,000,000. From April to July the Corporation created various companies for housing construction, the authorized capital in each company being as follows, although considerably more money was spent for this purpose:

$500,000 in the Bristol Realty Company, Bristol, Connecticut, for the accommodation of New Departure employees.
$3,500,000 in the Modern Housing Corporation, to build houses in Flint, Pontiac and Detroit.
$200,000 in the Lansing Home Building Company.
$200,000 in the House Financing Company of Detroit.

General Motors Building, Detroit

On June 12th, the stockholders ratified the increase in authorized stock to $1,020,000,000, the details of which have already been given at the beginning of this chapter.

From August on, the Corporation expanded rapidly. The directors authorized the purchase of the International Arms & Fuse Company plant at Bloomfield, New Jersey, for $1,175,000 and of T. W. Warner & Company, gear makers of Muncie, Indiana, for $5,000,000. Both deals were delayed in maturing. The Muncie property not being then obtainable, the Corporation took a lease as of September 25, 1919, with the option to purchase after January i, 1923, for $902,000 in Liberty Bonds and 31,238 shares of Debenture stock. This property became the Muncie Products division.

Various steps in the expansion program were:

The purchase of the Pontiac Body Company and its addition to Oakland.

The purchase of Domestic Engineering Company of Dayton, Ohio, for 35,451 shares of Common stock, given in exchange for 33,070 shares of Domestic Engineering Company stock valued at $9,000,000.

A merger was effected with Dayton Metal Products Company, whereby General Motors acquired all the outstanding shares of Metal Products (60,000 shares) in return for 25,338 shares of Debenture stock and 21,457 shares of Common stock in the Corporation.

The Dayton-Wright Airplanc Company, with assets in excess of $1,200,000, was bought for 10,960 shares of Debenture stock.

Through reorganizations and transfers all of these Dayton companies have passed from the picture, and their plants are now used for other Corporation products.

Of outstanding importance in the Corporation's history, also, is the decision taken on September 25th to buy a three fifths interest in the Fisher Body Corporation, which was effected by the purchase of 300,000 shares of Fisher Body Common stock at $92 a share. The purchase contract, among many other provisions safeguarding the Fisher interests, required these shares to be deposited in a voting trust. The importance of this alliance, which was later extended to the point of complete merger, can hardly be overestimated.[1] It gave General Motors first call on the production of the largest and best equipped body-building plants in the world. With the steady trend toward closed cars, the Fisher brothers, experienced from youth in the

FRED J. FISHER One of the founders of Fisher Body

difficult arts of body design and construction, had pushed forward until they occupied a foremost position in their line both as to quantity and quality. The Fisher name had become known far and wide, and the presence of a Fisher body on any car recommended it to the consuming public. By this one decisive step the General Motors Corporation wrote off future body difficulties by placing that business in the hands of outstanding specialists in the field of body manufacture. With advancing complexities in design and manufacture, the wisdom of this alliance has become increasingly manifest.

This date, September 25, 1919, registers the high-water mark of optimism during the Durant presidency. The Corpora-

---

[1] See Chap. XX: "Body by Fisher: The Motor-Car as a Style Vehicle."

tion had expanded its capital, brought under control many sources of supply, and on that date authorized the investment of up to $500,000 in the Common stock of Goodyear Tire & Rubber Company of Akron, which holding was sold soon afterward.

If General Motors is considered merely a manufacturer and seller of automobiles, the expansion from 1916 to 1920 contains a good many inexplicable elements; but the key to this period can be found in the word "motors." As will be shown in Chapter xv, this is the key to later diversification of the Corporation's products, especially in the direction of household, office, store, and farm equipment. The Corporation was no longer interested merely in motors that traveled and in the vehicles they propelled. Its scope had been enlarged to include motors designed for a wide variety of specialized and stationary uses, and in the commodities which housed motors in motorized refrigerators, motorized farm-lighting plants, motorized fans, and other equipment. The variety of its merchandise would grow from this time on, but always the growth would be in the direction of motor products. Expansion in the passenger car field stopped with the acquisition of Scripps-Booth which the directors were ready to cut adrift in 1919, and the short-lived experience with Sheridan in 1920-21. There would be further expansion in commercial vehicles, but in general the trend of General Motors' growth in the post-war prosperity was toward rounding out an established property and pushing it toward an ever-improved market position with the consuming public.

During the 1919 boom, with the market emphasis on quantity, General Motors reached its highest earnings to date, more than $60,000,000 net available for dividends, paid nearly $22,000,000 in dividends, and allocated more than $38,000,000 to reinvestment in the business, after setting aside some $30,000,000 for Federal taxes and miscellaneous items. But even that sum could not finance the driving expansion program under way. This prosperity was shared with employees through bonus awards in which were distributed the equivalent of 402,485 shares of present $10 par stock and 14,191 shares of

7 percent Preferred stock. The number of employees had taken another upward leap, reaching nearly 86,000, more than four times as many as the Corporation had in 1913. One influence of the World War on America may be read in these contrasting employment figures for 1913 and 1919. The period included two booms in which well managed industrial companies expanded staffs, drawing population from the land to the cities and concentrating employment as it had never been concentrated before in the United States.

The Corporation signalized the advent of 1920 a year destined to end quite otherwise than it began by increasing its authorized capitalization again, this time to 56,100,000 shares divided as follows:

200,000 shares of 6 percent Preferred, par $100
900,000 shares of 6 percent Debenture, par $100
5,000,000 shares of 7 percent Debenture stock, par $100
50,000,000 shares of Common stock, no par value, of which upwards of 20,000,000 were issued.
The senior securities issued stood in 1920 at slightly above $100,000,000.

To these gigantic proportions W. C. Durant had seen his brain-child grow from the original Buick Motor Company of $75,000. Conservative men began to wonder where this victorious march would end, and amazement gave way to some concern when falling grain prices in the spring of 1920 foreshadowed the post-war decline.

During the first three months of the year General Motors went along with its program of internal consolidations. Minority interests in Chevrolet of California were bought and steps were taken to complete ownership of Klaxon horn. Two companies were dissolved Samson Sieve-Grip Tractor and the Michigan Crankshaft Company.

The former name, Samson, is still one to rouse lurid reminiscences among the old-timers in the General Motors family. Under the Samson banner the Corporation entered the farm tractor field, expanding for that purpose the Janesville, Wisconsin, plant, later taken over by Chevrolet. In the full flush of his enthusiasm for motorized farm transport, Mr. Durant had not only spent lavishly on the Samson tractor; he had also tried to tame the Iron Horse, a small tractor for garden use, ordering large quantities of material for its manufacture. Always alert to new notes in sales campaigns, he undertook to revolutionize implement selling by organizing elaborate displays in specially designed show places on the Pacific Coast. These displays were beautiful examples of advanced merchandising, but events showed them to be ahead of their time. Indeed, the whole program of farm motorization so popular then was too advanced, being founded upon the assumption that the world could use practically unlimited quantities of foodstuffs at high prices. The slump in farm prices which set in with the post-war depression was to continue for years, reducing operations by nearly all manufacturers in that field.

By June, train-loads of motor cars standing undelivered in Western terminals gave a hint of what was coming. Livestock and grain prices were unsettled, and unemployment began to show itself. By late summer the decline reached a point at which General Motors could hardly look for earnings to flow in rapidly enough to meet the charges of its uncompleted expansion program. The pendulum which had been rising with but slight interruption since 1915 began a downward swing.

In a retrospective vein President Alfred P. Sloan, Jr., addressing the automobile editors of American newspapers at the Proving Ground at Milford, Michigan in 1927, graphically reviewed the 1920 situation as he saw it:

> In the spring of 1920, General Motors found itself, as it appeared at the moment, in a good position. On account of the limitation of automotive production during the war there was a great shortage of cars. Every car that could be produced was produced and could be sold at almost any price. So far as anyone could see, there was no reason why that prosperity should not continue for a time at least. I liken our position then to a big ship in the ocean. We were sailing along at full speed, the sun was shining, and there was no cloud in the sky that would indicate an approaching storm. Many of you have, of course, crossed the ocean and you can visualize just that sort of a picture yet what

> happened? In September of that year, almost overnight, values commenced to fall. The liquidation from the inflated prices resulting from the war had set in. Practically all schedules or a large part of them were cancelled. Inventory commenced to roll in, and, before it was realized what was happening, this great ship of ours was in the midst of a terrific storm. As a matter of fact, before control could be obtained General Motors found itself in a position of having to go to its bankers for loans aggregating $80,000,000 and although, as we look at things from today's standpoint, that isn't such a very large amount of money, yet when you must have $80,000,000 and haven't got it, it becomes an enormous sum of money, and if we had not had the confidence and support of the strongest banking interests our ship could never have weathered the storm.

On this occasion General Motors had strong banking connections. An agreement was entered into for the distribution of 3,200,000 shares of Common stock at $20 a share. Explosives Trades, Ltd., of London was interested by the du Fonts to the extent of 1,800,000 shares, the balance being underwritten by J. P. Morgan & Company.

The Explosives Trades passed a part of its allotment in this undertaking to Canadian Explosives, Ltd., a subsidiary, but before their subscription had been taken up, the American stock market showed such signs of weakness that doubt arose whether the Canadians would purchase at the figure. On top of the general situation, General Motors had weaknesses of its own, the net results of heavy commitments and slackening sales. These were complicated for the Corporation by the recurrence of the malady already noticed as present in 1910, namely, uncertainty over inventories. Estimates furnished the Finance Committee were approximate rather than definite, and even the approximations did not hold as the returns began to roll in from the wide-flung field operations. Again, it was apparent that centralized authority had not advanced to the point of giving the central office full information on plant commitments; there was still opportunity for plant executives to act independently enough to embarrass the Corporation. The initiative of plant managers, which the Corporation had always sought to preserve, and which had worked magnificently on the up-swing, brought complications on the down-swing. Discrepancies between estimates and actualities, in the matter of inventory commitments, ran into scores of millions. It began to look as if General Motors were in for another "squeeze," for another experience under banker control.

With the stock declining on the Exchange, a serious situation developed for Mr. Durant. His personal operations in the stock market had been large in volume. It was a common saying that he never sold General Motors but always bought it. On the 1920 decline he kept buying until his "Street loans" reached a critical condition. The explanation given by his friends for the more acute phase of his trouble is that he endeavored, by extending his buying on the decline, to maintain a price above $20 in order that Canadian Explosives would come in at that figure and take the 300,000 shares allotted to them. His battle with the falling market, however, reached proportions far in excess of the $6,000,000 involved in that particular transaction. When his situation became known to his colleagues, it was found that his commitments to bankers and brokers ran to several times that figure. His operations had been so huge and hurried that not even Mr. Durant himself could be quite sure what he owed. With each conference the sum kept growing, until it is said to have approached $35,000,000.

On a falling market, with public confidence low and General Motors in a position of rising inventories and decreasing earnings, the difficulties of its president became the difficulties of the Corporation. If Mr. Durant were to be sold out by his bankers and brokers, a possibility which became more and more imminent with every point decline, involving $2,450,000 loss to him, the forced sale of his pledged securities on a falling market would have meant a wild decline in the whole stock list, perhaps even a panic, and a severe blow to General Motors' credit. On March i, 1920, a ten-for-one stock split-up had been effected, and the new no par stock had already lost more than half its market value. The high and low figures for 1920 show the market nervousness high 42, low of 12¾.

From the standpoint of both the general welfare and the corporate credit, it was necessary for someone to finance a settlement with Mr. Durant's brokers and take his stocks out of the market. In the General Motors picture only the du Fonts had the financial strength to do this. Even with their resources and the strong incentives of their existing stock ownership, their "rescue" was an adventure of faith, when the general condition of the market and the country are considered. The transaction received the active aid of J. P. Morgan & Company and could hardly have been consummated without the assistance of that banking house. While general recovery was slow, it was steady from that point on until returning confidence ushered in the prosperity of the mid-'twenties.

The Durant stocks in jeopardy were taken over at a price which cleared his slate and left him a not inconsiderable margin. His friends might rate his fortune at this stage as far less than his deserts, considering his services and record through General Motors history, but on the other hand, the fact that something was saved for him when all might have been lost indicates that there was every disposition to deal fairly with him as far as the needs of a desperate situation would permit. He resigned the presidency on November 30, 1920, after four and a half years in office years of great achievement and bold expansion pushed at a pace so rapid that it brought difficulties.

There is a jaunty nonchalance, a cool courage, in the Durant character which on more than one occasion has let him rise from defeat. It showed itself in 1910 when he turned from General Motors to found Chevrolet and through Chevrolet to recover control of General Motors. Indeed, if one were attempting a psychological interpretation of so complex a nature, one might say that Mr. Durant never recognizes defeat. What might be considered by others an irretrievable misfortune, to be weighed in sorrow, is to him but a turn of the wheel on which he expects to swing round to the top again. So we see the founder of General Motors leaving its councils for the second and last time apparently without either remorse or wrath, and saying, as he put on his hat with something of a flourish, "Well, it's moving day."

The story runs that immediately after his eclipse Mr. Durant called together his "crowd," those who had followed him into

General Motors and out of it and back again. He told them that the powers left in control of the Corporation were "good people/' and that if the old crowd wanted to help "Billy" Durant, they should stay with the Corporation as long as they were wanted and serve loyally, because what he had left was in General Motors stock and, consequently, he would benefit as the Corporation benefited. A few days later one of his friends went to see him at a small office he had opened nearby, to say that he had been asked by the du Fonts to carry on. Mr. Durant advised him to do so. Then, going to the window, Mr. Durant raised the shade full length, and said with his head high, "It's a new day!" No better last line could be written for the end of a drama.

As the founder of General Motors leaves its history, it is pertinent to review his achievements briefly, not with the idea of passing judgment in either a laudatory or captious spirit, but rather to give point to lessons which the great army of General Motors employees and the public likewise may draw from his career in the Corporation.

First, from 1904 to 1908, W. C. Durant "made" Buick. Given a staunch car he had improved it steadily by a rigid insistence on quality. Although Buick was moderately priced, its chief insisted on good workmanship and high-priced materials. Although no operating man, in the sense in which the term is used today, he had high standards of what manufacturing should be. If he heard that Amesbury was furnishing expensive fittings for Packard, off he would go to Amesbury to buy equally good fittings for Buick, relying on quantity purchasing power to bring the price down to reasonable levels. But all in all he was a better salesman than manufacturer. With Buick, Mr. Durant proved, as he had proved in the carriage industry, that he was a master salesman, building up a sales organization which was the marvel of its day, and copied by competitors as far as they could do so. Many units of this early Buick sales organization are still dominant in the distribution of motor cars over wide areas.

In founding and financing General Motors through its first expansion period without first-class banking help, W. C. Durant demonstrated his skill as a promoter. He has been called America's greatest living promoter, and no exception has ever been taken to that. It is hardly too much to say that this quiet, soft-spoken man is the greatest promoter America has ever seen in action. General Motors is today the largest American corporation whose founding was the result of the promoting ability of a single individual, and as the trend is toward group rather than individual enterprises, his achievement in creating General Motors may never be equalled in any other field. Add to that record Chevrolet and United Motors, now merged into General Motors but brought into being separately, and you have a three-way success which is simply colossal. Of course, various elements in this energetic program proved to be disappointments, but enough survived to carry the whole through to a stupendous success. Napoleon remains a great soldier in the opinion of the world in spite of Waterloo. Both promoters and conquerors sometimes overstay markets.

One can call Mr. Durant a promoter without implying a derogatory note. There are times and situations in which the promoter is the indispensable man, the almost inspired servant of society. He may be moved chiefly by the desire for power and profits, and comprehend scarcely at all the social results of his activities; nevertheless, without his zeal in bringing labor and capital together along the line of his vision the world would have to wait longer than necessary for many boons. Of course, no one could foresee the full impact of motors on American life, but W. C. Durant did recognize some of the important futurities, and reacted toward them promptly. His early championship of good roads is a case in point, his keen acceptance of electric refrigeration is another, and a third was his instant recognition that this new industry could not afford to pay low wages, because the buying power of the masses must be kept active to absorb automobiles in quantity. On the personal side, he gave largely, paid generously for loyal or significant services, followed a liberal policy toward shareholders and in so far as he could, made sure that his friends and associates prospered along with himself.

The tap-root of the Durant nature appears to have been an incurable optimism which was the source of both his driving strength and his besetting weakness. He always hoped for the best, and never prepared for the worst in time to ward it off. In his long business career he said recently that he had been in business fifty-five years he wrestled many times with the business cycle without apparently becoming convinced of its periodicity. He could sell company stock, but he could never sell his personal holdings. Because Mr. Durant performed financing feats, and handled huge sums of money, the notion grew that he was a financier. Thus we find him hailed as one of "the Master Minds of Finance" in a Wall Street series during the flush period. The truth is that he was no financier in the sense that the term implies caution and conservatism. He could act boldly and daringly with money, and when circumstances favored the finance of courage he seemed a financier. But money is a two-edged sword, and when circumstances demanded the finance of caution Mr. Durant was lost.

Studied defences of Mr. Durant's responsibility for the vast expansion program of 1919 have been printed. From these it would appear that he declines to take full responsibility for some of the moves which brought the Corporation treasury low when the pinch in sales came. It is said that his plans for the office building in Detroit, and the concurrent expansion at Cadillac, were considerably more modest than the expensive projects put under way; that he yielded to the pressure of his associates. In the history of an industrial enterprise motives are more difficult to follow than results of record; a word, an opinion, sometimes even a gesture, may tip the scales this way or that. In this case all concerned were working at top speed, and it was inevitable that minds did not always meet when their possessors thought that understanding was complete.

There is an old saying that every man has three chances at riches in the course of the average business career. To put it another way, there are usually three complete swings of the business cycle in each quarter-century. It is human nature to "go broke" in the first depression one experiences, rally and survive the second by a narrow squeak, and then, having learned the lesson of experience, prepare for the third and emerge from it in full strength. As for men, so also for corporations. The special circumstances surrounding the automobile industry in 1907 kept Buick from feeling that shock severely, but a crisis came to General Motors in 1910. Ten years later came another; it found the Corporation better prepared to weather a storm though by no means as fully prepared as it would be ten years later.

With Mr. Durant's retirement, Pierre S. du Pont became president, continuing as chairman of the board. The new burden was one which Mr. du Pont did not seek; indeed, it was thrust upon him by the dire needs of the situation in which the Corporation found itself. In its 1910 trouble the bankers who rescued General Motors insisted on all possible legal protection as embodied in mortgage and voting trust. Although the 1920 situation of the Corporation was not as serious as its predecessor, nevertheless a receivership might have been asked for except for the high degree of confidence in which the new management was held by the banking fraternity. Mr. du Pont began his presidency by telling his associates that the Corporation was now operating under the most delicate and honorable relationship known to industry, "a receivership of our own," and in this spirit the Corporation carried on.

The totals for 1920 were relatively high because of the excellent earnings of the early months of the year, so the full force of the blow would not be of record until 1921. In 1920 General Motors earned $22,000,000 less than in 1919, and after dividends could reinvest only $14,000,000. The worst was still to come from the accounting standpoint, but by the close of 1920 the situation had been faced and the restoratives applied.

# Chapter XIV

## THE NEW ERA UNDER PRESIDENT DU PONT

WITH the presidency of Mr. du Pont, General Motors history entered a new phase. A calm, solid man, accustomed from youth to large affairs, and with the family habit of taking broad, far-reaching views, he brought to his new responsibility a poise the Corporation had never quite known before. He had a keen sense of fiduciary responsibility, and yet he was perfectly aware, as many conservative men are not, that by no means all the values of an organization can be listed on a balance sheet and appraised in dollars and cents. General Motors had had one experience of that sort, from 1910 to 1915, and it had ended abruptly with the Durant recapture through Chevrolet, as previously described, precisely because Mr. Durant knew the hidden values better than the bankers did.

On the other hand, Mr. du Pont was never driven by the devils of haste and expediency. The bright face of danger had no allurement for him. His attitude toward the vast property committed to his care was that of a conservator rather than a promoter. His faith in the future of General Motors and of the automobile industry in general held as firmly as that of his former associate, but he saw clearly that his job lay in reorganization, not expansion. General Motors was now a "made" property, and as such in need of long views and continuing policies. True, it had been made by desperate daring as well as by steady toil, but in 1920 it needed balance rather than daring. The emergency problem was to save the structure intact, certain that the swing of business would restore buying, and that General Motors would come up with the country if the weaknesses recently revealed were courageously met.

Mr. du Font's personality was not long in making itself felt. At the beginning he faced a difficult problem in reestablishing employee morale shaken by the sudden change of management. Remember that General Motors was a Middle Western creation. Important posts were filled by men who looked upon Mr. du Font's predecessor with adoration. Many of them had followed him into Chevrolet and back into General Motors. Now they were disposed to stick with the Corporation, realizing that General Motors was here to stay and that their futures were at stake. Nevertheless, accustomed as they were to the Durant procedure, they anticipated difficulties for themselves under du Pont management. Changes in personnel were necessary in many cases; and events indicated a need for a wholesale realignment of relations between the plants and the central offices. Whether that could be done seemed for a time doubtful; the situation was so tense, for instance, that in the case of one new manager there was grave doubt whether he might be allowed access to his office without interference from men who were being replaced.

This nervousness applied not only to managers but to labor staffs as well, and through them spread to whole communities. In no small degree the high morale of General Motors labor had been due to the dramatic leadership of the former president. To his old employees Mr. Durant had been both boss and friend, and while thousands of the later comers in his labor ranks had never seen him in the flesh, they knew him by hearsay. Years ago he had taken on for them the attributes of myth and story. He was their hero, their superman, doing the things they would like to do if they could. Particularly they had rejoiced when, like a Lochinvar out of the West, he had stolen the General Motors bride from the Eastern bankers in 1915. Now they began to grumble about absentee ownership and Eastern control, and wonder whether their jobs and houses would be worth anything a year hence. Even the substantial business men of the

PIERRE S. DU PONT
President, General Motors Corporation, 1920-23

Michigan cities with General Motors plants grew panicky at the thought of what might happen to them and their investments if, under the aegis of the du Fonts, plants were to be moved from West to East. Absurd rumors to that effect gathered headway; Chambers of Commerce grew excited and wrote in to plead consideration for their communities in the pending changes.

As soon as bandages had been applied to the sorest of the many sore spots on the corporate body, President du Pont took a statesmanlike swing around the General Motors circle. He visited plants and talked to groups of employees and associations of citizens. Everywhere he went he had the benefit of what the diplomats call a "good press." Local interests, from banks down to laborers, were reassured when they beheld this kindly, steady man and heard him tell them to be of good cheer. General Motors, he said, would stand by its investments in their communities as long as those communities stood by General Motors. There would be changes, of course, since not to change at the challenge of events was to risk corporate dry-rot, but the changes would benefit rather than injure the workers and the communities sustained by General Motors pay rolls. Opinion in Michigan completely reversed itself as soon as it heard that message: real estate values began to regain their buoyancy; labor, its morale; and plant executives, their stamina.

As on two previous occasions in the Corporation's history there were unfortunate accumulations to sweep away while reorganization efforts went forward. Liquidation of the expensive Janesville tractor experiment began. The Doylestown Agricultural Company, bought as part of the farm motorization program, went for $48,500, a loss of almost $100,000. The unfortunate Heany venture was finally wiped off the Corporation books when $1,205,000 in cash and securities was received from General Electric in return for General Motors interests in the electric light business.

An action most reassuring to the city of Flint was a General Motors subscription to $300,000 worth of capital stock in the Durant Hotel erected there and named for the founder of the Corporation. Meantime, the Corporation undertook a comprehensive study of its inter-relationships. As we have seen, General Motors' early growth was swift and chaotic. In that stage it was a hybrid rather than a unified organization. Attempts to organize it had been partially successful, but never wholly so, and for obvious reasons. Here was an enormous ganglion of highly active units, each of which faced problems peculiar to itself, each of which made its own approach to the public in the market place, and each of which had its own distinctive pride. Too much concentration of authority might very well reduce the initiative of plant managers and the efficiency of their staffs. On the other hand, since many of these units were dealing with one another, were buying similar goods and services, and all of them in the last analysis were taking their goods to the same market, the intelligent cooperation of the units was desirable up to the point of diminishing returns. Where to locate that point could be determined only by experiment, and preferably by cautious rather than rash experiment. Consequently various inter-divisional bodies were set up to integrate policies with respect to products, sales, purchasing, advertising, etc., thus initiating a program of close coordination but preserving decentralized administration.

One of the practical reforms of this period deserves especial attention since the danger which it corrected had twice played a destructive role by bringing General Motors into financial straits. This was a reasonable control of inventories. During the boom days when supplies were relatively scarce, plant managers had fallen into the habit of ordering in excess of need, sometimes to double or triple their requirements, merely to be assured that they would receive enough to let them fill their production programs. This had its obvious result. When business slackened, supplies were pushed out and sent forward faster than they could be used, and cancellations were not always applied in time to check the flow. Sometimes these cancellations did not hold, particularly if the order had been started on its production process. The new management met this by limiting the independent buying power of plant managers to the requirements of a four months' forecast. Control of inventory was the great lesson learned from the troubles of 1920-21.

Another weakness, in effect the other side of the inventory shield, was the lack of information regarding retail sales and stocks of cars in dealers' hands. None of the Corporation's automobile producers knew with even approximate accuracy how fast their cars were being taken out of the market by consumers; consequently they had no yardstick immediately available by which to regulate production. A first step was taken when General Motors encouraged R. L. Polk & Company to gather and make available to the entire industry, regular and frequent statistics of motor-car registration as fast as the figures could be gathered.

While real control of the production cycle, based on definite information, required some years in working out, the thought impulses in that direction began early in the reorganization. By 1923 each manufacturing division was obtaining from dealers ten-day reports on sales and cars in stock; by 1925 a degree of control over production programs had been achieved through a system suggested by Donaldson Brown, then vice-president in charge of finance. The essence of this improvement lies in control of production in line with retail demand as forecast on the basis of accurate and regular reports from the field, the final controlling calculations being worked out with reference to all calculable factors in the changing economic situation. Major errors of calculation have since been few and usually on the side of conservatism, so that at no time since this system was developed has General Motors been seriously threatened by unwieldy frozen inventories of either finished goods or parts in process. The decisions in this field are arrived at by consultations between headquarters and local managers with accurately gathered facts, usually bringing the conferees into agreement.

General Motors' outstanding contribution to industrial organization has been made along this and similar lines. Even a brief survey of the anatomy of industry reveals that, in general, industry is either autocratic or feudal in its set-up. Mr. Durant's rule over General Motors during his incumbencies may be accepted as a fair example of the second. Mr. Durant relied on trusted lieutenants and trusted some of them too far, their capacities sometimes being unequal to their responsibilities. In attempting to find a steady base somewhere between these different ways of expressing industrial authority which, of course, is a sine qua non of dependable production General Motors set up a system of liaison and control for which corollaries must be sought elsewhere than in industry.

This system has been likened in some respects to the staff-and-line organization of an efficient army, with the staff as the planner and coordinator of line activities, the former mapping the strategy of the campaign, the latter applying the tactics most likely to carry out the strategy. But if General Motors' organization has certain military aspects, it has others which are quite the reverse of military. Indeed, to the limited extent to which central control functions, it does so through a system of coordination seen to resemble the activities of a government rather than that of an army, while local managers, within the frame of consultative coordination described above, retain administrative jurisdiction over and responsibility for their divisions, each of which is an operating unit. Some concessions were made to the economic pressure of 1931-33, but these were not considered ideal relationships and were discontinued as soon as the need for them passed. The inter-divisional set-up of 1927 is thus described by Mr. Donaldson Brown:

Serving in the direction of crystalizing the important Corporation policies and making them effective, and to facilitate the adaptation of engineering improvements and operating methods, there are various so-called inter-divisional relations committees. They have suitable representation from the most important divisions, and are as follows:

| | |
|---|---|
| General Purchasing | Works Managers |
| General Technical | General Sales |

Institutional Advertising.

These committees meet separately once a month. The President of General Motors is a member of each one and besides, there is

> at least one other member of the Executive Committee. The work of the committees clears through various central office staff organizations, maintained so as to perfect the flow of information back and forth and to facilitate the orderly consideration of common problems of important policy and procedure... . Where there is a question at issue on the score of Corporation policy, the President of General Motors makes the decision or refers it to the Executive Committee. Cases of this kind are rare.[1]

The inter-divisional relations committees named above occupied themselves only with the policies in their respective fields, which required consideration from the standpoint of the Corporation as a whole, rather than from the standpoint of any single division.

If one thinks of these various committees as conducting investigations and taking the evidence of experts as Congressional committees do, and within a definite range of authority making decisions which are either put into effect immediately or referred to some higher committee for approval, then one will have a rough idea of how General Motors functions. An illustration, taken from the same address, will show how a committee works on a specific problem that of prices:

> The question of pricing product from one division to another is of great importance. Unless a true competitive situation is preserved, as to prices, there is no basis upon which the performance of the divisions can be measured. No division is required absolutely to purchase product from another division. In their interrelation they are encouraged to deal just as they would with outsiders. The independent purchaser that is buying product from any of our divisions is assured that prices to it are exactly in line with prices charged our own car divisions. Where there are no substantial sales outside, such as would establish a competitive basis, the buying division determines the competitive picture, at times partial requirements are actually purchased from outside sources so as to perfect the competitive situation.

---

[1] From Vice-President Donaldson Brown's address to the American Management Association, February, 1927.

After the date of Mr. Brown's speech the status of interdivisional relations was changed. The Operations Committee was no longer inter-divisional but did all the routine work of the Executive Committee. The roster of interdivisional relations committees was then reduced to four General Sales, General Purchasing, General Technical, and Works Managers. Several advisory committees were formed, as the Public Relations Committee, Advertising Committee, and the like. Somewhat further simplification in the direction of greater divisional autonomy occurred in October, 1933, when W. S. Knudsen became executive vice-president of the Corporation in charge of car and body manufacturing in the United States and Canada.

The acute financial pressure of 1921 had stirred the Corporation to another reform of lasting consequence both to itself and to industry at large. This was the better mobilization of its cash resources through the use of the fiscal transfer machinery which had been set up by the Federal Reserve System. In October, 1920, General Motors owed to banks $82,700,000. The strides made by General Motors in ten years from 1910 to 1920 can be measured by its borrowing power. In the former year it had to mortgage all its properties to borrow $15,000,000; in 1920 it could borrow more than five times this sum without recourse to unusual expedients. Yet during this stringency many of the divisions had more money than they needed. Clearly, one reform immediately necessary was the pooling of financial resources. This was accomplished in a way which has historical interest, as it was the first occasion in which the machinery of the Federal Reserve System was used to facilitate the fund transfers of a large corporation active over the whole United States, and the plan so set going has been studied and adopted in principle by many other corporations.

There were two elements in the problem. One was the rigidity of General Motors cash under divisional control, which forced Corporation borrowing when division funds were tied up by managers most concerned with being on the safe side in their own operations; the other was the time involved in transferring funds from city to city and from plant to plant, involving a "float" of ten million dollars at a time when every dollar could be used to advantage. Furthermore, the Federal Reserve System, having established means of practically instantaneous transfer by telegraph, desired to have those facilities tested in use.

Three plans were submitted by General Motors to the Federal Reserve officers, one of which was accepted as workable. It involved the automatic transfer of General Motors balances from its banks of deposit whenever the maximum agreed deposit to be retained by that bank had been exceeded. These remittances accumulated in eleven large banks as central reservoirs. In place of slow remittances by mail, telegraphic transfers shifted funds almost instantly.

This change involved visits to banks all over the country and many consultations with division executives to convert them to the idea of central financial control. As a means of meeting their natural concern to have abundant working capital promptly when needed, a system of disbursing funds to divisions upon requisitions was set up which operated as swiftly as the inflow.

As the plan became effective over the entire country, with all divisions and their banks cooperating through the Federal Reserve System, it was seen to have decisive benefits for all concerned. The Corporation benefited by reducing its "float" from $10,000,000 to $4,000,000, and also by increasing its potential borrowing power from $82,700,000 to more than $100,000,000. The banks cooperating in the plan benefited by having more stable General Motors deposits, and by having the interest rate on those deposits uniform for the entire country for several years after the plan went into effect. They were able to use General Motors deposits to better advantage than formerly. The Federal Reserve System benefited by demonstrating to the industrial world the efficiency of its service, the demonstration being so complete that many other large industries studied the plan and adapted it to their needs.

The financial stress of the post-war era soon passed. By June 30, 1922, the Corporation was out of debt, and for the balance of that year carried no bank loans, yet its cash had been so mobilized that it had a large potential borrowing capacity. The Corporation marched out of the post-war depression on a pair of seven-league boots. One of these was the internal reorganization already discussed. The other was the fruit of an energetic development of automobiles participated in by many companies but in which General Motors played no small part through the activities of its research department and those of affiliated companies. These improvements induced enough buying to make 1922 notable for the largest volume of business in units to date. Two years after General Motors had approached bottom in 1921 it was again riding the crest in 1923.

Whereas in 1921 the Corporation lost $38,680,770 the only deficit recorded in the history of General Motors, yet the following year saw a complete reversal with $54,474,493 net income available for dividends. Dividends were paid at the rate of $1 a share through 1921, but the quarterly dividend of 25 cents due February 1, 1922, was passed as a matter of precaution. Prompt recovery, however, permitted the payment of a special 50 cents dividend in December. During the 1921 slump the average number of employees decreased by 35,000 but 1922, with 65,345 on the pay roll that year, saw the beginning of another upward swing in employment.

Among the timely acts of the period were:

Sheridan at Muncie, Indiana, and Scripps-Booth at Detroit, were discontinued. During its short life Sheridan was produced at the former Interstate plant.

The Samson tractor plants at Janesville, Wisconsin, were switched to automobile production.

General Motors Building Corporation was formed to hold and operate the $20,000,000 office building in Detroit.

Saginaw Malleable Iron Company was acquired.

The outstanding shares of the Klaxon Company were acquired by exchange.

The Janesville Electric Company was sold to Halsey, Stuart & Company, acting for power interests.

Modern Housing Corporation shares were reduced from 17,500 to 1,750, $100 par.

The Corporation subscribed for 8,000 shares of the General Motors Acceptance Corporation at $125 a share, in the sum of $1,000,000.

Another sale of this period, indicative of the Corporation's efforts to forsake the real estate and home-building field where possible without distress to employees, was the sale by Harrison Radiator of 875 shares in the Lockport Homes Company.

Perhaps more significant for the Corporation than any other activity of the period was the drive which began in 1921 to improve the quality of its cars. Buick and Cadillac were in no need of overhauling, but the low-priced cars Chevrolet, Oldsmobile, and Oakland revealed weaknesses which flowed naturally from the post-war conditions. Then the cry had been for more and more production, with the result that engineering improvements had been postponed or overlooked entirely. As trade slackened, weak mechanical units began to affect both sales and reputation. One by one these problems were grappled with firmly. Under President du Pont began the long struggle to rehabilitate Chevrolet and establish it as the Corporation's leader in quantity production and earnings, which has finally made that subsidiary the world's leader in the low-priced field. Oldsmobile and Oakland, in their respective price ranges, were put under similar treatment. This farsighted policy proved immensely beneficial in after-years, bringing the Corporation to new levels of earnings and efficiency. Indeed, it may be held as the chief determining factor in lifting the Corporation to the impressive totals it reached from 1925 onward. Moreover, the lessons learned and remedies applied from 1920 to 1923 were to prove invaluable in the greater stress of 1930 to 1933.

On May 10, 1923, Pierre S. du Pont retired as president. The two and a half years of Mr. du Pont's presidency saw the Corporation rise swiftly from a perilous position to new levels of profits and financial stability. Dead wood had been cut away, and the weak properties brought to increased efficiency. One by one the Corporation's great producers had been taken in hand, tuned up, and placed in a position of market leadership. Progress had been made toward integrating a key control at headquarters. This eased interplant relations without reducing too greatly the responsibilities of plant managers, whose interest was to be further roused by the creation of the Managers' Securities Company a little later. Employee morale had been achieved by the maintenance of high wages, generous bonuses, and the highly successful conclusion of the first Employee Savings Fund program.

With the Corporation in this favorable condition, President du Pont, certain that the situation he went in to remedy had been fully cured, laid down the responsibilities of the presidency to return to his other large affairs. He was to remain chairman of the board for six years more. It is not too much praise to say that he saved General Motors by taking the helm in bad weather and effecting such reforms that it could not fail to reap the rewards of many of the policies he formulated. The long-range policies and forward views which Mr. du Pont took are still effective in Corporation thought and practice, and will continue to be influential there for many years to come. On leaving the presidency he attributed a large measure of the Corporation's successful development to Mr. Alfred P. Sloan, Jr., vice-president in charge of operations, who succeeded him in office.

Mr. Sloan entered the Corporation as a director on November 7, 1918, incident to the merging of United Motors, of which he had been president, and at first had charge of the operation of the United Motors group. With Mr. Durant he had been an important figure in the creation of United Motors, bringing into that organization the Hyatt Roller Bearing Company, with which he had long been associated. Combining technical training with executive experience, he was widely known throughout industrial America even before he assumed a high place in General Motors councils, and this acquaintanceship later brought into the General Motors family several of its present units.

That Mr. Sloan also possessed sales insight and had been a force in working out the coordinating reforms described in this chapter may be gathered from a tribute reported by B.C. Forbes

ALFRED P. SLOAN, JR.
President, General Motors Corporation, 1923

in the New York American, September 21, 1927:

> When he was selected president, in 1923, to succeed Pierre S. du Pont, the latter made this public announcement: "The greater part of the successful development of the Corporation's operations and building up of a strong manufacturing and sales organization is due to Mr. Sloan."

As General Motors integrated its policies and undertook a systematic supervision of its extended operations under Mr. Sloan as vice-president in charge of operations, he proved his fitness for the higher office so well that, when Mr. du Pont laid down the reins, there was never any question who would be chosen to succeed him as president. In him the Corporation secured an operating head possessing a combination of technical training, executive experience, and financial sense.

By 1923 General Motors had gained an important part of a trade already of gigantic proportions and destined to grow to even greater heights. America was just beginning to realize what the motor car meant to it economically and socially. The magnitude of the ever-growing industry is revealed in these statistics drawn from the 1923 census of manufactures:

> Three hundred and fifty-one factories were operated by the automobile industry in the United States and the District of Columbia 54 being in the state of Michigan.
> More than 318,000 persons were directly employed in the manufacture of vehicles in that year and 3,105,000 were benefited indirectly by the industry.
> A billion dollars was being spent on highways and more than $142,000,000 in the transportation by rail of completed vehicles.
> Eighty-five percent of the gasoline consumed in the United States during that year was used in cars and trucks.
> More than 10 percent of the products of the steel and iron industry were used in the production of motor cars or of their parts and accessories.
> One hundred and sixty establishments were engaged in the manufacture of tires and inner tubes while a total of 528 manufactured rubber products.

Of the thriving automotive trade General Motors already enjoyed a large share. After fifteen years of growth, punctuated by two stressful periods, General Motors in 1923 had reached unquestioned financial stability, had discovered the formula by which decentralized operations could be correlated and controlled within limits of efficiency, and had acquired the good-will of millions of satisfied buyers.

Opel Six at Shanghai, China

# Chapter XV

## ROUNDING OUT GENERAL MOTORS

AFTER the initial speed of accumulations had been checked and balanced by the banker reorganization of 1910-15, General Motors expansion came from two sources. By far the most impressive elements in the growth of the Corporation since 1923 have been written into the record by the internal growth of various divisions, chiefly the great car divisions and the accessory divisions supplying them, accounts of which will be found elsewhere under divisional heads. Another source of growth has been the purchase and adaptation of manufacturing units and service organizations which it seemed eminently desirable and necessary to control in the interest of industrial competence.

Strong reasons began to develop soon after 1923 impelling the Corporation to fortify itself wherever possible for a future which obviously would be considerably different from its past. Evidence began to accumulate that the industry as a whole was approaching stability. The phenomenal popularity of the motor car in its early stages had provided such large profits that many errors of judgment had been written off almost without notice. That era of "easy come, easy go" had passed, and with its passing the business of manufacturing motor cars became concentrated in fewer hands.

This "sweating-out" process had been going on all through the life of General Motors, and had in fact threatened various of its units at one time or another. The more than 300 manufacturers of 1908 had been reduced to 189 by 1914 and to 143 by

1923; these figures including promotions and other short-lived efforts. Of these hundreds, twenty-five manufacturing units would do more than 98 percent of the business in 1931.

The continued progress of the leaders created the impression that the entire industry was proceeding at the earlier rate of growth, whereas the truth was, as Mr. Sloan told the automobile editors in his Proving Ground address in 1927, in discussing the so-called saturation point, that "The industry has not grown much during the past three or four years. It is practically stabilized at the present [1927]. What has taken place is a shift from one manufacturer to another."

The trend from 1923 to 1927 meant a basic shift of the industry's attitude to the market. Henceforth the motor cars would have to be sold, instead of merely demonstrated. They must be advertised more and serviced better. Research and improved methods of manufacturing would have to be constantly speeded up to produce better and better cars for a market ever growing more sternly competitive. Precision methods, long applied in factory production, were equally needed in other branches of the trade in the calculation of logical production programs, in the accounting systems of dealers, etc. The industry had come of age and, with maturity, competition brought new responsibilities for management. Such increase in volume as developed would have to come from three sources: replacements, population growth, and the increased use of automobiles. Moreover, the share of business received by any producer would have to be secured by hard work and careful planning rather than through luck, boldness, or blind reliance on general prosperity.

Consequently, it was time to insure unquestioned control of certain essential supplies and to adopt, with relation to those sources once they were in hand, such improvements in their processes as would bring their goods into the Corporation's productive cycle as efficiently as possible.

A number of these additions merely registered the Corporation's completion of ownership in properties partly owned before and already an essential part of the production chain. In this category was Brown-Lipe-Chapin Company of Syracuse,

New York, which had been making differentials for General Motors cars almost from the latter's birth, and in which General Motors had been influential for years by reason of a large stock ownership. On December 19, 1922, the famous Syracuse company became a completely owned division of General Motors.

Armstrong Spring Company of Flint, though under various corporate ownerships during the period, had been making springs for General Motors for many years, working closely with the chief car divisions. It was purchased by the Corporation in December, 1923, becoming a wholly owned division.

In the field, perhaps the outstanding development of 1924 was the purchase and initial improvement of the Proving Ground near Milford, Michigan. Within easy reach of the Corporation's automobile plants at Detroit, Pontiac, Flint, and Lansing, the site was selected after an intensive search for a tract where surfaces and contours would provide the greatest variety of tests for motor cars. Here were heavy grades, sand and gravel hills, marshy lanes, and muddy bogs. Since the purchase, concrete and gravel roads, and a wide range of other road surfaces have been added and shops erected, where cars are taken down and examined after being put through their paces on long, stiff journeys over these varied highways. At the Proving Ground, which is really a great outdoor laboratory, all General Motors models are tested with the utmost rigor before they are placed in production by the various car divisions.

In the first year of General Motors' history, W. C. Durant used to tell his racing and test pilots to drive until their cars broke down under the strain, in an effort to locate weaknesses and indicate where improvements should be made. Now, sixteen years later, this idea materialized in a mammoth undertaking in which that exacting work could go on under scientific supervision, as part of a broad research program. The Proving Ground is a large-scale enterprise in fact-finding for the protection of the buying public. Into the corporation field in 1924 also came the new Ethyl Gasoline Corporation to market the "knockless" fuel which General Motors chemists had evolved through years of alert and adventurous research. The story of Ethyl will be told in detail elsewhere;[1] it is one of the characteristic triumphs of the open mind in industrial research. To market this new fuel a $5,000,000 corporation was formed, and $1,150,000 was subscribed in 1924, with Standard Oil of New Jersey and General Motors each having a 50 percent interest.

Fisher Body, with large earnings and the prospect of more, expanded its capital stock to 2,400,000 shares of $25 par value to take the place of 600,000 shares outstanding. Shortly after, Fisher began, with the acquisition of Fleetwood Corporation, an important growth cycle of its own, narrated in Chapter xx: "Body by Fisher."

Negotiations had been in progress for some time with the Hertz interests of Chicago looking toward the acquisition of an interest in the Yellow Cab Manufacturing Company of Chicago. On August 13, 1925, the directors approved the offer of that company relative to the transfer of the General Motors Truck Division to a new company, the Yellow Truck & Coach Manufacturing Company, a holding company in which General Motors Corporation acquired a majority interest. In addition to General Motors Truck, with large plants at Pontiac, Yellow Truck & Coach Manufacturing Company owns all or part of several other companies.[2]

October 21, 1925, marks the proposal by Vauxhall Motors, Ltd., of Luton, England, of a plan whereby General Motors entered the field of British automobile manufacture under excellent auspices.[3] This plan, which was consummated on November 24th, required a total outlay by General Motors of 510,000, or roughly $2,500,000 at par of exchange, for which the Corporation secured a controlling interest in an established British company with a possible production of 25,000 cars a year. Since the McKenna duties increased the tariff on American cars entering British markets, and preferential duties were in force in other parts of the Empire, the advantages of this coalition have been amply manifested.

---

[1] See Chap. XIX: "Research: the March of the Open Mind"

[2] See Chap. XXII: "Commercial Vehicles."

[3] See Chap. XVIII: "General Motors Across the Seas."

Another expansion of General Motors Acceptance Corporation capital occurred in December, the Corporation subscribing for 45,000 shares of GMAC stock at $125 a share, or $5,625,000. Instalment sales had grown tremendously in volume with the growing confidence of the country, and the country was approaching the point at which more than half the automobile output would be bought on credit.

General Exchange Insurance Corporation, with a capital of $500,000 and surplus of $1,000,000, was formed by General Motors Corporation to handle the vast insurance business arising through sales financed by General Motors Acceptance Corporation. The timeliness of this innovation was soon demonstrated, for within two years GEIC made profits of $676,000.[1]

After making a step toward retrenchment by authorizing the liquidation of the Lancaster Steel Products Company in 1926, the directors authorized two other large investments. Yellow Truck & Coach Manufacturing Company, in order to provide for expansion of its operations, sold additional capital stock, for the proceeds amounting to $14,000,000. General Motors underwrote the entire issue and secured its pro rata share. The other was the inauguration of a plan to buy up to $35,000,000 worth of Common stock before the close of 1930 for the second Managers Securities Company.

Following this important move the Fisher group was also greatly expanded by the construction and acquisition of new properties and companies making necessary supplics among them the Flint plant, the largest in the Fisher ensemble. Also, the Fisher structure was simplified by the purchase of the affiliated Ohio company. The "Golden

"Financing and Insuring the Buyer." 'Twenties" were busy years for Fisher, with four incorporations or acquisitions in 1923, two in 1924, two in 1925, five in 1926, two in 1928, and one in 1929. The complete Fisher set-up will be found in Chapter xx.

---

[1] A full account of General Exchange Corporation and its successor, General Exchange Insurance Corporation, will be found in Chap. XXVII:

August, 1926, marked the merger of selling efforts on certain important lines of the Corporation's electrical products in the formation of the Delco-Remy Corporation of Delaware, which was followed presently by the transfer of the starting, lighting, and ignition business of Delco, from Dayton to Anderson, Indiana, the seat of the present Delco-Remy division.

On November n, 1926, General Motors Acceptance Corporation stock was increased to $25,000,000, since it was clear that the country had overcome its hesitation of the spring and was confidently going forward, giving every promise of availing itself of time-payment plans to the full in automobile purchases. Another $10,000,000 increase in GMAC stock was ordered in June, 1927, and later GMAC acquired the General Exchange Insurance Corporation at its net asset value as of June 30, $2,176,702, the original cost to the Corporation having been only $1,500,000.

During these busy and prosperous years interests were also acquired in Bendix Aviation of Chicago; Fokker Aircraft Corporation of America, the latter becoming the General Aviation Corporation; Winton Engine of Cleveland, Ohio, makers of marine and stationary engines for railway use; McKinnon Industries of St. Catharines, Ontario; the Guide Lamp Company of Cleveland, Ohio; North East Electric of Rochester, New York; and others.

All through this period also there was an unceasing flow of changes in plant utilization. The union of Delco and Remy at Anderson, Indiana, has resulted in a large industry, although not all Delco activities were moved. There remains in Dayton the Delco Products Corporation, making Lovejoy Hydraulic shock absorbers, small motors, and a wide range of other appliances. North East Electric at Rochester, New York, bought in 1929, soon became Delco Appliance Corporation, and to it were transferred the Delco Light business and other electrical lines chiefly in the household appliance and radio fields. This shift of functions toward a more efficient alignment, both geographically and technically, has become more marked since the year 1930.

General Motors thus has become a large producer of motor-driven household equipment. The Frigidaire Corporation manufactures electric refrigerators, beverage and water coolers, refrigerating apparatus for households, hotels, restaurants, and apartments, air conditioning equipment for homes, offices, shops and railway cars. Delco Appliance Corporation makes electric light and power plants (Delco Light), electric motors and blowers, gas-producing units (Delcogas), and Delco oil burners, vacuum cleaners and water pumps, and electric fans. Small motors are made by Delco Products at Dayton. Gas refrigerators for household and apartment use are produced by the Faraday Refrigerator Corporation, also at Dayton.

The process of growth by purchase and development still goes on, though purchases of other properties are naturally at a reduced rate. A recent acquisition, in 1932, was the Packard Electric Corporation of Warren, Ohio, manufacturers of automotive cable products. The name has early connections with the industry, as the Packard car originated there. It will be noted that the Corporation in its twenty-five years of history has withdrawn from several activities for substantial reasons, and the motives making for abandonment have, of course, been especially strong from 1930 to 1933. The most important of these recent changes was the removal of Muncie Products activities from Muncie, Indiana, and the transfer of Brown-Lipe-Chapin activities to Buick and other plants.

With the exception of 1918, the war year in which materials could not be secured to satisfy the public demand, and in 1921, when the post-war slump reduced production by one quarter, the automobile industry had marched steadily forward in the quantity and value of its products for a quarter of a century. In the ten years from 1912 to 1922, the annual production had risen from 378,000 units with a wholesale value of $378,000,000 to 2,646,229 units with a wholesale value of $1,793,022,708. In 1923, the totals bulged to 4,180,450, valued at $2,592,033,428. A definite periodicity marks automobile production through the whole decade of the 'twenties. The trade began a succession of three-year cycles, apparently as a result of its own circumstances and enthusiasms, as follows:

| | | |
|---|---|---|
| 1921 Black year | 1924 Black year | 1927 Black year |
| 1922 Goodyear | 1925 Goodyear | 1928 Goodyear |
| 1923 Banner year | 1926 Banner year | 1929 Banner year |

The descriptions are, of course, merely relative, as new heights were being scaled all through the period, so that the good year of 1925 produced a larger volume than the banner year of 1923, and the good year of 1928 brought more business than the banner year of 1926. The figures on units manufactured, in the United States and Canada, and their wholesale value for the period are:

| | **Units** | **Value** |
|---|---|---|
| 1921 | 1,682,365 | $1,261,666,550 |
| 1922 | 2,646,229 | 1,793,022,708 |
| 1923 | 4,180,450 | 2,592,033,428 |
| 1924 | 3,737,786 | 2,367,413,015 |
| 1925 | 4,427,800 | 3,015,163,562 |
| 1926 | 4,505,661 | 3,214,817,491 |
| 1927 | 3,580,380 | 2,700,705,743 |
| 1928 | 4,601,141 | 3,162,798,880 |
| 1929 | 5,621,715 | 3,576,645,881 |

Owing to conditions which may be considered special, General Motors entirely escaped the 1927 reduction in volume common to the industry. Net sales for the Corporation rose by more than $200,000,000 over the 1926 figure, reaching $1,269,519,673, or, on the basis of units sold, nearly 44 percent of the entire output for the United States and Canada in 1927.

The trend is upward but not constant; after every banner year as noted, there is a dip, and then begins a two-year rise to new highs. To those aware of this swing of the pendulum it seemed likely that 1926 would be another banner year, unless there should be a general slackening of trade. For five years there had been a steady appreciation in urban real estate and security values, but farm produce had never recovered fully from the post-war price slump, and the condition of the farmers could not be considered good. The disparity between the farmer's dollar and the urban dollar was marked enough to cause some disquietude; only in automobiles and tires, owing to manufacturing improvements, could the farmer buy to advantage compared with the pre-war ratio. Moreover, while automobile prices had dropped steadily from 1912 to 1917, and again from the war peak of 1918 to 1924, they had shown a rising tendency in 1924 and 1925. Students of automobile economics were asking themselves whether the upward swing could continue; the Stock Exchange drop of March, 1926, enough to call forth a reassuring interview from President Coolidge, indicated a gathering tension.

The country, however, gathered courage and went on. The year 1926 rolled up the record figures of nearly $3,215,000,000 in wholesale value of the automobile industry's output, and the automobile cycle swung through another three-year period to even larger figures.

Economic historians will be debating the long continuance of the prosperity of the 1920's for many years to come. All precedents indicated that the boom should have reached its zenith before 1929. Evidently there were new factors in the equation. Among them were: increased efficiency in industrial production, growth of export trade, the nation's increase in gold holdings, and the wider distribution of wealth. Still another, of special moment to the automobile industry, was the enormous extension of instalment selling.

The amazing totals of 1928 and 1929 can only be explained by boom conditions confounding all calculations. We have seen how firmly Mr. Sloan, with all available data on hand, considered in 1927 that the industry as a whole was approaching stabilization, in which continued demand would be largely for replacement purposes, with some allowance for growth due to increasing population and gain in motor-mindedness. Yet in the next two years the automobile industry swept aside all these conclusions of conservatism under the lash of a demanding market.

# Chapter XVI LATER HISTORIES OF PASSENGER CAR DIVISIONS

THE early histories of the four of the five passenger car divisions of General Motors have been carried down in previous chapters to the time they joined General Motors of New Jersey, and that of Pontiac from its origins in Oakland down to the present time. This chapter presents briefly the stories of Buick, Cadillac, Chevrolet and Oldsmobile after they were merged into General Motors.

These compressed histories necessarily omit many interesting details and pertinent facts, as each of the car divisions is in itself an important, large-scale industrial operation deserving of more extended treatment. All have compiled historical material in considerable amount, some of which has been circulated widely in pamphlets and through press releases.

## BUICK

When Buick became the foundation stone of General Motors in 1908, it had already won leadership in the medium-priced field, and in volume of business it ranked first in the American automobile field. Ever since it has been one of the chief pillars of Corporation strength. From an output of only sixteen cars in 1903 Buick has reached a total production of more than 2,500,000 units, more than half of which are still in operation, including models of every year except the first. Its largest year, in unit sales, was 1926, when 280,009 Buicks were sold.

During its life-span of thirty years Buick has held closely to the medium-price range. In only six of those years has the average Buick the middle-priced car of all its models for the year exceeded $2,000. Only one model has been sold below $1,000.

One result of this consistent price policy has been the creation of a very large body of buyers who continue to order Buicks year after year. While many makes of cars have found new price levels in the history of the industry, a few rising and more dropping in the scale, Buick has kept its original position, compensating the customer as costs declined by adding to the value of the product year by year.

When Buick began, it was a manufacturing policy, common to the entire industry, to sell a bare car at a listed price and then furnish, as extra equipment at set prices, many appliances necessary to comfortable motoring which today's buyer looks upon as essential. For instance, the famous Model F two-cylinder Buick sold at $1,250, F.O.B. factory, but unless the customer bought and paid for a top he would ride in the rain; unless he bought lamps he would ride only by day; unless he bought a windshield he sat exposed to the blasts of Boreas. Extra equipment lists often included such essential elements as tires, horn, and gauges. Consequently a comparison of automobile prices, historically, means very little until manufacturers reached the point of selling their cars completely equipped. What one gets for his money is quite as important as what he pays.

Through this development from the skeleton car of the old days to the fully equipped car of the present, Buick was usually to be found somewhat to the fore in the matter of shifting extra equipment to standard equipment, rather than following the policy of pushing the price down and keeping the charges for extras up. For that reason a year-by-year statement of additions to Buick equipment is an interesting exhibit of the stages through which the American automobile industry advanced toward the present practice of equipping a car completely.

In the evolution of the automobile, both as a machine and as a style vehicle, Buick has always been consistently progressive, originating many improvements and applying others as soon as they developed reliability. For that reason, if one desires to see the steps by which the automobile has reached its present appearance, he can find some of the stages of that progress in the comparison of Buick specifications. One can see the automobile growing through additions like these: 1903-04, folding celluloid wind-breaker; 1905-06, side curtains; 1907-08, Presto-lite headlights; 1911, first closed car, a limousine, folding seats in tonneau; 1912, mohair top; 1913, clear vision windshield, electric lights, nickel trimmings instead of brass; 1914, first sedan.

On the mechanical side the progress can be charted by a few of many thousands of transitions: 1903-04, safety cranking device; 1904-05, disc clutch, steel drive shaft; 1907-08, aluminum crankcase, magneto, spark and gas control sector on steering wheel; 1909-10, pressed steel frame; 1911, oil splash system; etc., etc.

Bob Burman at wheel of "Old 41" famous Buick racer, 1909

Buick pioneered in glass fronts, fore-doors, internal expanding brakes and four-wheel brakes, and in the development of the rear axle assembly with differential.

In order to test its cars at high speed, Buick built a private track on its Flint property, where such speed and endurance were developed that its cars were soon entered in the more important races and hill-climbing contests. The sporting interest has already been referred to as developing automobile enthusiasm in the public. While amateurs might start these competitive events, the manufacturers, noting both their news value and test value, provided the competition, spending large sums to secure famous drivers. Buick's most famous racing team consisted of seven pilots, among whom were four of the first rank Bob Burman, Louis Strang, and the Chevrolet brothers, Louis and Arthur.

Among the more famous events of the many won by Buick during its first eight years were: hill climbs at Eagle Rock, Mt. Washington, and Pike's Peak'; the Vanderbilt Cup road race of 100 miles; the French Grand Prix; the 595-mile Elgin, and the 482-mile Los Angeles-Phoenix; track victories at Brooklands, England; St. Petersburg, Russia; and the Indianapolis Speedway. The Russian victory brought to Buick the Imperial trophy.

At the Indianapolis race appeared two odd cars, like big June bugs, with the name "Buick" on the side of each body. These two "Buick Bugs" were piloted by Bob Burman and the Chevrolets, Arthur and Louis. In the next two years they broke more records than any other team. In 1911 at Jacksonville, Florida, Bob Burman driving a "Buick Bug" established two straightaway records regardless of class: 50 miles in 35 minutes and 52.3 seconds, and two days later, 20 miles in 13 minutes and 11.92 seconds. Both records stood for more than ten years.

Victories like these were the result of careful organization and training, but if one retreats into the past as far as the car which is still affectionately called "old Model F" he finds amateurs at the wheel, and an informality like that of a country fair. When his little two-cylinder Buick first won the Eagle Rock hill climb, cutting the existing record almost in two, Walter L. Marr was asked how his small motor could develop the power to defeat much larger and more highly powered cars. He replied, quickly: "It's a Valve-in-the-head' motor." The phrase caught on and has been retained.

With the complete professionalizing of the sport, and the evident trend toward specially designed racing cars, Buick's interest in racing ceased while its great racing team was still at the top of its success. Although the records established by the non-stock racing cars of the present are being bettered year by year, some Buick records still stand in the stock classification.

From 1910 to 1920 Buick production rose from approximately 20,000 cars to more than 120,000 a year. Only three of those years show a recession from the preceding one, that of 1911 reflecting the conditions which brought on the banker control of General Motors, and that of 1918 being explained by the difficulty of securing material in a war year. During that decade the closed car came into its own, each year showing a larger percentage of production until the open car, except in sport models, has become a rarity.

Mr. Harry H. Bassett became president of Buick in 1919. His was a personality which endeared him to men of all positions both inside and outside the organization, his interest in the welfare of his workmen being particularly keen[1] He had been with the Remington Arms Company for fourteen years when, in 1905, Mr. C. S. Mott persuaded him to enter the Weston-Mott Company at Utica. The Weston-Mott plant at Flint being established in the meantime, in the next year Mr. Bassett went there as assistant factory manager. By 1913 he was made general manager, later also vice-president. On consolidation of Weston-Mott and Buick in 1916 he became the assistant general manager of Buick, the same year it was made a division of General Motors. In 1919 he became general manager, also a vice-president of General Motors, and then assumed the presidency of Buick Motor Company. In 1924 he became a member of General Motors' Executive Committee.

In Mr. Bassett's first year with Buick, capacity was increased to 750 cars per day, and in 1924 to 1,200 cars per day. The millionth Buick appeared in 1923. Upon his death Mr. E. T. Strong, who had been sales manager, became president in 1926 and was succeeded by Mr. I. J. Reuter in 1932. In October, 1933, Mr. Reuter resigned and was succeeded as general manager by Harlowe H. Curtice, who had been president of AC Spark Plug Company since 1929.

In 1928 Buick Motor Company celebrated its Silver Jubilee, the twenty-fifth anniversary of the organization, producing at that time its Silver Jubilee model.

During 1933 Buick's functions have been enlarged by the addition of the former Armstrong Spring division of Flint and

---

[1]See Chap. XXVIII for an account of his influence on General Motors Institute of Technology, etc.

the transfer to Flint from Syracuse, New York, and Muncie, Indiana, of differential and transmission manufacture.

## CADILLAC AND LA SALLE

In 1909 General Motors Company acquired the entire Common capital stock of the Cadillac Motor Car Company consisting of $1,500,000, par value, paying therefore $5,169,200 in General Motors Company Preferred stock and $500,050 in cash or a total of $5,669,250.

Cadillac: the first car equipped with an electric self-starter

In 1910 battery ignition was placed on Cadillac. The same year the first "big" order for closed cars (150 bodies) was given to Fisher Body Company. In 1912 the first electric self-starter was installed in Cadillac as standard equipment, as related in Chapter XIX: "Research: The March of the Open Mind." Mr. C. F. Kettering, the inventor of the starter, drove to the opening of the General Motors building at the Chicago Century of Progress Exposition in 1933, in one of these 1912 Cadillacs. For this contribution, Cadillac in 1912 won for the second time the Dewar trophy, awarded by the Royal Automobile Club of London for the greatest advance in automobile construction during the preceding year, its first victory having been awarded for interchangeability of parts five years earlier.[1]

Cadillac introduced, in 1914, the 90-degree V-type eight cylinder engine, the first eight-cylinder high-speed automobile power plant built in this country. This achievement of Cadillac was honored by the Smithsonian Institution which placed one of the first Cadillacs of this type in its museum in Washington. Every Cadillac built from 1914 to 1927 was powered by this same type engine. Even the bore and stroke, piston displacement and S. A. E. horsepower rating remained the same. In this period refinements in the power plant included thermostatic carburetion, the compensated shaft, and a positive system of crankcase ventilation. Other honors which came to Cadillac are treated in Chapter XII:

In 1917 Mr. Henry M. Leland resigned as president. He remained active in business and civic affairs until shortly before his death in 1932, at the age of eighty-nine. He was succeeded by Mr. R. H. Collins, who was president of Cadillac until 1921, during a period of rapid expansion. Mr. H. H. Rice then became president, remaining until 1925 when Mr. Laurence P. Fisher assumed the position.

Soon afterward Cadillac began a $5,000,000 expansion program. In 1927 the La Salle v-8 was introduced to a market which absorbed nearly 60,000 La Salles in the next three years; a sales record unparalleled for a new car in its price and quality class.

Both Cadillac and La Salle appeared in 1928 with non-shattering glass, the first cars to leave factories with that safeguard. Later in the year they were also equipped with the famous synchro-mesh transmission, in its day a decided innovation.

Synchro-mesh, the silent gear shift, was brought to perfection by Cadillac. Some ten models of this "silent shift" transmission were tried experimentally on twenty-five cars, which were driven 1,500,000 miles on the General Motors

---

[1] See Chap. VII: "Cadillac: The Triumph of Precision."

Proving Ground. Synchro-mesh was then introduced to the public, on Cadillac and La Salle in 1928, and later became standard equipment for General Motors cars.

Synchro-mesh marked a step forward in the direction of safety of operation with the driver in complete control of the car at all times, since it allows him to take advantage of the braking power of the engine in either high or second gear. Direct acceleration, a desirable feature in traffic, is also hastened by synchro-mesh. This transmission is designed to eliminate the clashing of gears and provide silent shifting at all car speeds.

Synchro-mesh is thus explained from the mechanical standpoint:

> The transmission consists basically of a small friction clutch which causes the two spinning members of the gears to revolve at identical speeds. When the driver moves the gear shift lever from the neutral position into second or high, one or the other of the two clutches, actuated by the lever, is engaged long enough to synchronize the gears. Since the gears are revolving at the same speed they engage with complete ease and without the possibility of clashing.

After three years' experimental work, Cadillac announced America's first sixteen-cylinder automobile, January, 1930. Its favorable reception led to Cadillac completing the series of multiple cylinder power plants by bringing out the v-12 later in the year. More than 450,000 Cadillacs and La Salles had been produced by the end of 1932.

Cadillac's evolution has been toward size and quality. From the one-cylinder engine which made its early reputation, Cadillac has risen to sixteen cylinders in its largest models. Even its one-cylinder engine was a precision job, giving more power than its pioneer competitors because of closer machining. The degree of precision has been increased steadily, reaching a high stage of refinement. In an industry which as a whole leads all others in combining large output with close work, Cadillac ranks as a leader. A master-set of precision gauge blocks, guaranteed to be accurate to the one millionth part of an inch, lies carefully guarded in a velvet-lined case as a talisman of measurement accuracy in the Cadillac plant.

More than a dozen similar sets, accurate to three millionths of an inch, are regularly used in the tool and gauge department to check other master gauges. There are more than 27,000 Cadillac and La Salle clearances which must be accurate to within one thousandth of an inch; there are 37,000 that must be accurate to less than two one thousandths of an inch; there are 800 operations carried out with such vigilance as to allow only one fourth to one half of one thousandth of an inch.

As one passes through the Cadillac plant he sees materials being tested with cameras which multiply their structure one thousand times; he sees diamond-pointed weights determining their mass hardness; he sees torsion machines and vibratory machines subjecting them to every kind of strain; his eyes fall on row upon row of scales and gauges in the hands of skilled workers; he witnesses each individual part of the car both before and after assembly being inspected; and he leaves with the feeling that he has been in a temple of precision where quality dwells.

Cadillac now offers eight-, twelve and sixteen-cylinder cars and La Salle on two wheel-bases. Cadillac v-8 appears in sixteen body styles, Cadillac v-12 in sixteen, while Cadillac v-16's are built only to order and limited to 400 a year, buyers having a practically unlimited choice in body design and finish. La Salle, with eight cylinders, is produced in seven models.

Cadillac's pioneering achievements were summarized effectively in the Founders Section of the Automobile Trade Journal of December 1, 1924, as follows:

> First to produce a car with parts sufficiently standardized to be interchangeable.
> First to introduce a complete electrical system of cranking, lighting and ignition.
> First to develop and incorporate thermostatic control of engine cooling.
> First in the United States to adopt Johansson gauge system of fine measurements. First to build a truly high-grade car in quantity production, and to sell at a moderate price.

> First American car to win the international Dewar trophy, awarded annually to the automobile making the greatest engineering advance. Won twice by Cadillac.
> First in the United States to develop v-type, high speed, high efficiency engine.
> First in thermostatically controlled carburetion.
> First to build an inherently balanced v-type eight-cylinder engine the vibrationless v-63.

To which might be added at this date:

> First to build a sixteen-cylinder engine for stock use, and first to install the silent synchro-mesh transmission.

Cadillac's record in the technical history of accurate craftsmanship as applied to large production, which is the most distinctive feature of America's contribution to machine practice, is unsurpassed by any automobile manufacturer.

## CHEVROLET

Chevrolet's spectacular entry into the General Motors family has been described. It was then a lusty child, seven years old, but already giving evidence of gigantic stature to come.

The seal had been set on decentralized operations by the sale, in 1916, of the Detroit property opposite the Ford Motor Company in Highland Park. If this, planned as the seat of a great plant, had been used as the site of a great central factory, as anticipated, Chevrolet's activities would have been concentrated there instead of being spread over the country, with Chevrolet manufacturing going on in six American cities and assembly in ten.

Chevrolet received its present corporate form on December 24, 1917, when it was incorporated in New Jersey as a sales company for $10,000, all issued and all owned by General Motors Corporation. Henceforward Chevrolet would rank as a division of General Motors, the largest on the roster.

After General Motors absorbed the Chevrolet Motor Company of Delaware on May 2, 1918, Chevrolet embarked on a policy of acquiring its affiliates in which local capital had been interested. Chevrolet Motor Company of St. Louis was purchased on

August 1st, and Chevrolet Motor Company of Canada on November 1st. This policy also brought Chevrolet of California under ownership in 1920. Meantime, assembly and sales, outlets were greatly expanded. A new assembly plant was built in St. Louis. As evidence of the driving power toward expansion, 500 retail stores for direct distribution of product to customer were authorized, and of these forty-eight had been opened by August 1, 1920. On December 22nd, following, Karl W. Zimmerschied became president of Chevrolet, succeeding W. C. Durant.

W. S. KNUDSEN
Executive Vice-President, General Motors Corporation

General Motors of Canada took over all of Chevrolet's Canadian operations January 1, 1921. Various steps in retrenchment were taken to meet the adverse conditions of the post-war slump. In March, 1922, Mr. Zimmerschied retired and was succeeded in authority by W. S. Knudsen, who, however, ranked only as vice-president until 1924. Mr. Knudsen's administration has seen the advance of Chevrolet to first place among quantity producers. Almost immediately Chevrolet began to march. On October 1, 1922, it took over from General Motors the Central Products group in Detroit, comprising the present gear, axle and forge plants; also the Janesville, Wisconsin, plant of the former Samson Tractor Company which was converted into an assembly plant. New assembly plants were built at Buffalo, New York, and Norwood, Ohio. Body-building activities of the St. Louis Manufacturing Company were transferred to the Fisher Body Corporation in 1923. A pressed metal plant was placed in production in Flint in June.

When Mr. Knudsen became president of Chevrolet on January 24, 1924, this great subsidiary of General Motors had manufacturing plants in four cities Detroit, Flint, Bay City, Michigan; and Toledo, Ohio six assembly plants and sixteen regional sales organizations blanketing the entire country.

The onward march continuing, new sales zones were opened in many of the leading cities of the country. These additions raised the total of zone offices to fifty-two, where it remained until reduced to thirty-six in 1932.

Manufacturing facilities of the company were also increased by these additions:

1925 - Acquisition of the Bloomfield, New Jersey, plant for export boxing operations.

1926 - Acquisition of General Motors Truck Company plant, No. 7, as of March 31, 1926, to increase the production of front and rear axles. This plant had been originally the home of Northway Motor & Manufacturing Company, one of the early purchases of General Motors Company of New Jersey. A new building was erected in Hamtramck, Michigan, a suburb of Detroit, for the experimental department of Chevrolet. 1927 Large expansion program completed. On September 1, Chevrolet took over the Grey iron foundry at Saginaw.

1928 - Production began in the new assembly plant at Atlanta, and construction started on the Kansas City assembly plant which entered production early in 1929.

Including Canada and export, Chevrolet built 1,188,053 cars and trucks in 1928, becoming the world leader in production for the second consecutive year. Production on the four-cylinder National model was discontinued in December, and the six-cylinder International model introduced on January i, 1929. This change set a new price level for six-cylinder cars, and in view of its novelty the public reaction was eagerly awaited. The "six" was favorably received from the start and in 1929 Chevrolet recorded its highest output 1,333,150 cars and trucks, a quantity more than 400 times as large as those produced in its first year, 1912, when the total fell just short of 3,000. The six millionth Chevrolet left the factory on July 1, 1929; by the end of the year 6,504,583 Chevrolets had been produced in the seventeen years of the company's life.

Only one considerable property has been added to Chevrolet since 1930, when the Martin-Parry Commercial Body Company was purchased, for the purpose of building truck bodies of types in general demand to be distributed through Chevrolet zone offices. Reference is made to Chevrolet truck sales in Chapter XXII: "Commercial Vehicles."

While total automobile sales for 1931 were off 29 percent, Chevrolet's decline was only 9 per cent. All through the depression Chevrolet's proportion of total automobile sales has been rising.

For its second year as a "six," Chevrolet took rank as the world's greatest producer of motor cars. For four of the last six years 1927, 1928, 1931 and 1932 it has led the entire American automobile industry in number of units sold in this country.

On April 1, 1932, an important realignment of functions between and among four of General Motors great producers brought to the Chevrolet organization increased responsibilities, Pontiac manufacturing, service and accounting being put in charge of Mr. Knudsen in addition to his former duties, while Mr. Reuter was placed at the head of Buick and Oldsmobile.

On October 16, 1933, Mr. Knudsen became executive vice-president of General Motors Corporation at Detroit and chief executive officer there, with general supervision over all car and body manufacturing operations in the United States and Canada. M. E. Coyle, formerly vice-president, succeeded to the presidency of Chevrolet.

With its vast production plants and its strategically located assembly plants, with its own highly developed engineering staff and the resources of General Motors behind it, Chevrolet has made its way to sales leadership, and in this Anniversary year of General Motors, is the best-selling motor car in the world.

## OLDSMOBILE

As the oldest of the General Motors car divisions, Olds Motor Works takes a just pride in its history. In 1931 it prepared a history of the company which contains this summary of its long career:

> Olds Motor Works was not only the first organization to produce passenger cars commercially, but it also pioneered many of the important advancements made in the industry. The introduction of the assembly line and of quantity production steps which revolutionized all existing automobile manufacturing methods and which established a system that the majority of manufacturers today are following were instituted by Olds Motor Works in the very early days of its career. Olds Motor Works produced the first two-cylinder and four-cylinder cars revolutionary designs for that time. The early achievements of Oldsmobile established public confidence in the automobile and made possible its further development. Olds Motor Works has continued its pioneering activities without cessation during the thirty-four years of its existence, nearly every year bringing forth some new and valuable development. The story of the Olds Motor Works is, in reality, a cross-section of the history and progress of the automotive industry.

General Motors bought Olds Motor Works in 1908, three years after the latter's removal from Detroit to Lansing, and immediately reorganized it. Production ran up to 1,690 cars in

1909, 150 of them being "sixes" of a model developed in 1907 and first sold in 1908. Thus, within a year after General Motors acquired a property which at the moment seemed to many without a bright future, Oldsmobile regained a good market position, standing as a leader in the four-cylinder field and being a pioneer in the six-cylinder field.

The celebrated Oldsmobile Limited's, with their 5-inch by 6-inch motors and 42-inch wheels, appeared in 1910, when 310 were sold. This most distinctive automobile of the day gave Oldsmobile the prestige of extreme luxuriousness. In all, 825 of these huge six-cylinder cars were built in the space of three years. Although highly successful from an engineering standpoint, the market for them was necessarily restricted, and the company's market reliance was the special "four." Sales of the latter model dropped to 1525 in 1910, reflecting the difficulties into which General Motors fell. The next four years were years of reorganization and readjustment to market needs. Not until 1915 was Oldsmobile again a market leader from the standpoint of quantity production.

During this interval Oldsmobile began its course, still continuing, as a manufacturer of cars light in weight but high in quality. As compared with its larger predecessors, the new Oldsmobile was planned to have equal finish and material, size only being sacrificed. Consumers have since become accustomed to high quality in small cars; but the program was then novel and received such firm support that Oldsmobile production leaped ahead, rising from nearly 8,000 in 1915 to more than 22,000 in 1917. After the lag of a war year in which supplies could not be obtained for a large passenger car schedule, Oldsmobile scored another marked advance in 1919 with 39,000 units. Of this total more than 11,000 units were "eights." The first Oldsmobile "eight," and the first offered to the American public in its price class, appeared in 1915. When the United States entered the World War, Olds Motor Works was one of the first automobile plants to get into production on government work. During 1917-1918, 2,000 kitchen trailers were produced. A new motor plant was built for the manufacture of Liberty Motors, although never used for that purpose. This plant was completed and considerable machinery installed by the time the Armistice was signed in 1918. A swift change in plans was necessary. By January 1, 1919, machinery and tools had been moved into the new plant from the Northway Motor Company at Detroit, and production started on the eight-cylinder engine which up to that time had been made at the Northway plant under the supervision of the Oldsmobile engineering department.

On May 18, 1921, A. B. C. Hardy became president and general manager. In years of service he was the eldest of the General Motors executives. From 1892, when he was made superintendent and secretary of the Wolverine Carriage Company, to May, 1921, when he came to the Olds Motor Works, he held practically every office of responsibility in the carriage and automobile industry of Michigan. In 1898 he was elected president of the Durant-Dort Carriage Company, and in 1902 he organized Flint's first automobile concern, the Flint Automobile Company.

Another round with carriage manufacturing, this time in Iowa, and Mr. Hardy was ready to concede that the day of the horse-drawn vehicle was nearly over. In 1914 he became vice-president and general manager of the Little Motor Car Company of Flint, and then of the Chevrolet Motor Company, later going to New York as assistant to Mr. Durant. In 1918 he was made assistant to the president of the General Motors Corporation.

While at Chevrolet he had shown ability in plant reorganization, now needed at Oldsmobile after the weaknesses caused by the interruptions and turmoil of the war period had been revealed by the post-war slump. After the black year of 1921, which registered a 40 percent drop in output for Oldsmobile, the company resumed its upward march with a 21,216 production in 1922. In the first twenty-two years of the century Oldsmobile had produced upward of 200,000 cars.

Following the post-war period there was a noticeable trend toward the "sixes." Oldsmobile had been building six-cylinder cars since 1908, but this type of automobile was relatively high-priced, enough so to be out of reach of the majority of automobile buyers, and "fours" had been Oldsmobile's dependence in reaching the masses. Work was now begun on a six-cylinder car to sell below $1,000. The first model of this light Oldsmobile u six" was introduced in September, 1923, and Oldsmobile left the four-cylinder field.

In 1926 Oldsmobile engineers designed another six-cylinder car, which brought a new high production record. In 1928 Oldsmobile sold more than 86,000 cars and in 1929 more than 100,000, which included the Viking model. From 1927 to 1929 twelve large buildings were erected and several additions were made to the Lansing factories. Oldsmobile brought out its first "straight eight" in 1932.

Olds Motor Works has been one of the chief factors in the growth of Lansing from 2,000 in 1890 to 79,000 in 1930. Employment rose from 3,700 in 1920 to 7,500 in 1930, the latter number including employees of the Fisher Body plant, located at the Oldsmobile factory, and devoted wholly to the construction of Oldsmobile "six" bodies.

Mr. I. J. Reuter succeeded Mr. Hardy as president and general manager. Mr. Reuter went to Opel in Germany in October, 1929, returning to Olds in 1932. In October, 1933, Mr. C. L. McCuen became general manager of Olds Motor Works with which he had been associated in an engineering capacity since 1926.

# Chapter XVII

## GENERAL MOTORS OF CANADA, LIMITED

GENERAL MOTORS OF CANADA, LTD., like many other organizations in the automobile field, had its beginnings in the carriage industry. The founder of the business was Robert McLaughlin, whose father came to Canada from the British Isles in 1835, and in 1867, the Y ear f Confederation of the scattered Canadian provinces, produced the first McLaughlin carriage at his primitive forge in the cross-roads village of Enniskillen, Ontario. Transportation was rudimentary it was eighteen years before Canada had a transcontinental railroad. The McLaughlin vehicles therefore found a ready market, particularly since they were built with thoroughness and craftsmanship.

His reputation established, Robert McLaughlin moved in 1878 to Oshawa, and began making carriages for the wholesale trade. A destructive fire late in 1899 caused temporary removal to Gananoque. The next year, the company reoccupied new and larger premises at Oshawa, which became the nucleus of the present large and modern plant in that city.

In the meantime, his two sons George W., and R. S. McLaughlin became associated with their father in the carriage business which, before it entered the motorized field, produced over 270,000 horse-drawn vehicles carriages, buggies, and sleighs.

The trend in transportation led to the formation of the McLaughlin Motor Car Company, Ltd., in 1907, with R. S. McLaughlin as president, Oliver Hezzlewood, vice-president, G. W. McLaughlin, treasurer, and Robert McLaughlin, the elder, in an advisory capacity. A survey of the entire automotive industry convinced the new organization that Buick was the ideal car to meet Canadian conditions. Thereupon the company made a fifteen-year contract with Buick Motor Company for sole manufacturing rights of Buick in Canada. The McLaughlin-Buick before long became Canada's standard car. At the close of 1914,

ROBERT MCLAUGHLIN

General Motors Corporation owned $500,000 of the capital stock of the McLaughlin Motor Car Company, Ltd., and McLaughlin Carriage Company, Ltd., had $503,000.

The McLaughlins gave up the carriage business in 1915 to make way for Chevrolet Motor Company of Canada, Ltd. Under them Chevrolet met with instant success in the Dominion.

On November 8, 1918, the McLaughlin and Canadian Chevrolet companies were merged into General Motors of Canada, Ltd., of which R. S. McLaughlin became president and G. W. McLaughlin, vice-president. The father was no longer active, but took a keen interest in the new organization until his death in 1921.

On December 19, 1918, General Motors Corporation bought the remaining interest in McLaughlin Carriage, Chevrolet of Canada, and McLaughlin Motor Car companies for 49,000 shares of Common and $550,000 cash. Capitalization of General Motors of Canada, Ltd., was increased to $10,000,000 soon afterward.

## GENERAL MOTORS OF CANADA, LIMITED

General Motors of Canada, Ltd., is a manufacturing company, with authorized capital stock of $10,000,000, par $100, of which $6,940,000 has been issued. Its sales and service subsidiaries are:

> Cadillac Motor Car Company of Canada, Ltd.
> Chevrolet Motor Company of Canada, Ltd.
> McLaughlin Motor Car Company, Ltd.
> Pontiac Motor Company of Canada, Ltd.
> Olds Motor Works of Canada, Ltd.
> General Motors Products of Canada, Ltd.

Mr. R. S. McLaughlin, now one of the older General Motors directors in point of service, heads the various subsidiaries of the Canadian company. Mr. McLaughlin was first elected to the board in 1910, and again in 1918, since which time he has served continuously.

General Motors Products of Canada, Ltd., is a selling organization with eight zone offices in the cities of Vancouver, Calgary, Winnipeg, Regina, Toronto, London, Montreal, and Moncton. Parts depots are also maintained in the principal cities, and there are 1,000 strong individual dealerships in Canada.

General Motors products are financed in the Canadian market by General Motors Acceptance Corporation, with headquarters in Toronto, and branch offices in the eight zone cities mentioned. Insurance service is furnished by General Exchange Insurance Corporation, also at Toronto. Manufacturing operations are centered in four cities.

The works at Oshawa are the largest ; having a floor space of 2,200,000 square feet and a production capacity of 750 cars a day. Employees, at the peak period, numbered 7,000.

The works at Regina, Saskatchewan, were opened in 1928 for assembly of Chevrolet and later, Pontiac. They have a floor space of 370,000 square feet and a capacity of 30,000 cars a year.

At Walkerville, engines are produced for Buick, Chevrolet, Oldsmobile, and Pontiac automobiles and front axles for Chevrolet, providing employment for as many as 2,000 workers. An associated factory, though independently operated, is that of the McKinnon Industries unit of General Motors at St. Catharines, Ontario, where generators, starters, differentials, spark plugs and other parts are made.

Domestic shipments of cars from the Oshawa plant amounted to 9,915 in 1916, and had grown to 22,408 in 1920. The high point for domestic shipments was reached in 1928 with 75,000 cars. In the seventeen years of production from 1916 to 1932, more than 725,000 Canadian made General Motors cars were manufactured. During 1932, sales of General Motors of Canada passenger car lines were 39.3 percent of all the sales in Canada, and commercial cars accounted for 41 percent of all commercial car sales.

R. S. MCLAUGHLIN

Export shipments from Oshawa have been made to as many as sixty-five overseas countries in a single year. In 1932 and 1933, Buicks for the British Isles were all shipped by General Motors of Canada, Ltd. (McLaughlin-Buicks).

Cars are manufactured, not merely assembled, in the plants of General Motors of Canada, Ltd., even bodies for most of the lines being made at Oshawa. It has been the policy of General Motors Corporation to provide as much employment as possible in the countries where sales opportunities are enjoyed. Canada is in a remarkable position in this respect, inasmuch as General Motors cars are built there almost completely "from the ground up," not simply put together with parts imported from the United States. In the early days, strictly speaking, there was very little automobile manufacturing in Canada, but in recent years there has been a steady development until today the Canadian automobile industry may be regarded as a manufacturing industry in the true sense, with General Motors of Canada, Ltd., playing an important role. General Motors cars made in Canada have a high percentage of Canadian content, to which the Canadian motor-car user reacts very favorably. Naturally, too, General Motors cars exported from Canada to other parts of the British Empire in a finished state are highly British in content, and this also is helpful in view of the recent emphasis on the Empire slogan "Buy British."

The evolution from assembling to manufacturing in the case of General Motors cars in Canada has been accompanied by frequent alterations in the tariff as it affects the Canadian industry. These alterations group themselves into two or three periods of varying duration.

From 1907 to 1926 the Canadian customs tariff on all imported automobiles was 35 percent. The degree of protection offered by this impost was sufficient to give the Canadian industry a good start. Leading companies in the United States, including General Motors, established large plants there for assembling and for such manufacturing as could then be carried on, in view of the limited supply of Canadian parts and material. It was during this period that General Motors built the large Walkerville plant, which is now a part of General Motors of Canada, Ltd. At first, this plant supplied some engines and parts of axles. At present, its contribution to finished Canadian cars is much larger and the plant has sufficient capacity to supply engines for Chevrolet, Pontiac, McLaughlin-Buick, and Oldsmobile in whatever quantities required.

The next alteration in Canadian tariff structure was downward. In 1926, the customs duty on small cars was reduced to 20 percent and that on large cars to 27.5 percent. At the same time a "drawback" of 25 percent was allowed on material to be used for cars of 40 percent Canadian manufacture, and in the following year this "drawback" was only to apply on cars of 50 percent Canadian manufacture. The reduction from 35 percent to 20 percent duty was made effective in the face of general protests from almost the whole Canadian automotive industry, and the figures showing imports of small cars for the next three years were striking enough to justify those objections. Imports in 1926 were 28,000 cars; in 1927 they reached 36,000; the following year they totalled 47,000; and in 1929, 44,000.

The ultimate effect on manufacturing process of this particular tariff period was to increase the number of Canadian parts and the Canadian content of finished cars. At first it was difficult with the complete factory output to qualify for the customs "drawback" on cars of 50 percent Canadian content, but before many years had passed this was easily accomplished. With the idea of providing the entire motor trade of Canada, including General Motors, with necessary parts, General Motors Corporation in 1929 purchased McKinnon Industries at St. Catharines, Ontario, which, interestingly enough, was a survival of the old-time carriage trade, as was the case with so many existing motor-car industries. At that time, McKinnon Industries was converted to the manufacture of differentials and rear axles. Development was very rapid, and many other parts have since been added, including steering gears, shock absorbers, and ignition parts. Recently, also, the making of spark plugs for the trade and for General Motors cars was commenced under the direction of the AC Spark Plug Company. In addition to supplying the whole Canadian trade with many motor parts, this plant also produces the unit for the Canadian-built Frigidaire refrigerator.

Entering a more recent period of tariff history, it is noticed that the motor-car schedule was completely rewritten in 1930. In the case of small cars, the customs duty was set at 20 percent (up to $1,200 trucks included) ; on cars from $1,200 to $2,100 duty was 30 percent; and on cars from $2,100 up, the duty was 40 percent. Moreover, the "drawback" was eliminated on a specified list of parts, and was made 25 percent for most of the remaining parts and 60 percent on a very restricted list in the case of cars of 50 percent Canadian content. A noticeable effect of this change was a reduction in the number of car imports. While the change came at a time when sales generally were decreasing heavily, yet the reduction in the number of imports was more than proportionate.

Imports between 1930 and 1931 declined much more heavily than did Canadian production. In 1931, another customs duty alteration took place. On imported cars the valuation for duty was fixed at a flat 20 percent discount from the list price. Inasmuch as discounts on many cars had been claimed at a much higher figure, the effect of the change was to increase the amount of duty paid on most cars. At this same time the importation of used cars was prohibited, except in a few cases, such as settlers bringing in effects.

The immediate result was a further decline in all car imports. The year 1932 saw a reduction of more than 80 percent in units imported, while Canadian production at the same time declined only 26 percent.

In addition to customs duties, there has been, since the war years, an excise tax of varying rates on cars either imported or made in Canada. The excise tax until 1926 was 5 percent on cars not more than $1,200, and 10 percent on cars valued in excess of $1,200. In 1926 this was changed, to allow complete exemption from the excise tax in the case of cars valued at not more than $1,200, if they were 40 percent Canadian content (50 percent Canadian content in 1927 and thereafter). The excise tax was also eliminated in the case of cars imported from Great Britain if they were of similar British content. The same rates of excise tax apply at the present time.

The final outcome of this evolution in tariffs and manufacturing is that General Motors cars in Canada are very largely Canadian-built. Examining the Chevrolet automobile, for example, it is found that all parts of the Dominion contribute. Chassis springs are purchased from firms which make them at Gananoque, Ontario, and Oshawa. Front axles are manufactured by the General Motors plant at Walkerville, forgings being supplied by a Canadian company in the same city. Third member assemblies, including differential, are made by McKinnon Industries at St. Catharines. The body is completely built at Oshawa. Since the Oshawa factory, from its earliest days, had an extensive woodworking establishment for manufacturing bodies, no Canadian division of the Fisher Body Corporation was formed. Instead, the operation was amalgamated with the general manufacturing. All General Motors of Canada cars, of course, bear the Fisher insignia. Engines are built at Walkerville and radiators at Oshawa, material in both cases being supplied by Canadian companies. Factories in many cities supply castings, bumpers, glass, wiring, paint, trim, hardware, and wheels.

Evolving to its present proportions under a Canadian management of long standing, which has coped successfully with manufacturing and sales problems rendered more intricate by tariff changes, General Motors of Canada stands today as one of the leading industries of the great Dominion.

# Chapter XVIII

## GENERAL MOTORS ACROSS THE SEAS

IN THE early 'nineties R. E. Olds, as previously related, made the first sale of an American automobile for export, a chance sale resulting from a description of the Olds steamer published in the Scientific American of May 21, 1892. Since then the overseas trade in motor cars has become a leading factor in America's foreign commerce.

World trade in General Motors products is a business of great complexity and extent. The Overseas Operations are divided into three major divisions, Vauxhall Motors, Ltd., which manufactures and sells its products in the British Isles; Adam Opel, A. G., which manufactures and sells its products in Germany; and the General Motors Export division, which is the assembling and merchandising organization responsible for the distribution of all products from whatever source in the world markets outside of the United States, Canada, Germany, and the British Isles.

The Export division consists of four great territorial regions, each under the charge of a regional director, who is responsible to the general manager in New York. These major territories under the control of the regional directors are:

Europe, with six plants located in France, Belgium, Sweden, Denmark, Spain, and Egypt.

South America and South Africa, with three plants, in Brazil, Argentina, and Cape Colony, South Africa.

Australasia, with assembly plants and a complete body manufacturing plant in Australia and an assembly plant in New Zealand.

Far East, with assembly plants in Japan, India, and Java.

The regional directors of those vast areas maintain headquarters at the home office but spend most of their time traveling far and wide over their territories.

But a goodly portion of the planet lies outside of these wide jurisdictions, and the trade originating there is handled by General Motors Export Company from New York City. This far-flung task embraces such distant and diverse areas as China, Mexico, all of Central and South America except Brazil, Uruguay, and Argentina, vast sections of Africa, and also the Caribbean and Pacific Islands.

The sun never sets on General Motors. Its Japanese workmen in Osaka are putting on their sandals to go to work while its South African workmen are starting homeward from the factory at Port Elizabeth. Its salesmen know what it is to compete with the indigenous transport on wheels and otherwise some of it most unusual, as exampled by the carabao (water buffalo) pulling a sled in Malaysia; horse-drawn carts with seven-foot wheels in Argentina; coolie-carried chairs in a Chinese bridal party procession; camel trains of the Sahara and camel-drawn wagons loaded with wool in Australia. In these globe-spanning operations and contacts General Motors representatives have met all manner of men and all varieties of problems in management, transportation, and human nature.

This peaceful penetration of the near and remote parts of the world has appealed to men willing to go forth, adapt themselves, and carry their message with the zeal and conviction of the missionary of old. In so doing, they have come upon all the colorful scenes of this earth from the narrow, crooked lanes of Java lined with thatched huts, to overcrowded tropical harbors and the long reaches of desert and river, each environment calling upon their ingenuity, resources of patience, good-will, and perseverance. Their lot is one of hard work, against a background of novelty, with always the duty of learning the manners and customs of each foreign field and tactfully seeking to understand its people and their needs.

The life appeals to adventurous men who are at home anywhere. There are General Motors families who answer a call to move with little more concern than if they were moving to the next block. One such family contains four children, each born in a different country, each speaking the language of his or her early environment better than English. All over the world a mobile force continues an unremitting drive for sales.

This vast organization began humbly enough on June 19, 1911, when the General Motors Export Company was incorporated in Michigan for $10,000, under the banker management then in control of General Motors. The president of General Motors Company, Mr. Thomas Neal, became its president. Export started operations with a staff of only three persons, but, even so, it was in a sense the most complete expression of General Motors because it was the first selling organization in which all the various General Motors units joined to carry a common objective. Hitherto, foreign sales, chiefly of Buicks and Cadillacs, had been made by the divisions' staffs acting independently; henceforth for many years all foreign trade would pass through the Export Company. During Mr. Neal's presidency and up to 1916, Mr. O. A. Bennett was the guiding spirit of Export. Mr. Bennett became president in 1915.

At first the Company pushed foreign sales along mail-order lines and made no really intensive effort to develop sustained activities in foreign selling. Capital was increased to $100,000 in 1912. Soon afterward Export brought its more important activities to the natural seat of foreign commerce, New York City, where sales for the entire world were handled by a staff of fifteen or sixteen people working in a three-room office on Park Avenue.

World War needs developed large orders, but the entry of America so reduced production that the Export Company found itself with a limited allotment of cars at its disposal, which it had to allocate as judiciously as possible among its distributors throughout the world. When Mr. Bennett resigned the presidency because of ill health in 1916 (he died in 1917), he was succeeded by R. H. "Trainload" Collins, the vigorous sales manager of Buick. Mr. Collins held the office until Mr. J. A. Haskell became president to remain at the helm until 1922. Headquarters were moved to n Broadway in 1917, and to 224 West 57th Street in 1918.

On June 27, 1918, General Motors Export was combined with General Motors (Europe), Ltd., with a capital of $212,932.75. General Motors (Europe), Ltd., had begun life in 1909 as Bedford Motors, Ltd., of London, England, making Bedford Buicks, the chassis being Buick. It became General Motors (Europe), Ltd., in 1912, but down to 1914 handled only Buicks. Reference will be made later to General Motors, Ltd., London, which succeeded to the English representation, and imported cars and trucks into England, independent of the Export Company.

What may be called the modern development of General Motors Export dates from the end of the World War, when demand for rapid transport rose rapidly. Export sales advanced from 6,004 units in 1918 to 14,665 in 1919. In 1919, Overseas Motor Service Corporation was organized to service and distribute motor-car parts and accessories in the export market. These include not only well known General Motors products such as AC spark plugs, AC speedometers, Delco products (except in the British Isles; Hyatt also excepted from Continental Europe), but also parts and accessories of other makers.

Up to 1920 the backbone of the overseas business was Buick. Plans were made to give it wider distribution and push its sale. The Export Company, now due for permanent expansion overseas, made studies of the export organizations of several other large American concerns doing business abroad. It also established a training school in the Buick Building, 1737 Broadway, to provide intensive training for Company employees who were to serve in the overseas field, and to aid those in the home office to improve themselves in their knowledge of the conduct of the Company's business. In 1920 unit sales reached a volume of 29,772 (Export 1927 annual report). Need for more room led the Company to move its offices from 224 West 57th Street to the Wurlitzer Building at 120 West 42nd Street.

In 1921 Mr. J. D. Mooney became a vice-president of the Export Company. The next year he succeeded Mr. Haskell as president. The new president had been general manager of the Remy Electric Company, another General Motors division. The management outlined an intensive sales campaign, the fruits of which appeared in the years of the "Golden 'Twenties." For this campaign six sales divisions were set up: Caribbean, South American, Australian, African, European, and Far Eastern.

While the United States and Canada absorbed cars at a rate which demanded constant expansion at home, other fields could have only the attention they insisted on having; the real effort toward intensive cultivation of these fields followed shortly after the post-war slump of 1920-21 when the domestic market failed to absorb its customary volume of cars.

By 1922 the Company employed 300 persons, including field men, but still the Export division had not taken on definite shape. In that year the Export Company returned to the General Motors Building at 224 West 57th Street.

In 1923 the Corporation began to advertise in overseas motor and other publications, reaching the public in as many as fifty-five publications in sixteen different countries. The senior field personnel was increased by one third in the same year, and new zone offices continued to be added, situated in strategic positions to take immediate advantage of any improvement in political and economic conditions.

Outside of the United Kingdom and Ireland, which was controlled directly by General Motors, Ltd., with headquarters in London, General Motors overseas offices were located in Paris, Soerabaia, Calcutta, Manila, Melbourne, Sydney, Johannesburg, Sao Paulo, Copenhagen, Bombay, Shanghai, Honolulu, Wellington, Madrid, Mexico City, and Buenos Aires. General Motors cars were being sold in 125 countries, giving coverage almost world-wide. General Motors World, organ of the Overseas Group, reports that leading markets in 1923, the first of a series of record-breaking years, were Australia, United Kingdom, Mexico, Sweden, Argentina, Spain, Cuba, Belgium, and Japan.

Chevrolet, steadily making a better reputation for itself, was now a factor in the low-priced field, and it seemed advantageous to assemble Chevrolets abroad. Accordingly, in 1924, plants were established in London and Copenhagen. The London venture soon came under the handicap of a horsepower tax. However, as there seemed to be a place for a satisfactory truck of the Chevrolet model, commercial body-building was started at General Motors, Ltd. The trend in the British Isles was toward a smaller, lower-powered and lower-priced product. The search for a satisfactory passenger car to supplement the truck led to the Vauxhall purchase, completed November, 1925.

Copenhagen, on the other hand, operated successfully from the start and furnished inspiration for assembly activities elsewhere. The original Company of 1911, primarily a sales company, and known by its old name of General Motors Export Company, still functions as such, finding its territory in those fields not covered by the self-contained, locally-operated organizations. As an operating concern, the Export Company has a managing director and a sales organization and a place in line with other operating companies of the overseas group. But in general, reference to the Export Company herein means the corporate organization rather than the operating one. More will be told about the plan of organization.

The focal points of General Motors assembly, manufacturing, warehousing, and distribution overseas are:

New York City, New York; Luton, London, and Birmingham, England; Copenhagen, Denmark; Stockholm, Sweden; Antwerp, Belgium; Riisselsheim and Berlin, Germany; Puteaux, France; Barcelona, Spain; Alexandria, Egypt; Buenos Aires, Argentina; Sao Paulo, Brazil; Montevideo, Uruguay; Port Elizabeth, South Africa; Woodville, Australia; Wellington, New Zealand; Osaka, Japan; Batavia, Java, D. E. I.; Bombay, India.

The assembly plants at Buenos Aires and Sao Paulo date from 1925, also that at Antwerp, Belgium, the seat of General Motors Continental. General Motors assembly operations in Spain began in 1925 with a plant at Malaga, operations being later transferred to Madrid and recently to Barcelona.

Outside of the Export division staff, but part of the General Motors Overseas Operations group under Mr. Mooney's jurisdiction, appeared at this time General Motors, Ltd., London, England, whose beginnings have been previously noted. The plant at Hendon, eight miles from London, was expanded in 1925. A conveyor system took the crates arriving from the Oshawa, Canada, plant and turned out the car ready for road tests and distribution in the United Kingdom. The commercial body-building plant heretofore noted began operations in 1925, and straightway proved itself one of the most successful of the overseas operations.

In 1929 General Motors, Ltd., took over the selling and general distribution of all Vauxhall cars. The arrangement was continued until 1932, at which time all products of Vauxhall Motors, Ltd., were sold, as well as manufactured, by the latter company. In 1932 Hendon had completely reversed its early procedure; cranes that once swung crates inward now swung them outward, so that instead of being an import, unboxing plant it became an export, boxing plant for the Vauxhall products which had been developed in the meantime.

Vauxhall Motors, Ltd., manufactures the Vauxhall, an English car of long-established reputation, with plant and offices at Luton; also the Bedford truck, an outgrowth of the Chevrolet truck activities in England, the change in name having occurred in 1931. Vauxhall then employed 2,000 men and made 1,700 cars a year, with a possible production estimated at 25,000 a year. The Corporation's initial transaction in Vauxhall, requiring 510,000, was consummated November 24, 1925. The Corporation has since enlarged its interest, and now owns a substantial proportion of the shares of Vauxhall. Three new directors representing General Motors were added to the board. The management and personnel remained British, while Vauxhall gained capital, a selling organization, and the benefit of research. More recently Vauxhall began to manufacture low-priced models of low horsepower to meet special English needs. Sales activities of General Motors, Ltd., have been taken over by Vauxhall.

In the German development were two concerns, known as General Motors, G.m.b.H., and Adam Opel, A.G. During the period of German import restrictions, the former company acted as a sales office placing its orders with the Export Company or with either General Motors Continental at Antwerp, or International at Copenhagen. When the German import restrictions on motor cars were lifted, in 1925, a warehouse was secured at Hamburg, and a manufacturing base opened in Berlin. In April, 1926, the first Chevrolet truck came out of the assembly. This expansion was accomplished in the fact of discouraging conditions in Germany which led to a severe crisis the following winter.

Adam Opel, A.G., represents the largest single investment of General Motors overseas, having grown to its present proportions through a long history of diversified manufacture. The Opel workshop was founded at Riisselsheim, Germany, in 1862, for the manufacture of sewing machines. Opel began to build wine-making machinery in 1883, high-wheeled bicycles in 1886, low-wheeled bicycles in 1892, passenger motor cars in 1899, motorcycles in 1901, and motor trucks in 1907.

In developing their plant the Opels drew on American experience. Wilhelm von Opel, the present head of the works, studied advanced production methods in the United States in 1909 and on his return to Russelsheim directed a tremendous plant expansion. Fire destroyed the works in 1911, leading to the end of sewing machine production. Larger plants were soon erected and by 1912, 4,500 persons were at work in Opel factories. In 1924, the conveyor system of production was installed, following the best American practice. Another extension occurred in 1926 when Opel reached a production of 43,500 cars.

As an industrial layout, Opel approaches Krupp's in completeness. Its equipment includes 7,000 working machines and automatics, more than seven miles of standard gauge railroad, six locomotives and 1,800 tip cars, 30 lift-trucks on eighteen transport lines, 700 loading platforms, 40 elevators, and 30 cranes.

From 1926 to 1929, Opel enlarged its Russelsheim plant by more than a million square feet to a total area of more than 7,500,000 square feet, with a daily capacity of 500 motor cars and trucks and 3,000 bicycles. Shortly after General Motors acquired its interest in Opel, in 1929, "Blitz" was adopted as the trade-name for Opel trucks. Export sales for all Opel automotive products are handled by the General Motors Export division.

Opel car in African exploration Schomburgk Expedition

Opel is unique in the Corporation in total amount of car fabricated in one plant: body, radiator, shock absorbers, fuel pump, clutch, valves and universal joint. Exports in 1932 were only 4 percent below those of the year preceding, and Opel cars accounted for slightly more than 68 percent of German car exports. Geheimrat Wilhelm von Opel is chairman of the board of Adam Opel, A.G.

While the Export Company was distributing cars in Australia, a new tariff of nearly three hundred dollars on every car body imported pushed prices of General Motors cars skyward.

Faced with a complete loss of its Australian trade, the Export Company made a contract with Holden's Motor Body Builders, Ltd., which already did the bulk of bodybuilding for the Commonwealth, besides handling a large part of General Motors' requirements, practically all body work being done by hand. Cooperation between General Motors and Holden's brought forth a new plant, contiguous to the existing one at Woodville. In the milling department a progressive line for all mill operations was set up, including Linderman and other modern machines new to the Commonwealth. The General Motors staff, in an emergency, built a hydraulic press, obviating the former method of shaping body panels by hand. These and other machines, also wood kilns, nickel-plating and Duco departments, were beyond the experience of workers there, and it took some persuasion to change their attitude regarding labor-saving devices. Also, they had to be instructed in the use of the new machinery. Results were highly beneficial. Before the end of 1925 several thousand employees were added for the production of General Motors bodies exclusively. In this General Motors-Holden's enterprise the Corporation acquired a majority of the Common stock in 1931.

The expansion of overseas business depended on the coordination of the Export group and other Corporation groups, many of which had goods to sell in that market. Field personnel gradually imbued foreign dealers with the spirit of American merchandising methods, adopting the policy of "Volume turnover service." While the total number of automobiles a given territory could absorb in a year was a matter of economics, "the percentage of General Motors products to that total was a question of representation and salesmanship."

A world survey of the automotive industry was undertaken in 1925, with three main questions to be answered: Where were the motor cars of the world to be found? How many were there? Who sold them? Through 1926, the spotting of other assembly plants and warehouses on the world map continued. The plants in Australia at Brisbane, Sydney, Melbourne, Adelaide, and Perth date from this time, also those at Wellington, New Zealand, and Port Elizabeth, South Africa.

These formative years reveal some of the motives actuating the Corporation in extending its overseas interests and in creating its assembly plant structures. At first, of course, it had an eye to savings in freight and duty. Other advantages made themselves felt in time and proved of no less importance. There was the advantage of being able to ship cars and trucks abroad at a rate not possible under the old distributor system: heavy consignments could be taken, delivered, and financed at the points of distribution as part of ordinary procedure. Another advantage entered with the invention and development of Duco, by which it became possible for assembly plants to finish their own bodies, with reference to local demands in color, finish, and trimming.

At first the Color division of the Export Company confined its activities to complete cars being shipped abroad (Single Unit Packing) ; then the assembly plants took it up; at length the Color division took over the work of the color program.

Selection of optional colors for each model of car is made and finalized six months in advance of any color change becoming effective in overseas plants. Wide tolerance in colors is permitted, but those are selected which are fundamentally correct from the standpoint of harmony and good taste, and which in addition meet local preferences. The latter are many and important. Red is taboo in some countries because of its Bolshevist connotation. In Japan maroon cannot be used, because it is the color of the Imperial Household, and in China yellow is taboo because it signifies mourning for death.

Until 1925, a financial service was available to overseas distributors generally from the General Motors Acceptance Corporation only through the Foreign Department located in New York City and, in the British Isles, through the London Branch established in 1920.

In 1925, at Antwerp and Copenhagen, General Motors Acceptance Corporation began the establishment of branches in foreign fields to conduct a financing service for overseas distributors and dealers in General Motors products and also their retail purchasers. This service, so valuable to the Export organization, has since been expanded to include the following

General Motors Acceptance Corporation branch offices: Alexandria, Egypt; Batavia, Java, D.E.I.; Bombay, India; Buenos Aires, Argentina; Honolulu, Hawaii; Osaka, Japan; San Juan, Porto Rico; South Melbourne, Australia; Wellington, New Zealand. In addition to the above direct branch offices, subsidiary corporations have been organized to offer similar services at: Copenhagen by General Motors Acceptance Corporation, Continental; Mexico City, Mexico, by General Motors Acceptance Corporation de Mexico, S.A.; Russelsheim, Germany, by Allgemeine Finanzierungs Gesellschaft m.b.H. ; Sao Paulo, Brazil, by General Motors Acceptance Corporation, South America ; London, by Vauxhall and General Finance Corporation, Limited; Port Elizabeth, S.A., by General Motors Acceptance Corporation of Delaware.

Effective as of February 1, 1926, a definite plan of organization and management for the Export group of General Motors was adopted, on the principle of staff and line control. While there had been previous charting of activities, necessarily rapid expansion had precluded any continuous set-up clearly revealing inter-relations, lines of contact, and division of authority. The Export group had taken on four general functions: manufacture, sales, supply, and finance, for each of which a departmental manager was now appointed. The president of the Export Company was given charge of all operations of the Export division as well as the manufacturing operations; on him was placed responsibility for the entire Overseas Operations group to the Corporation, of which he is a vice-president. In general this relationship still holds. The vice-president of the Export Company who served as its general manager was designated also as general manager of the entire Export division (not including the manufacturing activities in Germany and England). To him the four departmental vice-presidents of the functions stated above look for authority, each becoming responsible for the control and development of his particular function in all operating units throughout the group. The respective operating units were placed under managing directors directly responsible to their regional directors, who in turn are responsible to the general manager of the Export division in

New York. At Opel and Vauxhall, each self-contained in a single territory, no such far-flung world organization exists, but the functional set-up internally is practically identical. The General Manager of each of these three major operating divisions the Opel, the Vauxhall, and the Export division reports to the president of General Motors Export Company, who is also vice-president of General Motors Corporation in change of overseas operations.

By the end of 1926 General Motors overseas had rounded itself out. The situation was then outlined as follows:

> It becomes quite apparent that we have come, for 1927, to the next phase of our operation, for we are faced very definitely with the necessity of consolidating the advances we have made, and of developing the machinery we have set up to the degree of efficiency that will bring the permanent result we are seeking. ... As a group, we have just about reached the limits of our growth; few markets of worthwhile potentiality are now left without individual companies, largely self-controlled, installed on the ground to carry the tremendous resources of General Motors out directly to the retail dealer and the public. We can no longer look for the impetus to increase in volume that the creation of a new assembly plant provides. We have established our line organization on practically every battle front. Any growth, any expansion hereafter, must come from within these operating companies, and as a direct result of their increased activity... . This crystallizes directly to a definition of our objective in 1927: "Ratio to total American volume, return on investment, and car inventory turnover."

Wherever General Motors cars go, other benefits go in their turn. Group life insurance is carried on all American overseas employees of the Corporation. General Motors Acceptance Corporation participated in the financing of nearly 50 percent of the Export division's 1926 volume, employing one hundred workers solely on overseas activities. Its wholesale and retail plans of financing were in operation in fifteen countries, with nine other countries added in 1927.

The year 1927 saw assembly manufacturing started at Batavia in Java and Osaka in Japan, followed soon after, early in 1928, at Warsaw, Bombay, and Stockholm. The three South

American plants in Brazil, Uruguay, and Argentina were placed in communication with one another by radio. In the development of the line and staff organization in 1927, the Export Company, as an operating unit, was put on the same basis as other individual plants in the Export division, although it remained the source from which the officers and directors derived their corporate status.

All through this period advertising has been pushed by methods well known at home and by others keyed to characteristics of foreign publics and especially to their degree of motor-mindedness. Motion-picture films are effectively used. The automobile show is one of the main reliances for putting General Motors products before the eyes of populous centers; the one at Brussels always constitutes a main event in Continental's advertising program. Sao Paulo, which is the third largest city in South America, was treated to a special General Motors exhibition, which proved successful as a result of painstaking preliminary and follow-up advertising, although, as pointed out in General Motors World, March, 1927, previous shows at Sao Paulo and Rio de Janeiro had not been productive of sales. This show succeeded in bringing the closed car into favor with the Brazilians is never before. Europeans, Asiatics, and South Americans have developed as keen an appetite for automobile shows as have North Americans.

Overseas the "show" frequently takes the form of a caravan. Local advertising precedes the appearance of the fleet of cars and trucks at the various points scheduled. To the inhabitants of undeveloped countries perhaps the most exciting of these are the caravans organized to push their way into new or partially exploited territories. Cavalcades of General Motors cars, gayly decorated, start from Antwerp, Copenhagen, Osaka, or Melbourne, or some other base, for a long tour of the hinterland. They carry, in addition to service men and electricians, show troupes, stage and motion-picture equipment, loud speakers, field kitchens, and cafeterias. They are really sales and advertising expeditions, painstakingly equipped to maintain themselves in the field and to entertain large audiences gathered with the cooperation of local dealers.

Perhaps the most imposing cavalcade was that which started from New York to show the new v-16 Cadillac to Europe, a feature of the journey being the ceremonial Cadillac visit paid to the ancient seat of the Cadillac family in France. In Palestine and other Near East areas the motor cavalcade has been especially effective in bringing scattered dealers into touch with new models and sales ideas.

Impressive endurance runs have been made by General Motors cars for the same purpose, the most famous one being the Cape-to-Cairo run by Chevrolet. After crossing Africa from south to north, this sturdy car went on across Europe to Stockholm, welcomed all along the line and visiting all the principal General Motors centers. Chevrolet, of course, is the great volume producer in the export field as in the home market.

Other special features of export publicity, such as the round-the-world trips of Buick and Frigidaire, have united the whole export organization in synchronized publicity campaigns on an international scale.

Direct-mail advertising is a method used by the Export company extensively, the sales organization in New York serving those countries not covered by assembly plants. In some parts of the world it may be necessary to begin a sales campaign by offering proof that the Corporation actually exists and makes cars that can be bought locally. Witness the experience of a sales representative:

> I had thought, previous to going to West Africa, that we were so large and important that we must be universally known. Therefore it was a bit of a shock to find that many people were in complete ignorance of our products. To sell these people the idea of General Motors as a great and unique institution was my primary job and I found that after telling them about our size and strength, our multifarious manufacturing, sales and financing activities, showing them pictures of the Proving Ground and Research Laboratories, it was very much easier to develop an interest in the actual selling of our cars and trucks.

While publicity is as much a part of a selling campaign abroad as at home, objectionable propaganda is taboo. The following incident may be taken as an indication of how General

Motors adapts itself in a foreign country to custom and tradition. In 1927 General Motors Japan took over premises at Osaka, where its predecessors in business had established a Shinto shrine. Though there was no obligation to do so, the company restored it to use and now maintains it. Each year a ritual service is held and the mother temple has named the shrine, "The White Chrysanthemum."

By May 2, 1927, the home offices were finally quartered in the General Motors Building at 1775 Broadway. The goal set for 1927 overseas operations was reached, and a policy of plant expansion began. It was stated at that time that u we shall have to work definitely toward identifying ourselves with each market." Further explained, General Motors as an institution had already assumed an international character.

> In many countries General Motors is coming to be looked upon as an integral part of the basic industrial life of those countries. There is not one American in the entire organization of Vauxhall Motors, Ltd., in London. The daily business in the office in Osaka is conducted by Japanese, and the native workmen are carrying on their jobs under native foremen. The tendency to employ local personnel, and to promote this personnel into a position of responsibility and authority as fast as it qualifies, is present everywhere. It is definitely General Motors' policy to award the job to the man who merits it, whatever his nationality.

Altogether there were nineteen manufacturing and assembly plants in the Export division at the close of 1928, in which year the employees of the Company averaged eighteen thousand in number.

General Motors Export volume was greater than the volume of any group of export products going out from the United States with three exceptions... . The General Motors Export division also did a greater volume of business abroad than the Corporation did domestically seven years before. General Motors had become the biggest automotive concern doing business in Europe.

Sales abroad of American-source products by the export organizations of General Motors for a second four years of the modern development were:

CARS AND TRUCKS

| Year | Number | Value |
|---|---|---|
| 1926 | 118,791 | $98,156,088 |
| 1927 | 193,830 | 171,991,251 |
| 1928 | 282,157 | 252,152,284 |
| 1929 | 256,721 | 243,046,031 |

Since many overseas representatives usually come back to the United States in the spring, a regional conference of the Overseas group, attended by representatives from twenty-eight different countries, was held at Shawnee-on the-Delaware, Pennsylvania, May 22-30, 1929.

Before 1929 closed, the story of another pioneering effort came out of the Near East. There are twenty languages in General Motors Near East territory. The Parts division translated the Chevrolet Parts Book into each of the chief languages used in its territory: French, Italian, Arabic, and Turkish. The Turkish and the Arabic versions represented a monumental task. Arabic lacked an automobile vocabulary or even the root words upon which to build. The translator's book was praised by the Arabic press and found favor with government officials, such as those of the Egyptian customs. Constructive helps to overseas dealers in General Motors products are extended in three different forms: Service School, Dealer Accounting, and Retail Sales School. When dealers failed to send their service men to the plant school, the Rolling Service School was established and first took the field on wheels into Brazil in 1926. By 1930 most of the Overseas group had established similar schools on wheels.

General Motors 5 stake in world trade had become enormous by 1928, when its sales overseas amounted to 15.6 percent of its total business. General Motors' share of all American and Canadian motor-car business abroad was 12.4 percent in 1922, 24.3 percent in 1926, and 47.1 percent in 1928.

The peak passed. Danger signals began to fly abroad in 1929. By July the decline in the sales of automobiles abroad set in.

This was to continue steadily through 1932, and for the first two months of 1933. This decline was accelerated in the latter half of 1929 by the fall of international exchange and increasing disruption of world trade. In the Export division's affairs, those markets termed neutral, i.e., those where there was no major car manufacturing, as there was in England, France, Germany, and Italy, felt the adverse effects most severely, in reduced buying power as well as in increased competition from other than American products.

The sales account of Export in these three years of decline shows:

CARS AND TRUCKS

| Year | Number | Value |
|---|---|---|
| 1930 | 164,112 | $155,728,304 |
| 1931 | 125,606 | 110,525,817 |
| 1932 | 77,159 | 64,722,593 |

The attainment of these relatively satisfactory sales in 1930 to 1932 can be attributed to a substantial increase in the sales of the Corporation's German and English-source product in the world markets. In 1930, the first year in which these sales were included, they accounted for 20.7 percent of the Corporation's overseas volume, in 1931 this ratio was increased to 31.0 percent and further increased in 1932 to 47.7 percent.

As early as 1926, difficulties some already noted hindered expansion of the American export business. These might take the form of political alliances and loans, import quotas, and rising tariffs. The United States failed to import in volume sufficient to permit other countries to pay for our exports, and when American investors ceased to take foreign loans, a decided swing away from American goods began. By 1929 this trend threatened to become an avalanche which the Overseas Operations found it could meet best by capitalizing even more heavily on the manufacturing sources it possessed in Europe to protect its investment of approximately $68,000,000 in overseas business and to safeguard its distribution system with its thousands of employees and its 1,800 dealers in the event that its American cars, for one reason or another, should be shut out of these consumer markets.

There was also another good reason for foreign manufacturing. While American cars were being improved in quality, they were also being increased in size and weight, in contrast with European evolution toward small and light types, more economical to own and operate in countries where gasoline is expensive, and horsepower taxes are high. Differences in the tastes of foreign consumers could be satisfied by a flexibility of production not possible at home. Better geographic relation to some of the major volume markets also had its influence.

These considerations had led the Corporation to become interested in Great Britain as a manufacturing source, not only because the British Isles afforded an important and stable market in themselves, but also because the threat against American cars in foreign export fields, especially in Australasia, worked to favor British products. If the Corporation were handicapped in those fields, its investment would suffer. German manufacture was equally advantageous, both for the Reich and for many sections of Continental Europe.

With the arrested growth of American-source volume creating many investment problems, plans were made for a realignment of the Corporation's world-wide distributing structure. Reduced to a sentence, the problem resolved itself largely into one of "earning on the investment made by selling, in each market and from each source, the kind of cars the people in those markets want and can afford to buy." Such a policy demanded, in order to achieve the proper balance, as keen an interest in Vauxhall and Opel as in any of its American manufacturing companies.

Handicaps against American-made cars abroad reduced America's share of overseas sales from 52 percent in 1929 to 21 percent in 1932. In his report for the Overseas Operations group for 1932, the president commented as follows on the changing situation:

> The swing ... represents a change in the popular demand for motor cars which is of profound significance to General Motors and

> to all companies engaged in international trade. Its causes ... include the basic economic disadvantages which confront America in her present international economy; the growing sentiment of nationalism; the increased duty advantages and, in some cases, exchange advantages, which have come to prevail in favor of foreign products; the rather normal trend toward a smaller and more cheaply operated type of car arising out of the depression ; and most important of all, the more serious effects of the economic depression itself upon those "Neutral markets" throughout the world where the American car has always been strongest, as compared with the effects in the manufacturing markets, where the American car has never, actually, accounted for a very considerable proportion of the business.

With English and German production sources as its defence, General Motors was able to "soften the drop" in the total volume, and in the later months of 1932 these sources, including the products consumed domestically in England and Germany, were exceeding the American-source volume in General Motors' total overseas trade. Despite these bulwarks, however, General Motors unit sales abroad fell 38 percent below those for the preceding year, but at a decreasing rate as 1932 went along.

Of the total American-source business General Motors held its own in 1932 with 33.9 percent, while its German and English sources as against the volume from all foreign sources tended to increase their percentage, and amounted to 7.1 percent. Adam Opel, A. G., in 1933 marked up its first operating profit. In England, following Britain's departure from the gold standard, Vauxhall realized its greatest volume and profit, through putting out a low-powered car more in line with English demand.

The source-distribution of General Motors unit sales in the three Territorial divisions of the Overseas Operations showed the following percentages for 1932:

| | *American Source Percent* | *English Source Percent* | *German Source Percent* | ***Total*** |
|---|---|---|---|---|
| Export Division | 77.1 | 10.1 | 12.8 | **100** |

| | | | | |
|---|---|---|---|---|
| English Division | 2.7 | 97.3 | | **100** |
| German Division | 5.5 | | 94.5 | **100** |
| **Total Overseas Operations** | **52.3** | **20.4** | **27.3** | |

In the early months of 1933 the Export pendulum started its up-swing. Overseas shipments of General Motors American-source cars were 45 percent greater in the first six months of 1933 than in the corresponding period of 1932. June shipments were 133 percent greater in 1933 than in June, 1932. Sales of foreign source products by Opel and Vauxhall also advanced during this period.

General Motors' export selling reflects careful choice of distributors and dealers and a study of local needs; for each country, having its own peculiar methods of doing business, must be handled separately, although a casual survey might class it with some other country on the basis of like volumes of sales. In most territories General Motors products are sold on an exclusive basis. Building up the reputation of products is found to be slow work, and the quality of the retail outlet no less important than the quality of the product. Withal, it must be recognized that the first consideration of the car buyer in many overseas areas, and especially in Europe, is economy of operation. The world contains eighteen times the population of the United States yet only 28 percent of the motor cars are outside the United States, which has a car per 5.1 persons as against one for about two hundred people over the planet. America's opportunity to sell motor cars abroad seems still large, but there are limitations. An extract from a brochure, The Export Organization of General Motors, 1929 revision, sets forth some of these limiting factors overseas, where

> ... the motor car has not been generally accepted as an essential utility as in the United States. Tariff and cost of freight and handling add on the average 33 percent to the retail delivered price of the car, compared to the delivered price in the United States. ... A Buick owner overseas is in the price class of the Cadillac owner in this country

> .... Taxes are high and the cost of gasoline, oil, tires and other items make the operation of a car expensive by comparison with what it costs to operate in America.

The idea of the auto is an expanding factor, breaking the barriers. The new countries of European settlement which have developed exportable surpluses stand in the first rank as regards cars per capita of population; those of primarily British stock are sometimes spoken of as being saturated: United States, Canada, South Africa, Australia, and New Zealand. Their use of motor cars is not likely, for some time to come, to be approximated by countries on other standards of living, and under less favorable economic conditions.

By comparison with her neighbors to the west, the German Reich has only a half to one third the cars that might be expected. Latin Europe, except for France, is not even that well off in motor cars and, outside of Scandinavia, central and Eastern Europe are poorly supplied. Yet there are bright spots the world around where the automobile has come to be regarded as the necessity it is in the United States, although the hinterlands even of these countries have comparatively few cars. These areas are hopeful territories for cultivation in the future.

In February, 1933, a policy of more fully decentralized management went into effect. Four regional directors for Europe, for South America and South Africa, for the Far East, and for Australasia became responsible for all operations in their respective areas. They maintain headquarters at the home office in New York.

Through its far-flung organization, General Motors meets all the civilized peoples of the earth. Into many and varied social settings it has introduced its products through representatives carefully selected from the standpoint of character and understanding of the peoples among whom they are to work. In extensive manufacturing and assembly operations employing domestic labor, the Corporation applies both the factory technique and the labor policies developed at home. Care is taken to train native staffs, to the end that General Motors' overseas operations may become a part of the local scene to as large an extent as possible.

While the motor car is a tool which inevitably hastens social changes by enlarging the radius of human action and expanding the vistas of mankind, an attempt has been made to effect its entry into new scenes without affront to existing manners and customs. The result is that General Motors has escaped difficulties due to nationalist feeling, and can count citizens of all governments among its valued friends. As highways are gradually improved and lengthened, as international trade resumes, General Motors Export will press on to more and more mutually helpful contacts with all sorts and conditions of men who, whatever their surroundings, have in common the desire for rapid transport and the conveniences of a motorized civilization.

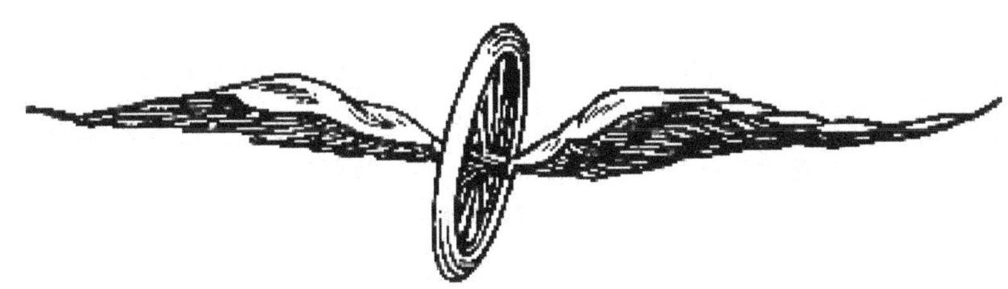

# Chapter XIX RESEARCH: THE MARCH OF THE OPEN MIND

THE automobile, as we have seen, came to pass as the result of a long evolution. At every stage the pioneers were handicapped by faulty materials, lack of capital, adverse public opinion and man's general ignorance of the forces of nature. Some of their efforts appear almost childish today, when the laws of matter and force have been tested and recorded in an imposing body of knowledge which grows day by day. But before we indulge in a pitying smile for their makeshift arrangements, let us recall that the book of science is by no means closed, and that at some distant day the very achievements on which modern man so prides himself will no doubt be considered inconsequential.

When the history of America is written centuries hence, it will be noted that in the first quarter of the twentieth century business began to subsidize scientific research. That resounding fact may then be considered more important than any of the political or military events of our era. The laboratory, where trained men work quietly amid controlled conditions, is a seedbed of social and political change as well as of scientific experiment.

Until recently, science was usually described as of two sorts: pure science and applied science. Pure science was held to be the fruitage of a disinterested search for truth into which no profit element entered and from which no practical benefit was expected. Work of this nature depended for support upon university or other foundations, or upon the private means of investigators. The purest of the pure scientists sometimes congratulated themselves on the prospect that what they discovered could never be of the slightest use to anyone. In this respect, at least, the pure scientists are defeated men, despite their triumphant contributions to knowledge.

There is a driving force in human nature which makes the pure science of yesterday the applied science of today. Even celestial research sometimes comes down to earth. Helium was discovered in the sun, through spectrum analysis; thereupon its presence on the planet was assumed, and after many years helium gas was found in such quantities on earth that it is today a commercial product and a factor in aviation and world politics. Years of pure scientific research on energy waves preceded the practical application of that knowledge through radio, with the result that everyman's home now shares in the results of what was once pure science.

Application of scientific knowledge to everyday life proceeded for several generations largely through the training which the colleges and universities gave in their professional courses. Once a man was graduated, he was cut off from laboratory opportunity and association with minds keyed to research. The tendency was for him to consider that scientific progress ended as of the day he lost contact with it. Of course there were notable exceptions to this generalization in all the learned professions, especially in medicine and chemistry. But a mind of unusual resiliency marked the man who, faced with practical problems in engineering day after day, continued to peer behind the veil which hid the unknown. Lesser minds might make minor innovations as they coped with novel tasks, but a long and steady drive on an objective which might never be found was usually beyond the resources of the individual worker or his employer. Meantime, the independent inventor and investigator struggled along by himself, without backing, living largely upon hope of developing something patentable. After securing a patent, he had to look far and wide for someone with capital enough to bring it to market. Great as the achievements of these men were, industries growing in volume year by year could hardly depend upon such random activities to produce the continuous improvements in technique upon which both earnings and market leadership depended. Consequently General Motors, in common with other large corporations, undertook the financing of scientific research.

It should be noted, however, that while this research concerns itself with long-reach problems involving many branches of science and which may be years in process of solution, the engineering staffs of the various subsidiaries attack with equal energy the problems involved in improving models from year to year. Not only do the divisional staffs commercialize research findings and ascertain their practical applications, but they also originate and put through important innovations on their own account, either independently or in concert with other staffs. Among the outstanding contributions of General Motors in the research field are the following, some of which have been the work of the Research Laboratories, some the work of divisional staffs, and others the result of joint efforts between the Laboratories and the divisions:

1909-10 Battery ignition (Delco and Remy).
1911 One-piece spark plug shell ( AC ) .
1911-12 Self-starter (Delco). First used on Cadillac.
1915 Tilt-beam headlights. First used on Cadillac.
1920 Cellular type radiator, of ribbon stock (Harrison).
1923 V-type fan belt (General Motors Laboratories) first used on Frigidaire and Chevrolet.
1923-24 Perfecting four-wheel brakes for quantity production (Buick).
1923-24 Ethyl "knockless" gasoline, developed by General Motors Laboratories.
1924 Harmonic balancer, to eliminate difficulties caused by torsional crankshaft vibration. Duco lacquer finish.
1925 Crankcase ventilation. First used on Cadillac. Thermo-control of water-cooling system (Cadillac). 1926 Balancing crankshafts in quantity production.
1927 Chromium plating. First used on Oldsmobile. Engine- driven fuel pump.

1928 Synchro-mesh transmission (Cadillac). 1930 Silent poppet valve mechanism (Cadillac). 193031 Carbureter intake silencer (AC).

1932 Automatic choke. First used on Oldsmobile. Supersafe headlights. First used on Cadillac.

1932-33 Fisher no draft ventilation.

1933-34 "Knee-action" front springs, and improved weight distribution.

General Motors' first move toward the creation of a research laboratory serving all constituent companies was taken by President James J. Storrow in January, 1911, when he conferred with Arthur D. Little, Inc., of Cambridge, Massachusetts, on the advisability of setting up a centralized testing and research laboratory and technical department. The Little company's plan was accepted on February 7, 1911. Until November of that year, the research organization was known as the Engineering Department; thereafter as the General Motors Research Department. After one year this department was placed on its own feet, the Little connection being changed from management to consultation.

Among the problems studied by this early research staff in Detroit were painting, lubricating and cutting-oil practices in various plants, tests of materials for purchasing departments, and investigation of new parts and accessories submitted to General Motors by outsiders. The original research staff consisted of nine technicians and their assistants. It was in this small laboratory, according to Arthur D. Little, Inc., that the electro-dynamometer was first made use of in the automobile industry.

The World War stimulated industrial research by lifting technical and scientific problems to the importance of life-and-death matters. Germany hitherto had led the world in the close cooperation of industry and science, a partnership encouraged by the government. Suddenly the rest of the world realized that it had been laggard in this respect. Everywhere in America the man of science took on a new importance, and nowhere did he find a warmer welcome than among alert industrialists. This enthusiasm outlived the war and continues unabated.

It was in this period that General Motors undertook seriously the organization and financing of scientific research, by taking over the Dayton Engineering Laboratories Company and organizing on that base, as of June 12, 1920, the General Motors Research Corporation, incorporated in Delaware for $100,000, with all the stock owned by the General Motors Corporation. In that way it took a short cut to a desired objective, as the Dayton Engineering Laboratories Company, long active in the scientific side of the automobile industry, possessed a competent equipment and tested staff.

CHARLES F. KETTERING

When one speaks of General Motors Research, the picturesque character and record of "Boss" Kettering leap to the mind. Charles Franklin Kettering was born on a farm in Ashland County, Ohio, near Loudonville, on August 29, 1876. He passed in turn through district school, Loudonville High School, the Normal School at Wooster, Ohio. Determined to work his way through all available institutions in the neighborhood, he was graduated from Ohio State University in 1904, at the age of twenty-eight, his education having been interrupted by outside work. After teaching school as a teacher he must have been full of surprises he entered the telephone business at almost its lowest rung, as installation man for the Star Telephone Company of Ashland, Ohio. To this day, by way of illustrating the need for efficiency in humble tasks, he is apt to tell his hearers how he discovered there is a right and a wrong way of digging postholes. It is of record that he almost bankrupted his company by installing what his directors thought was too large a switchboard, but events showed he was right, business flowing in to use the plant once it was installed. Looking far ahead is a Kettering habit and characteristic.

From telephones and Ashland he went to cash registers and Dayton, the city where he found opportunity to reach full stature. National Cash Register was then the last word in specialty manufacturing and sales promotion. Its general manager was E. A. Deeds, who recognized Kettering's genius, and promoted him to head of the Inventions Department. There Kettering developed and patented a number of improved cash registers, including one operated by electricity which did not come into full use until several years later.

With Colonel Deeds's backing, and that of other Dayton capitalists, the Dayton Engineering Laboratories Company was organized in 1911. While Delco, as it was called, stood ready to tackle any problems in electrical or mechanical engineering, its chief point of attack was Kettering's long considered and epoch-making idea of an electric self-starter for automobiles. He believed that the same power which switched on an automobile's lights and ignited its gasoline also could be used to start the engine, thereby eliminating the nuisance of hand-cranking.

Starting an engine by hand was difficult and often destructive of temper and clothes. It required, usually, a combination of physical vigors not always found in the same individual the strength of Ajax, the cunning of Ulysses, and the speed of Hermes. Ulysses had to adjust the spark and throttle just so; Ajax had to turn the engine over, sometimes over and over; and Hermes had to dart back like a flash to the controls to advance the spark and regulate the gas before the engine went dead again. Not everyone possessed all these qualifications; many an ardent motorist of today despaired of ever learning the difficult art of cranking, and the whole back-breaking operation was quite beyond the powers of all women save those of Amazonian proportions. This was a serious matter for the industry; serious, too, were the physical accidents incidental to cranking. The commonest of these was the broken wrist caused by the back-firing of the engine; but more serious hurts, sometimes even deaths, followed the forward charge of the car upon the cranker when, as often happened, the engine had been left in gear. It was an accident of that sort which gave Delco its first commercial opportunity.

Mr. Kettering says Delco's greatest obstacle was the closed mind he encountered nearly everywhere he went. First, the battery manufacturers said his idea of self-starting was absurd, for the simple reason that no electric storage battery could be built with enough capacity to turn a motor over. Why, they did not say; the fact that no such battery had ever been built was enough to convince them that it never could be done. They accepted their frame of experience as final. In effect, they said, "There ain't no such battery," and let it go at that. Finally Kettering had to explain, cajole, and bully them into making bigger and better batteries.

Meantime, Delco was trying to sell its self-starter idea to manufacturers. They, too, had the closed mind. The thing might work for a while, but it would soon exhaust the battery. Then how about the lighting and ignition? And what if it wasn't positive? "Positive" was a great word in those days. If a thing wasn't positive, it might be a bad buy; it often was. Finally Kettering got his chance, directly as a result of one of those accidents common to handcranking. An elderly friend of Henry M. Leland, founder of Cadillac, was driving a Cadillac on Belle Isle bridge when the motor stalled. Forgetting to throw out the clutch before cranking the car, he sustained serious injuries. Mr. Leland's grief over this accident drove him to encourage Kettering, and to give the Delco starting, lighting, and ignition system a chance. On February 27, 1911, the first Delco starter was installed on a

Cadillac car. It worked so well that in 1912 Delco was standard equipment on all Cadillac cars. Self-starting had arrived.

Other self-starters soon appeared on the market; in fact, more than one man was working along that line when Kettering was doing his pioneer work. Various systems of self-starting had been worked out before Kettering applied electricity to the task among them compressed air, springs, and acetylene gas. Other electrical manufacturers, notably the Halbleib brothers at Rochester in their North East plant, had worked out ways of doing the job about the same time as Kettering did. But in the ensuing litigation, the Delco patents held firmly enough to establish the priority of self-starting with Kettering and Delco.

With men of the Kettering stamp one thing leads to another with almost lightning rapidity. This has hardened into a code based on experience. Among the Kettering aphorisms, the value of time and the need for swift mental change are emphasized again and again. "Unless we progress from year to year we lose ground." "Today's dream is tomorrow's actuality." "Changes are born in men's minds and worked out in laboratories." Against a mental background of that kind, standing pat with the Delco starting, lighting, and ignition system, no matter how successful it was, would be merely standing still. The new idea came from a customer who had used his Cadillac to light his summer cottage in an emergency. He wrote in to buy a Delco system for domestic use. The result was Delco-Light, an electric power plant especially designed for farm and rural residences, another company being formed for its manufacture by the swift-moving Daytonians. From DelcoLight they went on to other electric-power adaptations for the rural market, establishing the Domestic Engineering Company. Another of their corporate creations was Dayton Metal Products Company. The Dayton-Wright Airplane Company, manufacturing airplanes during the war, was another important development established by this enterprising group of Ohio industrialists.

In 1916 W. C. Durant took over the Dayton Engineering Laboratories Company as part of his United Motors merger, and with that group it became part of General Motors in 1918. The associated companies in which Mr. Kettering was a leading factor followed the Laboratories into the General Motors fold within the next few years. Meantime Mr. Kettering had been commissioned to set up and direct research operations for the Corporation at Dayton, under the name of General Motors Research Laboratories. This was incorporated as General Motors Research Corporation, June 12, 1920, and transferred to Detroit in 1925, where it now occupies a specially constructed building directly back of the central offices of the parent corporation. General Motors Research Corporation is now inactive, the operation being conducted as a department of the Corporation under the title, "Research Section, General Motors Corporation."

The Research Building houses all the Laboratories' equipment under one roof. It contains chemical and physical laboratories, shops, drafting rooms, library, and all other necessary facilities. An eleven-story building, it is the third occupied by the Laboratories, the others being outgrown by the expanding enterprise.

To clarify the story of research, the Laboratories recently issued a booklet prepared by its engineers and scientists. It states:

> The greater part of the engineering and research, which makes General Motors cars better, is carried on by the engineering departments of its car manufacturing divisions. But, in addition, the divisions have the Research Laboratories to help them whenever and however possible.
>
> The Research Laboratories act in a consulting capacity to the divisions, and are continually searching for new principles and ideas of a more fundamental and scientific nature than would be possible for any of the individual engineering departments, which must be more concerned with immediate production problems.

How the various departments and sections of the Research Laboratories work out specific problems may be seen with reference to some of the more important contributions of General Motors research. Take, for instance, crankcase ventilation one of the meritorious achievements in motorcar advancement during the last decade. Complaints from motorists each spring showed a seasonal tendency toward balky motors and stretched timing chains. These early spring troubles attacked cars which had been conservatively driven, rather than those pushed at top speed with accompanying high engine temperatures. The affected parts, upon examination, showed a reduction in size beyond that to be expected from ordinary wear and tear.

In the Chemical section, undersized and heavily pitmarked parts were analyzed. The chemists found sulphate of iron in unusual quantity, and immediately recognized that sulphur was the villain in the piece. They set out to determine the relation of sulphur in the fuel to the weather conditions of winter and early spring. The explanation was eventually found. When a gasoline engine is cranked again and again in cold weather, as it frequently must be, water condenses on the cold cylinder walls and trickles down into the crankcase. This water comes from the burning of the fuel and air in the engine about one gallon of water for every gallon of fuel. During the process of fuel consumption, the sulphur in the fuel is being burned, forming oxides of sulphur which, united with the water, condensed in the engine to form sulphurous or sulphuric acid. This acid was corroding the metal parts, wasting away the chains, and eating the piston pins and camshaft. Even in diluted form, the acid slowly etched nearly all moving parts of the engine.

> It is one thing to find trouble; another to correct it. Water is a by-product of the process of getting power from fuel. It was not possible to keep out the water, nor was it practical to eliminate sulphur entirely from the fuel.
>
> The chemists reasoned that since they could not stop the acid from forming, they must prevent the acid from coming into contact with the metal parts of the engine, for the trouble arose only when acid touched metal... . Open-minded toward the application of the commonplace to the intricate, the chemists put this to the test. To keep the air in the engine continuously clean and free from acids, why not open the windows?
>
> So they put windows, so to speak, in the crankcase. And crankcase ventilation put an end to "spring fever" in the automobile.

This is a shining example of the work done in the chemistry of automobile progress. From first to last, from processing of raw materials to the last gallon of gasoline consumed before the car goes to the junk yard, the automobile is a chemical device. As the Laboratories put it:

> ... The materials entering into its construction, the fuel that makes it go, and the oil that lubricates it are chemicals. Iron, Copper, Lead, Tin, Carbon, and Tungsten are used in the electrical system. Iron and Steel make the chassis strong and durable. Babbitt and Bronze are used for bearings. Wood, Glass, Cotton, Wool, Nitrocellulose, and Rubber go into a body. Oxygen unites with the Hydrogen and Carbon of the gasoline and furnishes the power to propel the car. The Chemical Section of the Research Laboratories finds ways to improve and make better use of these chemical materials.

Fuel is so important an element in the motoring problem that it receives the undivided attention of a special section. There goes on the steady search for more economical fuel consumption. Fuel research is marching toward greater power and more miles per gallon, a fact which is reassuring in view of the frequent announcements of dwindling oil supplies and prophecies of future oil famines.

In the field of fuel research, an outstanding development has been the creation of Ethyl gasoline. One of the first barriers to progress in fuel research was the "knock." The knock had to go before fuel research could proceed further.

After years of research, a specific cure for the knock was discovered. The story of the discovery of Ethyl illustrates how a great laboratory feels its way along from one stage to another, applying the old method of trial and error but with greater speed and precision, until at last by process of elimination it evolves a specific remedy for particular ailments even though it may not have discovered the fundamental cause of the trouble.

Chemists noticed that ordinary combustion of gasoline produced a bluish flame, while knocking combustion produced yellow or orange flame. They thought that if the gasoline were darkened the knocking might be prevented, because darker color might absorb more of the radiated heat from cylinder walls. Trying one coloring matter after another, they concluded that certain coloring agents were effective on knocks because of their chemical properties and not because of their colors. Effective knock suppressors were found to be closely grouped from the standpoint of atomic weights. So it was decided to try the neighboring elements in the atomic scale.

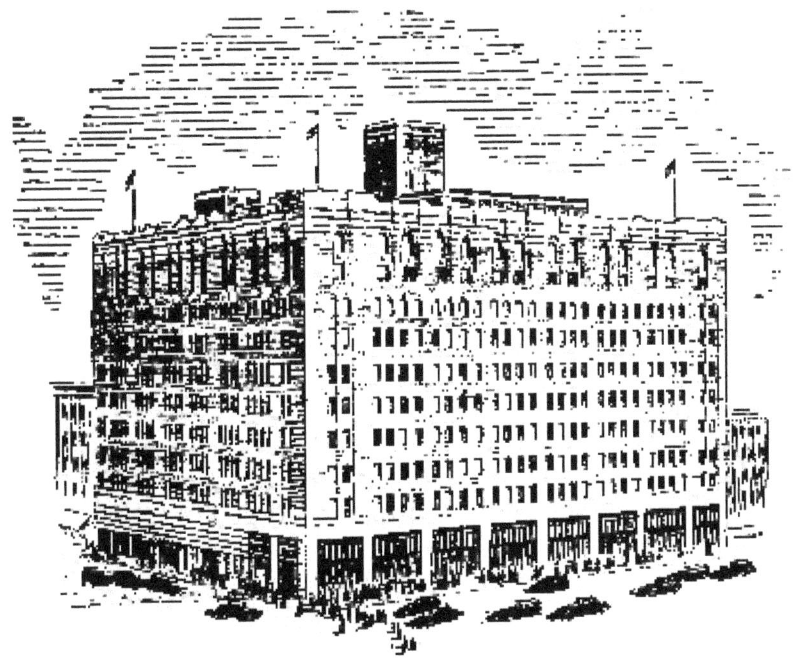

Research Building home of General Motors Research Laboratories, Detroit

One of these, tellurium, gave Thomas Midgeley, Jr. the key to the riddle. By eliminating other possibilities he concluded that the best available anti-knock agent was tetraethyl lead. Although composed of cheap elements, lead, carbon and hydrogen, the compound was rare, because it had little commercial use and was not produced in quantity. Other chemists aided in bridging this gap. Now it costs little to use knock-proof gasoline. In recognition of his success, Mr. Midgeley was awarded the Nicholas Medal by the American Chemical Society. The commercial rewards from this research activity have been large. In order to achieve a wide market for the new fuel in the shortest possible time, General Motors Corporation joined with Standard Oil Company of New Jersey to form the Ethyl Gasoline Corporation which was incorporated in Delaware, August 18, 1924, with $5,000,000 Common stock. Of this only $1,500,000 has been issued, of which General Motors took one half. Its investment has since been more than repaid by dividends, and Ethyl has become a household word to indicate anti-knock motor fuel obtainable on practically every highway. Under a liberal licensing policy, the Ethyl compound is now used by nearly all of the leading refiners. On the board of the Ethyl Gasoline Corporation, General Motors is represented by Donaldson Brown, Irenee du Pont, Charles F. Kettering, and Alfred P. Sloan, Jr., while Mr. Midgeley serves Ethyl as vice-president.

Six years after Ethyl gasoline was introduced, its effect upon the industry was described as follows by C. F. Kettering and Allen Orth in The New Necessity:

> The car manufacturers have not hesitated to take advantage of the possibilities offered by this new fuel. They have raised the compression ratio of their engines until now the average ratio is a little over 5 to I as compared to the 3½ to 1 ratio usually encountered fifteen years ago. Piston speeds have jumped from 1,000 feet per minute to over 3,000. The ultimate result is that from an engine of 250 cubic inches displacement, we now obtain over 80 horsepower instead of the 30 horsepower of 1915. Of course this cannot be attributed solely to the change in fuels. Engine designers have accomplished unheard of things during this period. They have done the seemingly miraculous with combustion chambers, valves, and the design of the rapidly reciprocating and rotating engine parts.

Both the Laboratories and the General Motors Proving Ground possess exceptional facilities for testing completed automobiles and the parts from which they are constructed:

> It is difficult to observe how the various parts of a car are operating when the car is traveling one hundred miles an hour over the road. The research engineer, confronted with the problem of finding out what might be happening under these and other driving conditions, had to devise other methods.

> The alternative was to make the road run under the car. The chassis dynamometer was designed to accomplish this. The rear wheels of the car rest on two large rollers, connected to a dynamometer or power-absorbing and measuring machine. Thus, the engine and chassis behave as if they were speeding down a road, yet the car stands still ; and the engineers are enabled to stand by the car, taking measurements and studying performance in a way that would be impossible otherwise.
>
> Weather conditions have to be manufactured to meet the demands of research, for a group of engineers, seeking facts about the operation of an automobile under varying temperatures, would find it both impractical and inconvenient to be traveling constantly between the Arctic Circle and the tropics.
>
> So the equipment in the Engineering Tests Section of the Research Laboratories includes a cold room 2,000 times as large as the average household refrigerator. The temperature inside this room falls as low as 50 degrees below zero.
>
> The room is large enough to hold two automobiles, and the engines may be operated with the air at any desired temperature. With this equipment, research engineers are checking the operation of carburetors, starting equipment and lubrication, as well as checking the action of materials and parts under extreme weather conditions.

The engine is viewed as a problem in mechanics and thermodynamics, the object of research being to find the best use of heat. The Laboratories define the automobile as a self-contained factory using gasoline and air as raw materials and producing chiefly water, carbon dioxide, and carbon monoxide, with power produced as a by-product. The engineers in this section follow the course of cooling air into the combustion chamber, into the exhaust pipes, and through the mufflers. At one end of the line is the spark plug, at the other end is the muffler, between them lie the cooling system and the engine. Along this line are also the gears, transmitting power from the engine to the wheels; consequently the Dynamics Section studies clutches, universal joints, propeller shafts, differentials, axles, and steering devices. These involve many varieties of gears.

> The gear is a mathematician's paradise. It entails the study of involute, cycloidal, epicycloidal, and hypocycloidal curves. There are spur gears, worm gears, helical gears, and spiral bevel gears, all with their own individual problems which bring up all sorts of technical questions, such as the pressure angle, diametral pitch, circular pitch, line of action, addendum, and dedendum. Few mechanical parts seem so simple when finished, yet involve so much detailed study as a gear.

Engineers cooperate with those of the manufacturing divisions in the continuous effort to improve the performance, quietness, smoothness, and durability of automobile power plants. This work is never-ending, as models are being redesigned all the time, and every change has to be studied with an eye to its effect on the entire car.

> The crankshaft, the backbone of the engine, furnishes an interesting example of the number of factors influencing the design of various parts. The power and speed of the engine, the weight of its moving parts, the number and arrangement of its cylinders, all play their parts in determining the size and strength of the crankshaft and the size of its bearings. Balance and smoothness must not be overlooked. To eliminate the roughness inherent in conventional reciprocating engines the crankshaft is usually counterbalanced to offset the forces arising from the up-and-down motion of the pistons and part of the connecting rod.
>
> The valves and valve gear bring up their own peculiar problems. The intake valves must open and close at the proper time to give a full charge of fuel to the cylinders, and the exhaust valves must be timed properly to rid the cylinders of the burned gases. These valves must open and close several thousand times a minute, noiselessly and without variation, and at red heat. The engineers must know materials which can withstand the heat and continuous hammering of the valve on its seat. They must calculate proper sizes, seat angles, lifts, and timing of the valves to allow a full charge of gas to be taken in, and to permit all of the exhaust gases to be expelled. The spring must be of correct size and tension, and the camshaft must give correct timing without noise.

An enduring element in this work has been the creation of machines for statically and dynamically balancing such parts as fly-wheels, crankshafts, clutches and propeller shafts. Here was achieved the "harmonic balancer. " This device balances out the torsional vibrations in the crankshaft, eliminating an annoying source of noise and undue wear as cumulative explosions give periodic blows to the crankshaft.

A code of General Motors Laboratories might be compressed into fault-finding and fact-finding. Its engineers are paid to find fault with everything about a car. It is the spirit of research to be dissatisfied with anything as long as improvement remains possible. Dissatisfaction by itself is not enough. Knowledge must be extended. New facts must be found, in order to help manufacturing divisions improve their cars. A Laboratory spokesman has stated the case in these words:

> One might think that engineers could design a mechanism perfect once and for all if they gave it study enough, but they can't. Everyone is changing. Change is the watchword of a progressive people and new experience brings new knowledge.
> Research is insurance the General Motors manufacturing divisions take out to help them keep just ahead of the calendar.

Sooner or later and usually with all possible dispatch, the economies developed through research affect prices, giving the consumer more value for his dollar. Research is one reason why the automobile dollar has brought more values in automobiles and tires, consistently, through the past fifteen years than the general merchandise dollar. During the boom days, when the cost-of-living dollar was worth only 68 cents as compared to the dollar of 1914, the automobile purchasing dollar was worth $1.28. For many years, automobiles and tires were practically the only goods which the farmer could buy to advantage, when farm prices were compared with those of manufactured goods. For this, quantity production and consistent research are jointly responsible.

Probably no piece of research has saved the consuming public more money than that which resulted in the widespread use of Duco and other quick-drying finishes for motor cars. The costs of painting and varnishing automobiles on the old system were high, because fifteen to thirty days were required to paint a car. On a heavy production program in a large plant, this meant that as many as fifteen thousand cars might be in process of painting at one time. "Twenty million dollars in automobiles sitting in warehouses waiting for the paint to dry!" A month might be required to finish a high-grade car. Regular varnish finish which the automobile industry had inherited from the carriage trade was a slow process. A black enamel finish could be completed in a few hours, but it was used chiefly for fenders, except on cheap cars which might have black enameled bodies.

In 1921, a 'Taint and Enamel" committee was appointed to consider the whole problem of painting and finishing cars.

> The industry absolutely had to have a finish that would decrease the time ... from days to hours; it must not require temperatures beyond those wood could stand; it must be as inexpensive as the varnishing process yet it must be applicable in all colors and last the lifetime of the automobile ... it must have all the advantages of enamel and varnish without any of the disadvantages.[1]

These requirements daunted the paint and varnish trade. Their representatives said: "You can't change nature." Just as the battery people years before had declared that no battery could be built strong enough to start a gasoline engine, so paint manufacturers, in 1921, considered as impossible whatever was outside of their experience. However, in various factories, cellulose nitrate was being sprayed on toys. Tests proved that this product dried even too fast for motor cars, but this demonstration neatly bracketed the research problem. Varnish was too slow; existing lacquers too fast. Between these extremes there must be some suitable chemical combination. What was it?

General Motors Laboratories worked out this particular problem with the help of other research staffs, notably that of the du Fonts. When the paint manufacturers upheld the old methods by saying they produced the best finish in the world,

---

[1] Kettering and Orth: The New Necessity, pp. 87-8 off

he thought, "What they mean is: it's the best finish they know anything about."

To follow Mr. Kettering and his many associates through all the mazes of their inquiry in chronological sequence would be too long a story. The riddle was finally solved after this manner. About a century ago, a French chemist discovered a brilliant, varnish-like coating by dissolving potato starch in concentrated nitric acid. A fellow countryman of his, using cotton instead of potato starch, called the resulting nitro-cellulose "Pyroxylin." Later a German chemist treated cotton with a mixture of sulphuric and nitric acid result, gun-cotton. After the Spanish-American War military and naval needs brought forth cellulose nitrate for smokeless powder, and incidentally revived interest in cellulose-nitrate lacquers. Low-viscosity cellulose nitrate today forms the basis of automobile finishing lacquers.

Another constituent of automobile lacquer is butyl alcohol. With efforts to make synthetic rubber about 1910, a microorganism was discovered which converted field corn into butyl alcohol, acetone, and ethyl alcohol in the ratios of 6:3:1. Acetone had commercial use in the manufacture of smokeless powder. The common source of acetone was charcoal, but when the World War increased the demand for smokeless powder there was not enough charcoal available for the acetone process. So the greatest of the naval powers fell back on the assistance of the humble microorganism which devoured corn and turned it into the chemicals mentioned above. Mars' tiny assistant refused to make acetone without producing twice as much butyl alcohol; the latter was a cumbersome by-product; for years it was thrown away as fast as made. The demand for acetone, continuing after the war, moved American distilleries, closed by the Eighteenth Amendment, to use their plants in its manufacture. One distillery, hoping that some day a use would be found for butyl alcohol, stored that liquid in a large swimming pool. Eventually they found that butyl alcohol was a good substitute for amyl alcohol, long used as raw material for making amyl acetate or "banana oil," an important ingredient of certain lacquers. So the cycle of investigation and experiment was completed by utilizing a by-product, then of almost no value, and which could be made in any desired quantity at low cost.

> Today practically every car has a lacquer finish a finish that can be applied in less than eight hours; one that is more durable, more easily applied, and more attractive than either varnish or black enamel. However, the use of this material has not been confined to the automobile. It has invaded the home and with it has come color furniture, radios and refrigerators are now lacquer-finished... . this development is a reassurance that nature is still on the job and always will be, operating through human agencies to meet the ever-increasing and broadening needs of people, providing a new substance to replace a disappearing one. There is no waste in nature. What we regard as waste is merely a dormant substance awaiting the hand of research to start it serving some practical and valuable career.[1]

When anyone says that a desired goal cannot be reached because the process is contrary to nature, he is referring merely to that small segment of nature which he comprehends. To explore that which is at present beyond comprehension is the whole duty of science.

Industrial science, as pursued under commercial auspices, must concentrate upon problems likely to have practical results within a reasonable time. Economics enters the scientific investigation at this point. There must be a nice balance between vision and practicality. General Motors has been fortunate in preserving that balance. One of the Kettering advices which has been taken to heart by the Laboratories organization is this: "Engineering must partake as much of economic horse-sense as it does of scientific principle."

The truth of this is revealed in the vast number of patents which never reach the market. Like all manufacturers, General Motors welcomes engineering ideas which come to it from outside, but, after these have been thoroughly tested from all angles, the verdict is apt to be disappointing to the inventors. Statistics show that out of 35,000 inventions submitted to a cer-

---

[1]Kettering and Orth: The New Necessity, pp. 92-3 ff.

tain manufacturer, only one out of 5,000 had merit. The inventors of the others apparently had wasted their time and money. Obviously, industry cannot wait for such random aid. If the tests through which outside inventions must pass are considered to be too severe, it must be remembered that millions are at stake in every change introduced into a manufacturing program based on quantity production.

Practical considerations are also behind the work toward standardization in which the Laboratories join through their association in the Society of Automotive Engineers. In the early days of the industry, all automobile parts were made by hand or on simple machines. The result was such variety that repairs were unduly expensive. Even the manufacturers had difficulty in securing standardized parts. According to J. K. Barnes, 800 different kinds of lock washers were made to fit three or four bolts of varying sizes. Automobile makers used 1,600 sizes of steel tubing; one manufacturer alone using 80 sizes. Standardization has now brought lock washers down to 16 sizes and steel tubing to 17 sizes in 13 thicknesses. Two hundred and thirty alloy steels have been reduced to 50 varieties. More than 200 parts and materials used in automobile manufacture have been standardized since 1910.

On the subject of standardization, Mr. Kettering has this to say as to the benefits automobile owners have received as a result:

> That word "interchangeable" is the sum and substance of all standardization. It is the key to mass production. Fabrication has become a matter of timed operations rather than of individual ingenuity in getting mismatched pieces together. There are specified tolerances that are rigidly adhered to. Pieces are no longer antagonistic. They are interchangeable they fit. And the owner nowadays, when a plug fails him, goes to an accessory shop or maybe to a corner filling station, and asks for a spark plug. In about the time it takes to replenish the water in the radiator, it has been installed and he is on his way again. The owner and the manufacturer both profit by this new form of simplicity. It has

> been estimated that the annual saving to the automobile industry alone amounts to three quarters of a billion dollars.[1]

There are constants in the automobile business, plenty of them; but, paradoxically, the greatest constant of all is change. He who stands still is lost. The public must be wooed through change as well as through quality; science

and art provide the material for beneficial change in such abundance that the timing of change becomes an all-important factor; hence the question which management faces is not so much, "Shall we change?" but rather, "How and when shall we change?" Mr. Kettering calls his research workers "economic-change men." They provide the management with changes from which it can choose those for which it considers the public is ready.

Not all beneficial changes can be introduced drastically as soon as they have been developed. While the public insists on change, it prefers evolution to revolution. Millions can be lost by being too far ahead of the times, just as they can be lost by being too far behind the times. The average man likes to move forward by short steps on firm ground rather than by long leaps into what he considers the dark. Scientists may be quite aware that the fancied darkness is really broad daylight, but there is nothing gained by arguing against a public prejudice in expensive advertising space. In a year or two, the public may have caught up, and then may accept gladly what would have been rejected before. The findings of the economic-change men of General Motors must consequently be winnowed through the sieve of public acceptances.

In acting as a general staff for the engineers of all the Corporation's manufacturing divisions, the Research Laboratories are not confined to the automobile field. As the Corporation grows in the diversity of its products, the Research Laboratories also tend to become more diversified. Various of its members have done effective work in refrigerants; others are likely to give keen attention to the future of aero-dynamics. General Motors

---

[1] Kettering and Orth: The New Necessity, p. 100.

has recently become a large factor in aviation, and the Research Laboratories henceforth may consider aviation as well as automotive problems.

Back of the whole research program is the progressive philosophy that one change leads inevitably to another, that new goods produce new wants, and that through innovations in the conveniences of life both man and society reach higher levels of comfort and efficiency. The director of the Research Laboratories has quoted approvingly this statement of Thomas A. Edison, and the Laboratories staff will do its part to raise what Mr. Edison calls the "thinking capacity of society," by applying its cumulative experience and specialized wisdom to more and more fields of applied science:

> ... the automobile has done more to make America a nation of thinkers than any other invention or agency. The great value of the automobile is not the fact that it has made it easier and quicker and cheaper to go to places, but the fact that it has inspired several million people to go. It has set their gray matter to work. It has revealed to them how petty and meaningless their lives were becoming
>
> ...Most of us view the automobile principally as a great business and manufacturing achievement. It is but it is a greater educational achievement. In the beginning we were a pioneer people a restless people. But when things came easier for us we began to lose our restlessness. The automobile is helping to restore it. And that is one of the most healthful signs of the present generation. Restlessness is discontent and discontent is the first necessity of progress. Show me a thoroughly satisfied man and I will show you a failure. The wheels of progress especially those of the automobile have worked results which may be called miracles. But their service has been to raise the thinking capacity of society.

This statement by the great Edison is one of the most penetrating tributes ever paid to the automobile industry; and, beyond that, it is a stirring declaration of the spirit which moves General Motors Research Laboratories a reasoned dissatisfaction with what is, a reasoned certainty that what it does will improve the lot of mankind.

## Chapter XX BODY BY FISHER: THE MOTOR CAR AS A STYLE VEHICLE

T.HE year 1908 seems to have been a lucky one for industrial beginnings, particularly in Michigan and the automobile trade. In that year General Motors was born and also Fisher Body, destined to become one of its larger subsidiaries and of such proportions that, standing by itself, it would be one of the nation's most important industries.

Six Fisher brothers, all craftsmen, were led to Detroit and fortune by the eldest of them, Fred J., who came to the hub of the automobile world from Norwalk, Ohio, in 1901, where their father, Lawrence Fisher, made carriages and wagons. The brothers, learning their trade under the father's watchful eye, were vehicle craftsmen of the third generation, the grandfather Fisher having been skilled in that art in Germany, though in this country known as a merchant in Peru, Ohio. His son, Lawrence, lived to a ripe age and, before his death in 1921, had the satisfaction of seeing six of his sons rated among the masters of their craft in America. These six brothers Fred J., Charles T., William A., Lawrence P., Edward F., and Alfred J., are today actively associated in General Motors, Fisher Body, and many other large enterprises.

Going to work in Detroit for C. R. Wilson Body Company, then the largest firm in the automobile body business, Fred J. Fisher rose to be superintendent. Within four years he and Charles T. Fisher, who had been associated with him at Wilson for two and a half years, formed Fisher Body Company, incorporating in Michigan in July, 1908, for $50,000, with slightly more than $30,000 paid in.

A few months later Louis Mendelssohn and Aaron Mendelson became associated with the Fishers, obtaining a substantial financial interest, and also became actually engaged in the business, the former as treasurer and the latter as secretary, remaining in these positions until the time Fisher became a part of General Motors Corporation.

The venture looked hopeless to many. The country was still shaken from the effects of the 1907 panic. Money was tight; general business slow. For those who have courage, however, a "bad" time is the best time to start a business then a bold man can rent real estate cheaply, and can get his pick of the labor market. If he knows his business better than his competitor, he is safe in starting at scratch, because his costs will be lower than those of his competitors.

The business prospered from the start, since these Fisher brothers the younger brothers soon joined the older ones in Detroit knew how to build automobile bodies superlatively well. There was plenty of room for improvement in body building. The older concerns had been builders of carriage bodies, and they brought the simple arts and machinery suitable for carriage bodies over into the automobile business with as few changes as possible. Hence, they were a little slow to realize that far greater strength and resilience were required for powered vehicles than for drawn vehicles. As a result, early automobile bodies soon became loose and noisy under road shocks. Even relatively simple open bodies did not give continuous satisfaction.

The Fishers kept their eyes on closed car possibilities from the start. They saw that motoring would remain a summer sport until drivers and owners could be comfortable in the winter months. Women would never be really pleased with the automobile so long as their gowns and hats were at the mercy of wind and weather. After pressing these points on car manufacturers for two years, they were at last rewarded in 1910 by an order from Cadillac for 150 closed bodies, the first "big order" for closed car bodies ever placed in America. Seeing more business ahead in this line, the Fishers organized the Fisher Closed Body Company in December, 1910, thus initiating the march of the closed car to a position where it now dominates the market, with approximately 95 percent of all American cars carrying closed bodies. In 1912 Fisher entered Canadian production with Fisher Body Company of Canada, Ltd., at Walkerville.

CHARLES T. FISHER
One of the Founders of Fisher Body

These plants turned in combined profits of $369,321 in 1913-14, $57,495 in 1914-15 and $1,390,592 in 1915-16, profits based on increasing quantities and decreasing unit costs. Indeed, one of Fisher's greatest difficulties was inducing manufacturers to place reasonable prices on closed cars, consonant with the prices Fisher charged for their closed bodies. At this period the company contemplated entering the automobile business in order to break the deadlock and bring closed cars to the public at fair prices. Finally one manufacturer broke away from the old practice; others followed, and closed cars became popular. The vigorous organization, pushing on, merged its three companies into the Fisher Body Corporation, incorporated in New York, August 22, 1916. Authorized stock was $6,000,000, par $100, 7 percent cumulative Preferred stock, of which $5,000,000 is shown as issued, and 200,000 shares of Common stock, no par, but valued at approximately $10 a share in an early statement. Starting with little more than $30,000, in 1908, Fisher Body had climbed in eight years to the point of owning land and buildings worth more than $2,000,000, and machinery worth nearly $1,500,000. Its capacity was 370,000 bodies a year, the largest in the industry.

In 1919, after three successful years in which the Fisher name had become established, the Fishers sold to General Motors Corporation a three fifths interest in the Fisher Body Corporation, 300,000 of the then outstanding 500,000 Fisher shares at $92 a share, for $27,600,000, on a series of five-year notes. General Motors agreed to purchase all its automobile body requirements from Fisher for ten years. The management remained with the Fishers.

Both parties reaped substantial advantages from this arrangement. General Motors placed its body business with the leading producer. On the other hand, Fisher Body solved a problem which had been growing in proportions as Fisher increased in size. With a $20,000,000 property to conserve, it became a matter of prudence for Fisher to safeguard its future by seeking an assured outlet. There were several ways in which greater security might be obtained. One way was for them to manufacture automobiles, a plan to which serious thought was given. Two other manufacturers were in the market for the Fisher connection when the General Motors union was being considered. General Motors was accepted partly because of the Corporation's stability, partly because its cars occupied a wider price range than those of any competitor. While Fisher was interested in quantity production, it was also interested in continuing to build high-priced bodies for cars with quality reputation. General Motors offered the connection under which the Fishers could best maintain their leadership which had been firmly planted in the public mind by their consistent building of quality bodies.

With this important transaction completed, Fisher expansion went on swiftly, the whole automotive world being then in an expansive mood. National advertising began to carry the craftsmen's name to the public. Fisher Body Ohio Company was incorporated October 17, 1919, in Ohio, to produce bodies in a Cleveland plant. The capitalization was $10,000,000 of 8 percent cumulative Preferred stock, par $100, and 100,000 shares of Common, no par. Control was in the Fisher Body Corporation but Cleveland capital was largely interested. The plant erected there, with 1,500,000 feet of floor space, was then the largest body plant in the world, with capacity for 7,000 employees. It is now the second largest of the Fisher plants, having been surpassed by Flint No. 1. Fisher Body Corporation eventually acquired the entire stock of the Ohio company.

To insure its threatened glass supply, Fisher made a $4,000,000 investment in National Plate Glass Company, acquiring control. This leading plate glass producer had been incorporated in Maryland, January 17, 1920, to acquire three large and well known glass factories: Columbia Plate Glass Company at Blairsville, Pennsylvania, organized 1901; Federal Plate Glass Company at Ottawa, Illinois, organized 1903; Saginaw Plate Glass Company, Saginaw, Michigan, organized 1900. These companies had a combined annual capacity of 11,000,000 square feet of plate glass, and under a ten-year contract with Fisher Body dated January i, 1920, they obtained an assured outlet for their product in the growing closed car trade[1].

General Motors made a further investment of approximately $4,500,000 in Fisher Body in the up-swing of 1923. Fisher prospered greatly in the next three years, a period marked by the acquisition of Fleetwood Body Corporation, dating from 1919, and the purchase or building of large properties in many parts of the country. On June 30, 1926, the outstanding minori-

---

[1] Part of Fisher interests in National Plate Glass Company have since been acquired by other companies.

ty interest in Fisher Body Corporation passed to General Motors Corporation on the basis of one and one half shares of Fisher for one of General Motors, 664,720 shares of General Motors Common being paid. In the following year General Motors completed acquisition of Fisher Body Ohio Company. Since 1926 Fisher Body Corporation has functioned as a division of General Motors, but remains under Fisher management. Fred J., Charles T., Lawrence P., and William A. Fisher are directors of General Motors. Mr. E. F. Fisher is vice-president in charge of production in the far-flung Fisher manufacturing activities, and Mr. A. J. Fisher is vice-president in charge of engineering, a responsibility embracing the creation and development of as many as seventy individual body types and styles annually for General Motors cars.

Before Fisher Body Plant No. i at Flint, Michigan, now the largest in the world, was taken over from Durant Motors, Buick bodies had been completed in Detroit and were hauled to the Buick plant by truck and trailer over the Dixie Highway. These long, box-like vehicles moving on the Dixie were familiar sights for many years to Michigan motorists. Fisher's decision to build Buick bodies in Flint started that city on another wave of growth, since the plant employed in normal times 5,500 men in its more than 2,000,000 feet of floor space. Large as this plant was when Fisher took it over in 1926, it was further expanded. Fisher Plant No. 18 in Detroit employs almost as many men as Flint No. 1. Wcrc it not for new handling methods originated to conserve floor space, labor and materials, still larger factories would be required for the huge production.

Fisher Body is the world's Largest user of steel for automobile body construction, its requirements reaching 335,000,000 pounds a year. On all models, from the lowest in price to the highest, the sturdy body is encased in steel, electrically welded into a single unit. Strength and safety body-tests applied to the lowest priced Fisher-equipped car have shown resistance to 12,000 pounds (six tons) diagonal pressure without appreciable damage to any part of the car or body structure.

Among the many advances achieved in these bodies is the extensive use of solid steel panels. Special steel had to be developed flexible enough and strong enough to endure tremendous strain. Years of patient work on the part of Fisher engineers and steel mill experts finally produced a metal which could be drawn large enough for a body panel under a pressure of 800 tons. The typical Fisher Body of today incorporates the strength of some 200 basic parts, on which more than 1,200 distinct operations are performed to make certain of their accuracy and quality. All the body parts for a given model are as interchangeable as machine parts.

Miniature Napoleonic coach, winner in one of the competitions of the Fisher Body Craftsman's Guild

The Fisher No Draft Ventilator system, applied in 1933 to all General Motors cars, provides a solidly built-in ventilator pane for each side window. The ventilator and the window can be regulated, independently of one another. Fresh air can be drawn in without causing drafts, and any individual passenger can get as much fresh air as he likes without disturbing his companions. Smoke and stale air are drawn out by the vacuum created by the motion of the car. After the division engineers designed the system, its subsidiary, the Ternstedt Manufacturing Company, was called upon to produce it in tremendous quantities, for the system was no sooner perfected than it was decided to use it on all models. Twenty-five hundred men were put to work, days, nights and Sundays, and in thirty days had all the necessary machines, tools, dies, plating units, punch presses, polishing machines, etc., designed and produced and were turning out 2,000 sets or 6,000 "ventipanes" a day.

Another advanced contribution is safety glass. All windshields and ventipanes are made of laminated glass which under impact may be broken without shattering. Polished plate glass is used in all Fisher Bodies, in such volume that the National Plate Glass Company, which they then owned, became one of the world's largest glass manufacturers.

The metal fittings which are used in Fisher Bodies are made by the Ternstedt Manufacturing Corporation of Detroit, a Fisher subsidiary which takes its name from the late Alva K. Ternstedt. Ternstedt became nationally known to motorists in 1920-21, when Fisher introduced the Ternstedt window regulator, the first dependable built-in and concealed window control for closed cars. In developing Ternstedt to its present proportions, Fisher merged into it three companies: International Metal Stamping Company, Shepard Art Metal Company, and England Manufacturing Company. In the Ternstedt organization, Fisher gathered, for the first time in the history of the industry, a complete staff of appointment engineers, designers, artists, and modelers, prepared to fill the modern demand for artistic harmony of design in body hardware, interior fittings, and similar appointments.

One of the important elements to which the Fishers have always devoted unsparing attention in the building of bodies is the quality of quiet. When they began to build bodies, closed cars were the rich man's toy, ridiculed as "show cases," and seldom driven outside of cities. Even open bodies rattled and creaked. The central problem of body-building all along has been to increase strength and quietness without increasing weight. Progress to the present degree of excellence has called forth the best efforts of a large staff of engineers.

Finally, there is the decisive quality of style. Its pursuit begins of course with initial designs. After these have been discussed with the car manufacturer and a choice made, a one quarter size clay model of the car is built in the Art and Color section by skilled sculptors. This model is also carefully considered and changed until it suits the fastidious tastes of all concerned. Then another model is made, and if found satisfactory, a full-sized wood and clay model is constructed in the exact dimensions of the complete car, the clay being rubbed down to the smoothness required for a perfect finish. This is then finished inside and out according to color schemes worked out by the Color section. In the lacquered body, the upholstered interior with its hardware trimmings, one sees an exact reproduction of the car as it will appear on the sales floor. Months pass in these elaborate preparations for manufacturing the automobile as a style vehicle.

Color has dominated automobile styling since about 1925, when the Duco finishes made it possible to meet a public demand for color which had already been roused by manufacturers in other lines. In his book, *... and then came Ford*, Charles Merz says of this period,

> ... The invention of pyroxylin finishes had changed the raiment of the fashionable car from dark blue or black to hues as delicate as the pale tint of an Easter egg or as ruddy as a sunset. "The tendency toward a more varied color-scheme was apparent in 1925," the New York Times observed, "but with the beginning of 1926 the trend toward color was seen to be a stampede."

The color obsession led to bizarre effects in motordom for a while; later tendencies are more restrained and refined. More than 800 colors were taken on by the industry after 1923, and as a result one of the chief tasks has been to simplify color schemes. This has been done through analysis of the spectrum, breaking up the rainbow into its nine steps; white, black, red, yellow, orange, green, blue, indigo, and violet, each hue with its accompanying tints and shades. By this means "off colors" are sorted out and thousands of possibilities reduced to 450. Of these 150 are unsuited for automobile use. The remaining 300 are possibilities. By the use of Maxwell's color wheel, to which colored discs are adjusted with certain areas exposed, it is determined just what proportion of the various primary colors is required to produce a desired blend. By this means blended colors are fixed so accurately that each can be reproduced faithfully on each new shipment of cars. All these refinements come to market in General Motors cars with Fisher bodies.

From antiquity down to the present, style and decoration have been influential in vehicle design. Excavations in the strata of buried civilizations bring forth evidence that the chief's chariot was as elaborately ornamented as his shield and helmet. All through the coach-and-four period, a man's carriages indicated his rank and fortune. The coaches of royalty and aristocracy received the attention of noted artists as well as of skilled craftsmen. In recognition of their beauty, Fisher adopted as its emblem a silhouette derived from the designs of two coaches used by Napoleon Bonaparte, one at his coronation and the other at his marriage to Marie Louise. These are among the finest examples of coach work of the pre-automobile days and the emblem drawn from them is one of the most significant symbols of quality in modern workmanship. It represents the artistry of an unhurried age which Fisher, in spite of its mechanical prowess, holds as a standard before its staff and extends as a pledge of quality to the users of Fisher Bodies on General Motors cars.

The growth of Fisher Body has been phenomenal. In fifteen years after its founding, the muster roll of Fisher employees had grown from 300 to 40,000. Fisher built 105,000 bodies in 1914, and more than five times that number in 1924. In its twenty-five years of existence, Fisher has produced more than 9,000,000 automobile bodies impressive as emphasizing the recognition of Fisher craftsmanship.

## FISHER BODY CRAFTSMAN'S GUILD

No review of the growth and character of the Fisher Body Corporation would be complete without attention to the Fisher Body Craftsman's Guild.

The purpose of the Guild is to foster the ideals of true craftsmanship among boys of high school and college age

A June 1, 1933, summary of the Fisher Body Corporation, division of General Motors, shows the following properties and companies operated:

| | Location | Year Acquired or Incorporated | Nature of Activity | Square Feet of Floor Space | |
|---|---|---|---|---|---|
| Fisher Body Company.. | Detroit, Mich. | 1908–1916 | | | |
| Fisher Body Corporation | " " | 1916–1927 | | 3,524,699 | Approximately 62,000 acres timberlands in northern Michigan |
| Fisher Body Corporation Division of General Motors ........... | " " | 1927 | | | |
| Fisher Body Service Corporation ....... | " " | 1912 | Service Parts | 173,856 | |
| Fisher Body Corporation Division— | | | | | |
| Fleetwood Unit... | " " | 1919 | Body Plant | 1,373,275 | |
| Pontiac Unit .... | Pontiac, Mich. | 1922 | " " | 1,393,004 | |
| Flint Unit No. 1.. | Flint, Mich. | 1926 | " " | 2,127,909 | |
| Flint Unit No. 2.. | " " | 1923 | Body Assembly Plant | 266,443 | |
| Lansing Unit .... | Lansing, Mich. | 1923 | " " " | 622,234 | |
| Buffalo Unit..... | Buffalo, N. Y. | 1923 | " " " | 150,692 | |
| Tarrytown Unit.. | Tarrytown, N. Y. | 1925 | " " " | 376,924 | |
| Fisher Body Company of Cleveland— | Cleveland, Ohio | 1919 | Body Plant | 1,525,217 | |
| Norwood Division. | Norwood, Ohio | 1923 | Body Assembly Plant | 182,167 | |
| Fisher Body St. Louis Company— | St. Louis, Mo. | 1922 | Body Plant | 884,362 | |
| Janesville Division. | Janesville, Wis. | 1923 | Body Assembly Plant | 257,790 | Inactive |
| Kansas City Div... | Kansas City, Mo. | 1929 | " " " | 133,573 | |
| Oakland Division. | Oakland, Cal. | 1923 | " " " | 163,920 | |

| | | | | | |
|---|---|---|---|---|---|
| Fisher Lumber Corp.— | Memphis, Tenn. | 1924 | Wood Working Plant | 658,960 | |
| Acquired Fisher Hurd Lumber Co. .......... | " " | 1926 | | | Approximately 160,000 acres timberlands in Southern states |
| Acquired Pritchard Wheeler Co. ... | " " | 1926 | | | |
| Acquired sawmill at ............ | Ferriday, La. | 1926 | | | |
| Acquired sawmill at ............ | Wisner, La. | 1926 | | | |
| Formed Fisher Delta Log Co.... | Memphis, Tenn. | 1924 | Logging and Log Buyers | | |
| Fisher Body Company of Atlanta ......... | Atlanta, Ga. | 1928 | Body Assembly Plant | 130,285 | |
| Fisher Body Company of Seattle ......... | Seattle, Wash. | 1928 | Wood Working Plant | 193,505 | |
| Ternstedt Manufacturing Company— | Detroit, Mich. | 1920 | Hardware | 1,244,503 | |
| Acquired England Mfg. Co. ...... | " " | 1921 | | | |
| Acquired Shepard Art Metal Co... | " " | 1923 | | | |
| Acquired International Metal Stamping Co.... | " " | 1922 | | | |
| Fleetwood Body Corp... | Fleetwood, Pa. | 1925 | Body Plant | 230,465 | Inactive |

in the United States and Canada. In form and spirit it is a revival of the picturesque workers' guilds of the Middle Ages. Its membership today numbers 750,000 boys, in all parts of the United States and Canada, and in the three years since it was organized (1930), the Guild has earned international recognition.

Fourteen leading personalities in academic and practical engineering training serve on its honorary board of judges. An advisory board, made up of the heads of secondary public school systems and of leaders in manual arts teaching, gives the Guild the benefit of its members' experience in dealing with boys from twelve to twenty years of age.

In more than 2,000 high schools, in nearly 600 major cities of the United States, the Guild is now an accepted and approved educational activity for boys. The Y.M.C.A. endorses and encourages it. The only recognition which the National Headquarters, Boy Scouts of America, has ever extended to another activity among boys is given to the Guild. Daniel Carter Beard, National Commissioner, Boy Scouts of America, is honorary president of the Guild.

The President of the United States, presidents, premiers, ministers of public instruction, such personalities as General Baden-Powell, founder of the world Boy Scout movement, the Governors of all the states in the Union, and the Lieutenant-Governors of all the provinces of Canada, commend the ideals and purposes of this institution.

The Guild for the last three years has provided an annual competition in craftsmanship, in which any boy in the United States or Canada over twelve years old and not more than nineteen may enter. The principal awards for excellence in this competition have been a series of four-year university scholarships valued at $5,000 each. This year's competition (1933-34) provides twenty-four university scholarships, ranging from $500 to $5,000 and with a total value of $51,000.

The subject of competition is the reproduction of a miniature model of the Napoleonic coach. This tests skill, perseverance, and patience. It also offers the widest opportunity for skill in metal-craft, wood-craft, paint-craft and trim craft.

The Guild today has ten of its alumni, winners of first awards, in American and Canadian universities. Five others are fitting themselves to enter universities. In addition, the Guild has distributed 3,220 other substantial awards for excellence in craftsmanship.

The president of the Guild for the United States is Mr. William A. Fisher, president of Fisher Body Corporation; for the Dominion of Canada, Mr. R. S. McLaughlin is president.

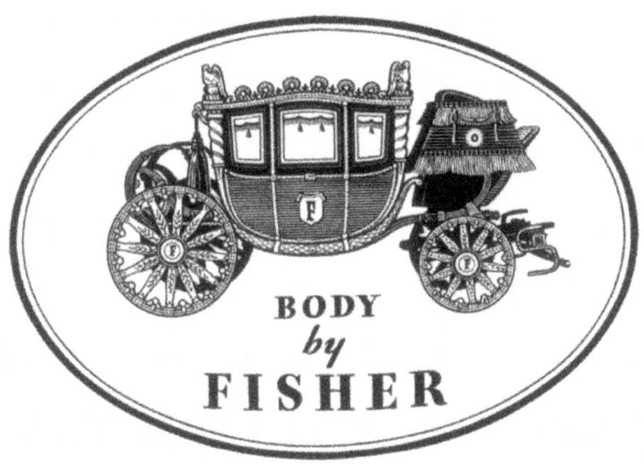

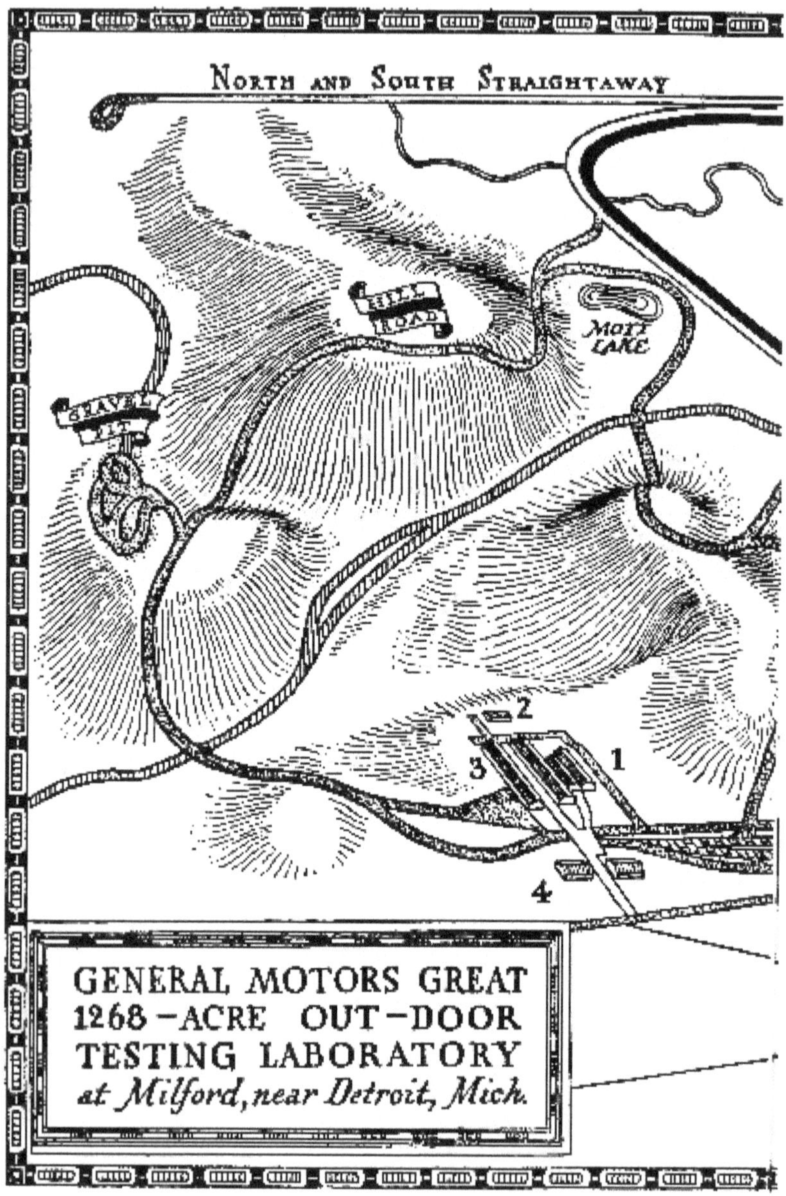
North and South Straightaway
Hill Road
Mott Lake
Gravel Pit
1
2
3
4
GENERAL MOTORS GREAT
1268–ACRE OUT–DOOR
TESTING LABORATORY
at Milford, near Detroit, Mich.

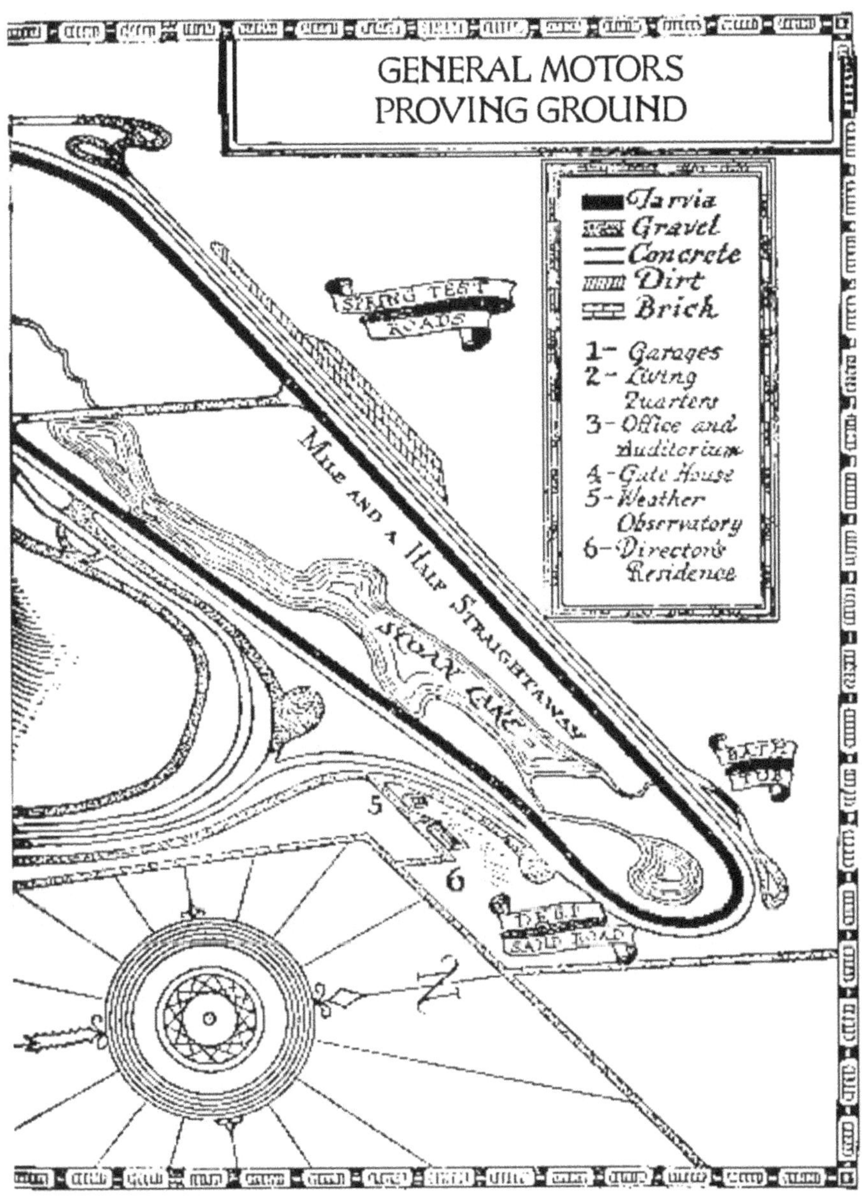
GENERAL MOTORS
PROVING GROUND
Tarvia
Gravel
Concrete
Dirt
Brick
1- Garages
2- Living Quarters
3- Office and Auditorium
4- Gate House
5- Weather Observatory
6- Director's Residence
SPRING TEST ROADS
MILE AND A HALF STRAIGHTAWAY
SLOAN LAKE
5
6
N

## Chapter XXI FRIGIDAIRE AND ELECTRIC REFRIGERATION

FRIGIDAIRE entered the General Motors family as a result of one of W. C. Durant's lightning intuitions. The Murray Body Company, long established in Detroit as automobile body builders, organized the Guardian Frigerator Company in 1916 as a pioneering project in the then almost virgin field of electric refrigeration. The head of the firm told Mr. Durant that all was not going well with Guardian. There seemed to be a little family friction, since the Murray boys had none of their father's faith in the enterprise. This is scarcely to be wondered at, as the whole concept of electric refrigeration in small units for household use was still untried, and Guardian had scarcely crept beyond the stage of rather discouraging experiments.

Mr. Murray took Mr. Durant to the fourth floor of an old factory building, where he found several refrigerators in various stages of manufacture, and a most inadequate machine equipment two lathes, a shaper, a hand milling machine, drill press, tool grinder, and hand tools. Three quart-bottles sufficed for the distillation of the sulphur dioxide used as the refrigerant. Only fifty Guardian refrigerators had been made and sold during the year; the concern was losing money, and the elderly Mr. Murray could see heavy losses present and to come.

It was Mr. Murray's idea that Mr. Durant should put in new money and let Guardian continue, but this did not suit Mr. Durant's purposes. Negotiations held fire for some time. Even before 1918, when he made a personal investment in Guardian, Mr. Durant told a gathering of General Motors executives and their financial friends that here was a babe-in-arms destined to rapid growth and likely before many years to earn enough to pay the salaries of all the staff at General Motors headquarters. His listeners looked at him with amazement, yet it came to pass within ten years.

The naming of Frigidaire is thus described by Mr. Durant. He recollects that he wrote to various of his friends describing the new article of merchandise and asked for suggestions. Lists of names came in from many sources and Frigidaire, the creation of an Ohioan, was selected. It has since become a household word in all parts of the world.

Guardian became the Frigidaire Corporation on the eighth of February, 1919, taking the name previously given to the product. On March 21st following, the directors of General Motors authorized the purchase of the capital stock of Frigidaire for $56,366.50, which represented the amount Mr. Durant had invested in it, his intention from the first having been to link it with the Corporation as soon as it showed promise. Manufacturing continued in Detroit until 1920, when the business was turned over to the DelcoLight Company of Dayton, Ohio, created out of Domestic Engineering Company.

The rapid growth of Frigidaire dates from this change of base and management. Delco-Light, making individual electric light plants for homes and farms, had already mastered the difficult art of specialty selling. In the redesigning of Frigidaire, Delco-Light was fortunate in having available the services of C. F. Kettering, now General Motors' vice-president in charge of research, and also vice-president of Frigidaire Corporation. Already Mr. Kettering had developed Delco equipment for automobile lighting, starting, and ignition. For years he had been keenly interested in electric refrigeration for rural homes, as an adjunct of Delco-Light plants. Ever since Frigidaire's move to Dayton he has taken a leading part in its engineering councils.

Selling at a relatively high price $750 and up in a skeptical market, Frigidaire had to be pushed with determination until costs could be brought down by quantity production. The early years developed these figures in units sold:

| | |
|---|---|
| 1921 | 365 |
| 1922 | 2,000 |
| 1923 | 4,700 |
| 1924 | 20,000 |

These jumps in production reveal the influence of mass production methods, lower prices, improved sales methods, and the development of overseas markets. Frigidaire had to carry the novel message of electric refrigeration directly to the consumer, breaking down the household habits of centuries. As the product improved, the market became more receptive, and sharp price cuts kept public favor from lagging. Sales mounted with dizzy speed. Meantime the factories at Dayton, operated on triple shifts, were being constantly enlarged and improved by the installation of special mass-production machines and systems. Today Frigidaire occupies fifty-three acres of factory space.

By 1926 Frigidaire had so far outstripped Delco-Light in sales that a separation was advisable. The present Frigidaire Corporation was formed, and in 1930 Delco-Light was transferred to the North East plant at Rochester, New York, and rechristened Delco Appliance Corporation.

In 1926 Frigidaire abandoned the wooden cabinet in favor of sheet metal built on a wooden frame. Porcelain was applied first to the inside of the refrigerator, but its superiority over paint for inside finish soon brought porcelain to the exterior of the new metal cabinet Frigidaires. In the present factory layout is located the largest porcelain enameling plant in the world.

Export sales multiplied amazingly, taking one seventh of total production, and employing an overseas staff of 2,000 persons. Every civilized country in the world had its staff of Frigidaire salesmen, and many lands which would hardly be considered civilized in the ordinary sense of the word have seen Frigidaires installed at the instance of aggressive representatives. The export trade is still predominantly commercial, while in the United States the bulk of installations is for household use.

Household refrigeration was the first concern of Frigidaire, but in 1923, at the request of ice cream manufacturers, the company undertook to supply them with special cabinets for their retail outlets. Since that time, partly as a result of its ever-improving compressors, Frigidaire has entered appropriate machines in many other fields of retail business groceries, meat markets, restaurants, dairies, and florists' shops.

Electric water coolers for office buildings and factories were added in 1926. Frigidaire builds six types of individual water coolers and three tank types, and cooling equipment for draught and bottled beer and other beverages. The entire Frigidaire line consists of 155 different models in 19 various lines. Among these are 12 models of household refrigerators, de-humidifiers and several types of air conditioners for domestic and commercial uses as well as complete air conditioning equipment for railway cars.

The last two lines represent Frigidaire's entry, in 1932, into the new and growing business of air-conditioning, a logical move in view of the Corporation's long experience with electric refrigeration and its advance from small wooden refrigerators for household use to the cooling of large rooms for fur storage, and other extensive installations. Frigidaire engineers have already installed air-conditioning equipment in hundreds of homes, offices, and stores, sometimes under difficult conditions due to the fact that the buildings were not properly designed for the purpose. Their experience indicates that air-conditioning offers possibilities fully equal to those already realized in electric refrigeration.

Frigidaire air-conditioning equipment was selected to provide the requisite conditions and controls for the comprehensive allergic tests conducted at Johns Hopkins University, Baltimore, Maryland, by Dr. Leslie N. Gay, to determine whether high altitude conditions favorable to hay fever and asthma victims could be reproduced in homes and offices for the relief of sufferers. In summarizing his results, Dr. Gay reported:

> The experiments in the treatment of pollen hay fever and asthma with air-conditioned atmosphere represent a new method of attack. Complete relief was given to patients suffering with symptoms of hay fever, whether they occupied the room for several hours or for longer periods of time. Striking relief was given to patients suffering with pollen asthma within twelve hours after admission to the room. For individuals who can make provision for such atmosphere, whether in their homes or offices, great relief can be offered.

A new refrigerant at a lower cost, Freon, odorless, noncorrosive, non-inflammable and non-toxic under ordinary conditions, lately developed in the research department of Frigidaire, is expected to be a factor in all the branches of temperature control.

At the peak of production Frigidaire employs more than 11,000 persons in its Greater Dayton plants, uses thirty-five carloads of steel a day, consumes more silver solder than any other manufacturer in the world, and is one of the world's largest copper consumers. Eighteen porcelain furnaces operate continuously, 300,000 pieces being given three burns each and consuming 400 tons of porcelain frit each month. Four production lines creep through the Moraine City plant, and there are two conveyors, each a mile and a quarter in length, carrying materials to the thousands of workmen stationed along the assembly lines. The impressive technique of material movement worked out in the automobile industry had been applied fully by Frigidaire in its building of electric refrigerators. Refrigerators are given tests for temperature, in which thermometer readings are taken inside the food compartment while the cabinet stands in a room at 90 degrees, a running test of three hours, and a noise test in a room arranged to amplify sound hundreds of times.

The public has purchased more than half a billion dollars' worth of Frigidaires. The 2,500,000th Frigidaire left the factory at Dayton on July 8, 1932, for a cruise around the world and was displayed by a Frigidaire dealer in every port of call.

Within fourteen years, the period of General Motors' association with Frigidaire, electric refrigeration has come to be standard equipment. The evolution has been from wood to porcelain, from complexity to simplicity of operation and repair, from small units to large, until in modern air-conditioning one finds entire buildings under temperature control. Costs have steadily been reduced, dropping from a minimum price of $750 in 1921 to $96 in 1933 for a much superior but smaller refrigerator. This represents a turnover for quantity production and intelligent sales effort unsurpassed in any field of American enterprise.

Frigidaire is General Motors' largest producer in the non-automobile field. Mr. E. G. Biechler has been president and general manager since 1926 when the present Frigidaire Corporation was incorporated.

## Chapter XXII COMMERCIAL VEHICLES

MENTION has been made of the steam "drags" built in England and France from 1821 on, some of which were used alternately in towing both goods and passengers in trailers. In America, the first power road vehicle especially designed for goods transport was a huge steam traction engine begun in New York City in 1858, shipped West, and assembled there to haul goods from the Missouri River to Colorado. It broke down on its first trip seven miles after starting, the spot being marked by a monument at Nebraska City, Nebraska.

Better luck attended the self-propelled "steamers" built for fire-fighting purposes by the Amoskeag Mills at Manchester, New Hampshire, one of which was bought by the City of Boston after it had been rushed there to help fight the great fire of 1872. Hartford, Connecticut, bought a steam fire fighter, the Jumbo, in 1876. Many other less successful attempts to develop self-propelled fire-fighters are on record, one having been built by the famous Captain Ericsson as early as 1840.

Gasoline delivery wagons began to appear shortly after the light gasoline cars proved their superiority over steam and electric vehicles in the Chicago race of 1895. The Langert Company of Philadelphia entered one in the Cosmopolitan race of 1896. An interesting early variant was a sightseeing stage built by C. S. Fairchild of Portland, Oregon; this carried eighteen passengers and was powered by a kerosene engine. Among the pioneers in commercial vehicles were:

Charles E. Woods of the Woods Motor Vehicle Company of Chicago, builder of light electric delivery wagons.

Hiram P. Maxim, then connected with the Pope Company at Hartford, Connecticut.

L. F. N. Baldwin, who converted a Boston horse van into a steam wagon with a 6 horsepower engine and a side chain-drive.

Alexander Winton, whose Winton delivery wagon was the first gasoline commercial vehicle produced for sale in any quantity, eight being under construction in October, 1898.

Charles E. Duryea, also in 1898, with a three-wheel gasoline delivery wagon, weight 1,000 pounds.

A. L. Riker, who built for B. Altman & Co., New York City, in 1898, electric delivery wagons which were the first to be operated regularly by a large metropolitan store.

Early motor fire engine, then largest in the world, owned by Hartford, Connecticut. From Cosmopolitan Magazine, 1896

Electric vehicles, with the Pope and Whitney millions behind them, had the better of their gasoline and steam competitors in the late 'nineties, not only in the matter of city goods deliveries, but also in cabs for hire. Electric cab service began in New York City in 1897, and soon after was extended to other cities under the same financial auspices. Electricity held the advantage for some years against the challenge of steam and gasoline; but its prestige was weakened in 1900 when Altman's, after three years trial of electrics, switched to gasoline delivery wagons. However, only three gasoline cars were exhibited in Madison Square Garden at the 1900 show, the first exclusive automobile show held in America. In that year Detroit Automobile Company, predecessor of Cadillac, introduced a most advanced gasoline delivery wagon, unusual for its aluminum, gun metal, and nickel axles. No market seems to have been found for it, however. The year 1900 saw also the founding of the Autocar Company of Philadelphia and, in Detroit, the first step in the evolution of General Motors Truck.

## GENERAL MOTORS TRUCK

In 1900 Max Grabowsky built and sold to the American Garment Cleaning Company a commercial vehicle, powered by a single-cylinder horizontal engine. In 1902 he organized the Rapid Motor Vehicle Company which put out 200 units in its first year. Moving to Pontiac in 1904, Rapid erected there the first building in America for the exclusive manufacture of self-propelled commercial vehicles. In 1908 General Motors acquired a majority of Rapid Motor Vehicle stock. Shortly afterward Reliance Motor Truck Company of Owosso, organized in 1905, also came into General Motors, and a sales company was organized to handle the output of both truck companies under the name of General Motors Truck Company. This company took over manufacturing operations in 1912, and moved Reliance to Pontiac in 1913.

Between 1900 and 1910 the gasoline commercial car developed a decided supremacy for all-round use, but the electric delivery wagon had made a place for itself in cities and for light merchandise. The rout of the steam cars, however, was complete. The light steam delivery wagon disappeared from the market by 1905, the heavy steam truck by 1910. Except for a few pioneers, like Rapid, which staked everything on truck specialization, most of the trucks produced from 1900 to 1910 were adaptations of commercial bodies to stock passenger car chassis. In the next five years of fast and furious evolution the commercial vehicle manufacturers had settled down more or less to a standard type "distinguished by a pressed or rolled steel frame, four-cylinder vertical engine mounted in front, water-cooled, with magneto ignition, three-speed selective gear transmission, shaft and worm-gear drive to rear-axle differential, seat back of hood and dash, left-side drive with wheel-steering and center-control levers."

Having led in establishing some of these fundamental principles of design in gasoline commercial vehicles, General Motors Truck in 1912 recognized the usefulness of electric delivery wagons for city use by building a full line of electric delivery wagons, which it continued until 1916. By that time refinements in the construction of light gasoline cars had reached a point at which the electric delivery wagon business might be expected to show diminishing returns. The small, light delivery wagon, adapted to a passenger car chassis, had shown its possibilities early. Both Cadillac and Oldsmobile had offered such delivery wagons in 1904, the latter carrying off first prize in its class against stiff competition in the Automobile Club of America tests. These two manufacturers soon after went into the production of larger cars, the light delivery wagon business going elsewhere. But in 1917, after General Motors and Chevrolet came together, Chevrolet began the manufacture of commercial units, a decision which may have had a bearing upon the abandonment of electric delivery wagon manufacture by General Motors Truck.

## CHEVROLET COMMERCIAL VEHICLES

The light-weight commercial car production of Chevrolet, which has since grown to tremendous figures, began modestly with the manufacture of 437 chassis and 40 light delivery wagons. The commercial chassis, one-half ton with open body, was priced at $595 f.o.b. Flint. A one-ton truck, known at the Model T, and powered with the FA four-cylinder motor, was introduced in 1918 at from $1,125 to $1,320. Prices on comparable units rose with the advancing price levels until the latter part

Pioneer police patrol wagon, Chicago. From Scribner's Magazine, February, 1913

of 1920 and then began to drop with increasing quantity production. In 1933 the halfton chassis, much more powerful than the original of 1917, and powered with a six-cylinder motor instead of a four, sold for $330, and the larger truck, capacity increased to one and one half tons, sold for $480. During the intervening period production of Chevrolet commercial units rose steadily year by year to a peak of 344,963 in 1929, when commercial cars represented more than 25 percent of total Chevrolet unit production. Through the entire Chevrolet truck experience, its commercial units have been roughly 17 percent of total production, more than 1,500,000 units having been manufactured since 1917.

In 1927 Chevrolet undertook on a small scale to build part of its commercial bodies, especially closed cabs; and in 1929 added a high-grade body known as the Sedan Delivery. In 1930 it introduced single and dual wheels as standard equipment on the heavy duty truck, and brought out a one and one half ton truck with 157-inch wheelbase.

Owing to the large production necessary, securing proper bodies for installation of Chevrolet commercial bodies had become a major problem, leading to the acquisition in 1930 of the Martin-Parry Corporation plants in Indianapolis, then one of the largest commercial car body builders in the world. The formation of the Chevrolet Commercial Body division followed, with emphasis upon advanced design, volume production, improved material, sales control, and GMAC financing. Since this acquisition Chevrolet's participation in the total commercial car market has steadily increased from 32.7 percent of the 1930 business available in the weight class in which it sells trucks to approximately 50 percent of the business available in 1933. The latter figure means that Chevrolet sold approximately as many cars in this class as all its competitors combined. In the account of the General Motors Fleet Sales Corporation in Chapter xxvi, reference is made to Chevrolet's sales to fleet owners and to the government, in both of which classifications it leads all competitors by overwhelming margins.

## YELLOW TRUCK & COACH

Yellow Truck & Coach Manufacturing Company, a Maine corporation, succeeded the Yellow Cab Manufacturing Company, which was incorporated in the year 1910.

Yellow Cab grew out of the Walden W. Shaw Auto Livery Company, operating taxicabs at loth and Wabash streets, Chicago. Mr. Shaw was president, and Mr. John D. Hertz, vice-president and general manager. Mr. Hertz described their condition, after a disastrous strike, as bankrupt, $97,000 in debt and with nothing but forty battered cars as assets. The Shaw Company and another taxicab company, the City Motor Cab Company, were merged into a Maine corporation on August 25, 1910.

After a year's work, the first specially designed taxicab was completed on Christmas morning, 1914, and it was manufactured in 1915. With its service record of 600,000 miles appended, this famous cab was shown at the Century of Progress Exposition in Chicago in 1933. Forty cabs of this model were built during the first seven months of 1915, and placed in service on Chicago streets on August 2, 1915. In 1916 a small factory at 310 East Huron Street, Chicago, with eighty workers, was turning out one car a day. Sole ownership of the company was acquired by the Walden W. Shaw Corporation, a New York organization, in 1916. Three years later the Auto Livery Company ceased operating taxicabs, confining its activities entirely to manufacturing commercial vehicles. Taxicab operation was continued by the Walden W. Shaw Corporation which in 1919 became Chicago Yellow Cab Company, Incorporated.

In evolving a sturdy taxicab, a nation-wide demand there had been uncovered. Orders for cabs began to come in from other cities, and production rose to twenty-five cars a day. In 1920 the Yellow Cab Manufacturing Company was formed to succeed the Shaw Livery Company, and a further separation of the Yellow and Shaw interests followed in 1921. The branching-out process which we have seen so often in automotive history began, resulting in the formation of Yellow Motor Coach Company in November, 1922, and Yellow Sleeve Valve Engine Works at East Moline, Illinois, within the following month. Yellow Manufacturing Acceptance Corporation was formed in December, 1923, to finance sales.

A merger of the General Motors Truck properties and the Yellow interests was effected in September, 1925, when Yellow Truck & Coach Manufacturing Company, formerly Yellow Cab Manufacturing Company, acquired all the stock of both General Motors Truck Corporation and General Motors Truck Company from General Motors Corporation in exchange for a controlling interest, now above 50 percent, in the Yellow Truck & Coach Manufacturing Company.

The Yellow Truck & Coach Manufacturing Company, through its subsidiaries, builds and sells trucks, motor coaches, and taxicabs of a wide variety, thoroughly adapted to the complex needs of the commercial world. The merger brought together General Motors Truck, which had been active in its field for seventeen years, and the leading manufacturer of motor coaches and taxicabs. This company continues to be one of the largest builders of commercial vehicles, with practically all of the leading coach operators in the United States using its equipment. A special type of omnibus developed for the purpose was selected to provide transportation within the grounds of the Worlds Fair of 1933. The company's products include a complete line of trucks from one and one half tons to fifteen tons, as well as trailers.

Manufacture of these vehicles is centralized at Pontiac, Michigan. The General Motors Truck plant in Pontiac which was completed late in 1927 is the largest plant in the world devoted exclusively to the manufacture of commercial vehicles.

Mr. P. W. Seiler has served as president of Yellow Truck & Coach Manufacturing Company since 1927.

# Chapter XXIII

## GENERAL MOTORS IN AVIATION

GENERAL MOTORS made its first effective contact with aviation during the World War, when its plants delivered more than 2,500 Liberty engines for airplanes.[1] More than 10,000 Libertys were on order from the government when the war ended. This experience, and the Corporation's habit of looking ahead, naturally led it to consider the future possibilities of aviation as a peace-time industry.

An enviable record in war-time production of airplanes had been established by the Dayton-Wright Airplane Company of Dayton, Ohio, a 1917 offshoot of the Dayton Metal Products Company, one of the several Dayton companies already referred to as being established in rapid succession in the Ohio city by Messrs. Kettering and Deeds and their associates. As the home of the famous Wright brothers, first to fly a heavier-than-air machine, Dayton had excellent reason for a keen interest in aviation. When the war demand for planes developed, the industrial leaders of the city promptly organized to take the lead in production. Neither before nor since have airplanes been manufactured at the speed and in the quantity at which Dayton-Wright produced them for the United States Government during October, 1918, when an average of forty completed planes a day were turned out. In all, Dayton-Wright produced more than

[1] As shown in Chapter XII: "The War Years".

3,000 of the famous DeHaviland (DH) fours, and 300 Standard JI training planes.

During this period the liaison between the United States Army Air Service and Dayton-Wright Airplane Company was very close. One of the latter's Hying fields was taken over by the Air Service, renamed McCook Field, and as such became famous in American aviation history.

Airplane production fell rapidly through 1919, but the long view held that eventually aviation would take its place as a major industry. General Motors purchased substantial interests in several Dayton companies, and acquired on September 25, 1919, all the outstanding shares of Dayton Metal Products Company, and Dayton-Wright Airplane Company. The latter was re-incorporated as the Dayton-Wright Company on December, 24, 1919, until recently a wholly-owned subsidiary of General Aviation Corporation. Various of its properties were turned over to other uses as the other Dayton interests of General Motors developed, notably Frigidaire, and Inland Manufacturing. Until the move to Detroit in 1925 General Motors Laboratories occupied one of the former Dayton-Wright buildings. The Dayton-Wright Realty Company, wholly owned by General Aviation Corporation, was formed in 1933 to hold title to the largest tract of land used as a flying field during the war.

General Motors' interest in aviation revived and a connection was made with Anthony H. G. Fokker. The American market for Fokker planes having been developed through importations, American manufacture was begun through the Atlantic Aircraft Corporation of Hasbrouck Heights, New Jersey, in 1926. Atlantic Aircraft Corporation, incorporated December 14, 1923, retained its separate status as a corporation and, until recently, was a wholly owned subsidiary of General Aviation Corporation.

A second manufacturing plant was established at Glendale, West Virginia, Fokker Aircraft Corporation being incorporated December 3, 1927, as a holding company covering both operations. From 1926 to 1930, 211 commercial airplanes of nine models were manufactured, the most popular being the Super

Universal of which eighty were delivered from January to October, 1929. There were delivered also forty-five service airplanes, all trimotor, to the Army and Marine Corps. With these, army pilots hung up many notable records among them:

1. San Francisco Honolulu, non-stop flight in a United States Army c-n trimotor piloted by Lieutenants Maitland and Hegenberger.
2. A one-stop flight by Major Bourne in the Marine Corps TA-I trimotor from Washington to Managua, Nicaragua.
3. In an Army C-II-A trimotor, Major Carl Spatz established the first endurance refueling record by remaining in the air over 150 hours.

GA-43 climbing. This low-wing, all metal monoplane is made by General Aviation Manufacturing Corporation at Dundalk, Maryland

## GENERAL AVIATION CORPORATION

Shortly after the formation of Fokker Aircraft Corporation, General Motors acquired 400,000 shares of the Fokker Common stock. In payment, General Motors turned over all of the capital stock of Dayton-Wright Company. The assets of the latter company consisted of McCook Field located in Dayton, Ohio, a large number of aviation patents, and additional cash assets of substantially $6,500,000. On May 24, 1930, the name of the Corporation was changed to General Aviation Corporation, and operations were consolidated in a plant at Dundalk, Maryland, leased from the Curtiss-Caproni Corporation. On August 29,

1931, the manufacturing operation was incorporated as General Aviation Manufacturing Corporation, wholly owned by General Aviation Corporation, which also owned Atlantic Aircraft, the Metalair Corporation, and the Dayton-Wright Company.

General Aviation Corporation was a holding company organized under the laws of Delaware, with authorized capital, as of June 21, 1930, of 5,000,000 shares of no par Common stock and $1,000,000 in $25 Preferred shares. In July, 1932, capitalization was reduced to 1,000,000 shares of stock without par value. In April, 1933, the capital stock was changed to $i par value per share. General Motors owns approximately 50 per cent of the 980,900 shares of Common stock issued by General Aviation, the balance being widely held over the country.

## NORTH AMERICAN AVIATION, INC.

In April, 1933, General Aviation Corporation sold its manufacturing interests and certain other assets to North American Aviation, Inc., in return for approximately 43 percent of the Common stock of the latter which, through various stock holdings, reaches down to transport and manufacturing companies in various parts of the country. In the merger between General Aviation and North American Aviation, ownership of General Aviation Manufacturing Corporation, Atlantic, Metalair, and Dayton-Wright (except as to real estate already noted) passed to North American Aviation. As the latter already owned the B/J Aircraft Corporation, another property located at Dundalk, Maryland, near Baltimore, this company has been merged with General Aviation Manufacturing Corporation. North American Aviation owns completely Eastern Air Transport and has substantial interests in two other transport companies Transcontinental & Western Air, Inc., and Western Air Express as well as in Douglas Aircraft, Inc., a manufacturing company. Atlantic Aircraft has since been dissolved. Through the merger of April, 1933, General Motors extended its aviation interests, until it is now one of the leading factors in both manufacturing and transport, since North American Aviation, Inc., is one of the three largest aviation groups in the United States.

North American Aviation was incorporated December 6, 1928, as an investment trust specializing in aviation securities. Its authorized capital stock was 6,000,000 shares of which 2,000,000 shares were offered for sale. On March 9, 1932, the stock was changed from no par to $5 par value. Later, North American Aviation became a holding company. Its shares were changed to $i each in 1933. In order to insure control, General Motors Corporation bought an additional interest in North American Aviation, making the total amount of North American Aviation stock owned by General Motors and General Aviation Corporation in excess of 51 percent.

## TRANSPORT

In the transportation field, Eastern Air Transport, Inc., formerly Pitcairn Aviation, Inc., which was formed in 1927, is a wholly-owned subsidiary of North American Aviation. Eastern Air operates mail and passenger lines between New York and Southern cities. On the fifteenth anniversary of the first regular air mail contract, established May 15, 1918, between Washington, D. C., and New York, the New York Times cited Eastern Air Transport's present competence as an example of progress in the period. The first air mail made 75 miles an hour for 200 miles once a day, while over the same route Eastern Air Transport now operates 18 passenger planes every hour on-the-hour over this first route and then proceeds for 2,268 airway miles more on regular schedule.

In Transcontinental Air Transport, Inc., North American Aviation also owns a substantial interest. T.A.T., incorporated in Delaware, May 14, 1928, operated the first combined air and rail service from New York City to Los Angeles and San Francisco in conjunction with the Pennsylvania Railroad. It ceased operations in October, 1930, continuing in existence as a holding company, owning 47.6 percent of Transcontinental and Western Air, Inc. T.A.T. has a small holding in Western Air Express Corporation.

In Western Air Express Corporation, North American Aviation, Inc., owns a controlling interest. Organized in 1926, Western Air Express operates mail, passenger, and express service from San Diego to Los Angeles to Salt Lake City, and from Cheyenne to El Paso and to Amarillo, Texas, via Denver and Pueblo, Colorado. Western Air Express Corporation also owns a 47.6 percent interest in Transcontinental and Western Air, Inc.

GA-43: front view

Transcontinental and Western Air, Inc., was formed in October, 1930, to take over as one transcontinental air route the lines formerly operated by Western Air and Transcontinental Air Transport, forming the shortest

coast-to-coast air route. T. & W. A. is the operating company with which Colonel Charles A. Lindbergh is identified as chairman of the Technical Committee. Colonel Lindbergh laid out in 1928 for T.A.T. the line now operated by Transcontinental and Western Air from New York to Los Angeles.

Over the three transportation systems brought together under North American Aviation control, 43,576 miles are flown daily on regular schedules.

Eastern Air Transport, Inc., flies 13,500 scheduled miles daily over the 2,493-mile airway, serving twenty-eight cities. This company operates ten round trips daily between New York and Washington, five of them fast non-stop express schedules and all flown by air liners carrying fifteen to eighteen passengers. Planes of Eastern Air Transport, Inc., and its subsidiaries have covered 18,000,000 miles. Transcontinental & Western Air, Inc., familiarly known as "TWA," operates 23,298 miles daily over a 4,401mile airway from New York to Los Angeles and San Francisco via St. Louis, Kansas City and Albuquerque and via Springfield, Missouri, Tulsa and Oklahoma City, serving twenty-three cities on this Mid-Transcontinental airway. Its planes have flown a total of more than 20,000,000 miles. A fast New York-Chicago service is maintained.

Western Air Express, Inc., is one of the oldest air transport lines in the United States. It flies 5,016 miles daily from Cheyenne to El Paso via Albuquerque, to Amarillo via Pueblo, and from Salt Lake City to Los Angeles and San Diego, a total of 1,762 airway miles. Twelve cities are served, and the company's planes have flown millions of miles on schedule.

These three systems have flown more than 40,000,000 miles, carrying more than 475,000 passengers and more than 10,000,000 pounds of mail.

## MANUFACTURING

The wholly-owned manufacturing plants of North American Aviation are located within easy reach of each other at Dundalk, Maryland, a suburb of Baltimore, where commercial transport planes, as well as military and naval planes, are manufactured. The plants are modern and close to flying fields, and front directly on the new municipal airport now being developed, which makes the plants available to both land and water craft.

## GENERAL AVIATION MANUFACTURING CORPORATION

One series in which General Aviation Manufacturing Corporation takes especial pride is the GA-FLB type Flying Lifeboat of the Coast Guard. Five of these staunch seaplanes are on duty on the Atlantic Coast, bearing the names of stars Antares, Altair, Acrux, Acamar, and Arcturus. After experiments with various amphibians, the Coast Guard decided that life-saving on the high seas made imperative the creation of a seaplane to fill the following requirements: An aerial "eye" capable of extended search, radio-equipped to maintain constant contact with surface, thus saving hours and possibly days of delay of search ; an aerial ambulance capable of a speed of 100 miles per hour, able to land in rough sea, equipped with hatches large enough to admit of stretcher cases and to be able to take off in rough water ; a demolition outfit to effect the destruction at sea of derelicts and obstructions to navigation within a few hours after the report of location; a high speed flying patrol for observation, landing and returning with rescued crews of distressed small craft and capable of taking aboard fifteen or more passengers from distressed craft and standing by for lengthy periods on the surface, maintaining in the meantime radio communication with surface craft until transfer can be made of its passengers.

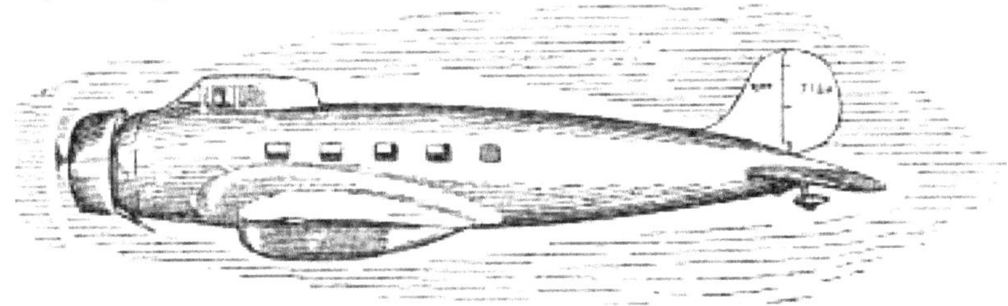

GA-43: side view

In the United States Naval Institute Proceedings for January, 1933, Colonel Harold C. Reisinger of the United States Marine Corps says:

> These specifications were turned over to the General Aviation Corporation of Baltimore, Maryland, and the flying lifeboat was constructed there. The finished product has by test and in actual service lived up to the fondest hopes of its most ardent advocates.

Captain Reisinger gives several stirring accounts of rescues effected by these ships near Cape May, which would have been impossible without them. Perhaps the most thrilling of the many rescues by the flying lifeboats was that of Paul Long, saved by the Arcturus of the Miami Station in a sea so high that its left wing was unavoidably damaged. Thereafter the Arcturus taxied thirty miles to shore through waves eight to fifteen feet high. The whole record of the two flying lifeboats on the Miami Station is one to arouse admiration and establish the utility of this craft in its special field. In fifty-three days the two ships at Miami received twenty-seven calls, and answered all of them successfully.

At the former B/J plant General Aviation Manufacturing Corporation is also producing military and naval airplanes. This plant is equipped with a wind tunnel with the aid of which the aerodynamic forces on a model can be accurately measured through a system of electrically driven balance arms. From these data, predictions on the performance and flying characteristics of the actual airplane can be made.

Following the decline in demand for commercial planes, B/J concentrated successfully on fighting craft and supplies for the national services. In May and June, 1929, orders were received for three experimental planes a single-seater Navy Fighter, a two-place Navy Observation plane, and a two-seater pursuit ship for the Army Air Corps.

Following the building of experimental models, B/J received production orders for twenty-five army pursuit planes known as YP-16, and twenty-seven Navy observation types called OJ-2 land and seaplanes, the former going to the army air base at Selfridge Field, Michigan, and the latter to the naval air station at San Diego, where they are being used on cruisers of the Pacific fleet.

In process of development in 1933 were a new two-place combined fighter, and long-distance scouting plane designated as XF-2-J, all metal except for covering on the wing and control surfaces; also an experimental single-seater fighter known as XF-3-J, embodying the latest features in aerodynamics and structural design, and powered with the newest military engine.

In addition to its construction work the company conducts research in aerodynamics, design and structures, its wind-tunnel services being in demand by other manufacturers lacking that equipment.

## DOUGLAS AIRCRAFT COMPANY, INC.

North American Aviation, Inc., also has a substantial interest in Douglas Aircraft Company, incorporated in Delaware, November 30, 1928, to take over the assets of the Davis-Douglas Company (the Douglas Company from 1921), a partnership active in Western aviation since 1920. Douglas has developed an extensive European business, building planes specially designed for foreign use. It manufactures commercial and military aircraft at Santa Monica, California, being one of the nation's leaders in military aircraft, and owns 51 percent of the Common stock of the Northrop Corporation organized in 1932.

Before General Aviation bought into North American Aviation, Inc., the latter had become the second in point of transport by 1932, and also held a good position as a producer of planes. Hence, with 43 percent of North American

Flying Lifeboat of the Arcturus Class GA-FLB

Aviation stock in its treasury, General Aviation becomes one of the largest factors in both the manufacturing and operating branches of aviation. Through its controlling interest in General Aviation, the General Motors Corporation is in position to cooperate fully in the anticipated progress of aviation, and expects to assist that development by placing its research facilities and manufacturing experience at the disposal of its aviation affiliates. For these reasons the recent enlargement of General Motors activity in aviation has been hailed approvingly by the press. It seems to be generally recognized that aviation has reached a stage at which scientific and business assistance on a large scale will mean accelerated progress. The New York Herald Tribune for April 27, 1933, thus sums up the Corporation's present standing in aviation:

> The broadening of this phase of General Motors activities, it is believed, will bring to the public and to the aviation industry, both from a manufacturing and a transport operating standpoint, the organization, technical efficiency and experience which

> General Motors has thus far given in main part to the automotive industry. It is hoped that through the acquisition, control and co-ordination of these individual units in the aviation field General Motors Corporation will play a part in that field comparable to its development in the motor industry.

Mr. Ernest R. Breech is president of North American Aviation, Inc.

# Chapter XXIV

## THE POINT OF VIEW OF GENERAL MOTORS

T,HE most complete statement of the point of view of General Motors in its broad relations with stockholders, employees, dealers, suppliers, and the public is contained in the speech which President Alfred P. Sloan, Jr., delivered to representatives of the automotive press at the Proving Ground on September 28, 1927, and previously referred to in this volume. The occasion was a three-day visit to the Proving Ground by the automobile editors of American newspapers, and this important gathering occurred at one of the most interesting junctures in the history of the automobile industry.

It was a time of rising trade and yet of considerable uncertainty, since no one could quite foresee the boiling business thermometer of the next two years, and there were plain indications that the industry as a whole had reached a condition of relative stability as compared to previous eras in which facilities for manufacturing motor cars had been generally too small to meet the ever-growing demand. Consequently, there was need to review the past, analyze the present in the light of lessons learned, and restate the principles and policies under which the Corporation would grapple with the future:

President Sloan's speech follows, in part:.

"You of course appreciate that this industry of ours the automotive industry is today the greatest in the world. Three or four years ago it passed, in volume, steel and steel products, the next largest industry. This means, expressed otherwise, that upon its prosperity depends the prosperity of many millions of our citizens and the degree to which it has become stabilized in turn has a tremendous influence on the stabilization of industry as a whole, and therefore on the prosperity and happiness of still many more of our citizens. Directly and indirectly, this industry distributes hundreds of millions of dollars annually to those who are connected with it, in one way or another, as workers. It also distributes hundreds of millions of dollars in the aggregate to those who have invested in its securities. The purchasing power of this total aggregation, as you must appreciate, is tremendous.

"I believe that if you questioned many of your readers as to the present position of the automotive industry, they would tell you that it is growing by leaps and bounds. I believe further you would sense uncertainty as to what is going to happen in the industry when the so-called state of saturation is reached. I do not know whether you appreciate it or not, but the industry has not grown very much during the past three or four years. It is practically stabilized at the present time.

"What has taken place is a shift of business from one manufacturer to another, and the announcements in the press as well as the general publicity of those manufacturers who have succeeded in increasing their business give, I think, the impression that this is true of the whole industry. If we could assume, for the sake of argument, that we will reach the point at which twenty-five million cars and trucks will be registered in the United States an assumption that from what we have accomplished so far is certainly perfectly reasonable then I think we could safely say that the replacement demand, plus the export demand which will increase for many years yet, plus the normal growth, would amount to something like four to four and one half million vehicles a year and would require the manufacture of a number of cars equal to or greater than has yet been produced in any year in the history of the industry... .

"I am sure that I do not need to elaborate what the automotive industry consists of, its influence on the prosperity of the United States, the influence that it has had in many other industries which contribute to its production necessities. General

Motors is an important part of this great industry of ours and as my contribution to your visit with us I would like to tell you in a brief way something about General Motors; how we are thinking, what we are doing, and our ambitions for the future.

"Let me deal here with what General Motors includes and with the responsibility that rests on its management.

"First. We have approximately 60,000 stockholders.[1] The market value of the securities that these stockholders hold at the present time exceeds $2,000,000,000 a very tidy sum. This enormous sum and the responsibility of acting as trustees throws a very great responsibility on General Motors' organization. In 1926 these stockholders received, in dividend disbursement, an amount in excess of $100,000,000. The purchasing power of this $100,000,000 has an important influence on our general business situation.

"Second. We have 180,000 people in our own organization on our pay rolls and directly concerned in our prosperity. I believe it can be conservatively stated that allowing four dependents to each worker, we have approaching three quarters of a million people whose prosperity and happiness is directly concerned with ours.

"Third. We have our dealer organization. There are in the aggregate something like 18,000 dealers. If we assume, and I think we have a right to assume, that each dealer would average twenty-five employees, I really think it is much higher than that, we have something like 500,000 dealers and members of their direct organization, with their dependents a total of over 2,000,000. I estimate these dealers are employing a capital of over $500,000,000.

"Fourth. Next we have over 4,600 suppliers of material.

1 haven't any idea how many workers are involved in the organization of those suppliers applicable to General Motors production, but as you can well appreciate, it is very large.

"Fifth. We have the enormous aggregation of people whose prosperity depends, in turn, upon the prosperity of those I have mentioned above their purchasing power, in other words. As a

---

[1] The number of stockholders increased to 372,000 from 1927 to 1933.

matter of fact, there are several cities of importance in the United States whose prosperity is absolutely linked up with the prosperity of General Motors.

"I have estimated that a very appreciable percent of the total population of the United States is directly affected by the prosperity of General Motors. Expressed otherwise, the happiness of an enormous group of citizens is dependent upon the establishment of sound principles of administration in General Motors and the development of sound thinking in the formation of its programs and policies.

"I mention all the above to give you some appreciation of what General Motors includes, and a better appreciation of what our whole industry includes because, naturally, the whole must be larger than any part. The responsibility is tremendous, and although it is, necessarily, more or less divided, yet the fact remains that upon a very limited number of individuals must rest the responsibility of formulating policies and principles upon which this vast enterprise revolves and upon which its future depends.

"Recognizing the responsibility that rests upon us, I want to deal next with certain things that we are doing that, in our judgment, have not only contributed much to our progress, but which, in my opinion, if fully appreciated, should add a sense of great security to those whose prosperity and happiness is so intimately linked up with us.

"To illustrate my point, I want to tell you a true story of what happened in General Motors in 1920.

"In the spring of 1920, General Motors found itself, as it appeared at the moment, in a good position. On account of the limitation of automotive production during the war there was a great shortage of cars. Every car that could be produced was produced and could be sold at almost any price. So far as any one could see, there was no reason why that prosperity should not continue for a time at least. I liken our position then to a big ship in the ocean. We were sailing along at full speed, the sun was shining, and there was no cloud in the sky that would indicate an approaching storm. Many of you have, of course, crossed the ocean and you can visualize just that sort of a picture yet what happened? In September of that year, almost overnight, values commenced to fall. The liquidation from the inflated prices resulting from the war had set in. Practically all schedules or a large part of them were cancelled. Inventory commenced to roll in, and, before we realized what was happening, this great ship of ours was in the midst of a terrific storm. As a matter of fact, before control could be obtained General Motors found itself in the position of having to go to its bankers for loans aggregating $80,000,000 and although, as we look at things from today's standpoint, that isn't such a very large amount of money, yet when you must have $80,000,000 and haven't got it, it becomes an enormous sum of money, and if we had not had the confidence and support of the strongest banking interests our ship could never have weathered the storm.

u Now, as a result of that experience, which was at the time the present administration came into the picture, we recognized that our first duty was to obtain a proper control over the operations of this big ship. We should not be satisfied to go along, unconcernedly, when times were good, with no thought of the future. We should first devise scientific means of administration and control whereby we should be able to project ourselves as much as possible into the future and discount changing trends and influences and, second, that we should be prepared at all times to alter the course of this ship of ours promptly and effectively should circumstances develop that required us to do so. This has all been accomplished and I feel at the present time, dealing again with this great responsibility that falls upon our management, that no matter what the future may bring forth or no matter what changes may take place, irrespective of how suddenly they may take place, we have at all times the organization and machinery to deal with them in such a way that the adverse effect upon the great interests that we represent will be reduced to a minimum.

"At the time specified, General Motors, and I think the same applies to other manufacturers, never had any regard for the number of unsold cars in the field. The sole idea was to make as many cars as the factory could possibly turn out and have the sales department force the dealers to take and pay for those cars irrespective of the economic justification of so doing I mean, irrespective of the dealers' ability properly to merchandise such cars. That was wrong and it is just as wrong in other industries as it was in ours. The quicker merchandise can be moved from the raw material to the ultimate consumer and the less merchandise, of whatever it may consist, is involved in the 'float,' so to speak, the more efficient and more stable industry becomes.

"To my mind, one of the strongest factors influencing continued prosperity, or expressed otherwise, one of the strongest influences operating against a sharp reaction in industry, is the fact' that American business today is largely conducted on that basis. As a matter of fact, the war taught us that lesson. The automotive industry is particularly fortunate in dealing with large units of relatively large value and of having direct contacts with the retailers from whom proper statistics can be developed. In General Motors, we receive reports from our 18,000 dealers three times a month. These reports inform us as to the number of cars they have on hand usually the models they have on hand. Also as to the number of used cars on hand as well as the number of forward orders booked for future delivery. Upon these reports the manufacturing schedules that involve our 180,000 direct employees, our 4,600 suppliers, and each and every one of our operations, at home or abroad, are developed.

"The movement of merchandise into the hands of the ultimate consumer is our fundamental index, and it should be the fundamental index of every business to the fullest degree possible. It is absolutely against the policy of General Motors to require dealers to take cars in excess of the number they properly should 'take. Naturally, once in a while in the closing out of a model, our dealers must necessarily help us. They appreciate their responsibility, and never object to doing so. Our policy in formulating our production schedules is to err, if we must err, on the conservative side. Naturally, errors of judgment will occur, as it is difficult to forecast the consumer demand, five months ahead, as we have to frequently, but we are able to come remarkably close.

"We publish each month, in order that the whole world may know, exactly what the movement of cars to the ultimate consumer is in order that those interested in the statistical side of American industry may take that for what it is worth as a measure of the general business trend.

"We have developed a system whereby we forecast each month for the current month and three succeeding months, every detail of our operations production, sales, overhead, profits, inventory, commitments, cash, and all the other elements that are involved in an operation like ours. We have developed this procedure to a remarkable degree of accuracy. I think it would be safe to say that all of these indices, with the procedure we now have, are forecast within a very close percentage. We are able, therefore, to look forward and provide for the future with the assurance that we know very closely at all times where we will stand four months ahead.

"After all, everything I have told you can be expressed in two words proper accounting, and I now come to the point where I want to outline to you what I believe to be a great weakness in the automotive industry today and what General Motors is trying to do to correct that weakness.

"I have stated frankly to General Motors dealers, in almost every city in the United States, that I was deeply concerned with the fact that many of them, even those who were carrying on in a reasonably efficient manner, were not making the return on their capital that they should. Right here let me say that so far as General Motors dealers are concerned, from what facts I have, I realize there has been much improvement during the past two or three years, but interested as the management of General Motors must be in every step, from the raw material to the ultimate consumer, and recognizing that this chain of circumstances is no stronger than its weakest link, I feel a great deal of uncertainty as to the operating position of our dealer organization as a whole. I hope that this feeling of uncertainty is unwarranted. I am sure that with a responsibility so great, all elements of uncertainty must be eliminated and that our dealers should know the facts about their operating position as clearly and as scientifically as we feel that we know the facts about General Motors operating position, just outlined.

## OUR DEALERS ARE PARTNERS

"We consider our dealers partners in our business. It is true they operate on their own account, but they are, nevertheless, partners in the sense that their prosperity is linked up with our prosperity, and all good partners should recognize the necessities of each to the other and they should cooperate so that all weaknesses can be eliminated. This is exactly what General Motors is doing in this connection. We have organized a subsidiary whose sole function will be to establish proper accounting systems wherever desired by our dealers. We will audit such accounts periodically in order that our dealers may have the assurance that their records are properly established, and that the facts that come to them are facts rather than fiction. We feel that with the great amount of specific knowledge we have, involving all phases of the automotive business, and with an organization that specializes in this particular branch of accounting, with nothing else to think of, we can, through evolution and with the cooperation of our dealers, place before them facts and figures that will indicate to them very clearly what they should do and what they should not do. I do not think there is anything that will contribute more to our complete stabilization than an accomplishment of this kind. I do not think there is anything that will establish greater confidence in the minds of the banking interests whose cooperation we must have in carrying on. Some time ago I saw it stated, and I believe it is absolutely correct, that if business, using that term in its broadest sense, were equipped with proper accounting, a very large percentage of the failures and losses incident thereto could be eliminated. We hope to be able, in due course of time, to place before our dealers 'bogeys/ I might say, showing the proper relationship of each expense item to the business as a whole with the result that if a dealer will conduct his affairs along the lines that we can ultimately outline to him, he will, in a sense, take the straight and direct course to a reasonable and fair profit.

"I have told my associates time and time again, that with this program of ours accomplished to the degree that I am hopeful that it can be accomplished, it will be the greatest achievement of General Motors.

"Every once in a while my attention is called to items in the papers that you gentlemen are publishing statements to the effect that this General Motors line or that General Motors line is going to be discontinued. That is unfair to your readers who have invested in those particular cars; it is unfair to the dealers handling those lines, and it is unfair to General Motors. It is unfair to you because you want to tell the facts. I will take a few minutes to tell you exactly what our policy is in this regard.

"It is our hope and ambition to develop a complete line of motor cars from the low-priced group to the high-priced group within the limitations of reasonable quantity production. It is our hope and ambition to make each of those lines of cars represent a greater value than anyone else can offer. It is our hope and ambition so to develop the confidence of the buying public in our policies and purposes, as to have it feel that whatever price car may be needed, the most outstanding value, from every point of view, is in the General Motors line. Much has been accomplished in. that direction, but no one appreciates more than I do that much more can and will be accomplished. It was in the development of this program that we added the Pontiac, and it was also in the development of the same program that we added the La Salle.

"Should we feel that our line of cars at any time is, for any reason, incomplete, we will add other lines to the end that, from the highest price group to the low price group there will be a General Motors car with reasonable difference in price, to fit the purse and purpose of all, and they will all be quality cars you may be sure of that. We will never make the fatal mistake of sacrificing quality for price. All that I have said means, expressed otherwise, that there is no possibility or probability of any of the present lines being discontinued. On the other hand, they will be expanded and improved and made more effective and more efficient as the ability of the General Motors organization makes this possible.

"As a result of the policy just outlined we have already made substantial progress; we have increased the proportion of General Motors' new cars registered in the United States from one in six to better than one in three. We have increased our retail sales during the past three years to the point where, this year, we expect to sell at retail, in excess of one and one half million motor cars. We have expanded our business in volume during the past three years to over $1,000,000,000 annually. Our organization recognizes that there is much to be accomplished, and I am sure that it will be accomplished. We are on our way.

"General Motors profits, like its business, have increased rapidly during the past two or three years. In 1926 they were something like $190,000,000. If the balance of 1927 continues along our forecast, the earnings of General Motors during the last six months should be as good as the last six months of 1926. I am in hopes they will be a little better. Now, this aggregation of profits is the largest that any corporation has ever made in times of peace and, as a matter of fact, has ever made in the whole history of industry with the exception of a single instance in times of war. Unfortunately, this has led to a false impression due to a lack of understanding of the facts, that General Motors must be making a very large profit per car. This is absolutely not true. General Motors profit per car is less today than it has been at any time except the one year in which we made no profit or, as a matter of fact, made a loss.

"The statement that I made as to the increase in our profit account and the statement I have just made as to reduction in profit per car, may appear to you to be somewhat contradictory. I think, therefore, that I should perhaps explain the statement a little more in detail. It is not realized that General Motors profits come from many different sources. As a matter of fact, not more than half the Corporation's profits come from the manufacture of motor cars in the sense that other manufacturers produce motor cars. Our motor-car operations are, I think, equally, if not more completely self-contained than those of competitors and yet, as I said before, they contribute less than one half to our profit account.

"You will agree with me, I am sure, that when we are charged with the responsibility that I so thoroughly outlined to you, when we invest our stockholders' money as trustees, we must do it in the firm belief that the capital is safely invested and that the return to the stockholders, as a result of the investment, will be fair and equitable ; otherwise we have no right to make the investment at all. The stockholders' investment during the past three years has increased roughly $400,000,000. This investment carries with it an obligation as to a return. Part of this money has been invested in the expansion of our motor-car operations directly in increasing capacity of existing lines and in the establishment of the Pontiac and the La Salle, but very large sums of money have gone in entirely different directions, viz., in expanding and establishing new activities entirely independent of the motor-car operations, and some even outside the automotive industry.

"We have $50,000,000 invested in the General Motors Acceptance Corporation, the functions of which I will mention to you later. This $50,000,000, being stockholders' money, is entitled to make a fair return consistent with a normal banking operation. We have tens of millions of dollars invested in accessory operations which contribute a reasonable return to the stockholders and, naturally, add to the profit account. We manufacture roller and ball bearings for all purposes including railroad cars; and electric light plants for farm lighting. Through our Fisher Body division we build our own bodies from beginning to end. We own timber tracts and saw-mills. We carry the operations all the way from the forest to the dealer. This takes capital, and that capital is entitled to a return, and the return increases the total profits.

"We operate the Frigidaire, having about $40,000,000 invested. We believe the electrical refrigeration industry is destined to be a very great industry. It goes without saying that that has nothing to do with the automotive industry, and its contribution to the stockholders' profits has nothing to do with the profits per motor car, yet the support of the Frigidaire Corporation to the stockholders' profits is very substantial, and it is increasing.

"We have employed at the present time, something like $50,000,000 in overseas operations; in assembly plants and in merchandising operations abroad. The stockholders are entitled to a return on this as well as they are on other investments I have mentioned. It is not necessary that we should do this. We could sell our cars f.o.b. New York for cash and have them distributed by others. In that way we would employ less capital, but our stockholders, of course, would not get the return. The advantage of doing it our way is that we can do a bigger and better business. We can reduce the price of the cars in overseas countries. We therefore sell more cars and we can make all the cars at a lower cost in consequence. We can also paint and trim the cars more to the liking of the people in those countries. We employ labor and purchase material in those countries, and in that way we make ourselves more a part of the industrial life of each country. What is the result? A good return on the additional capital employed for the stockholders; an increase in the volume of business, meaning additional profits; a lowering of the cost of all cars produced, hence greater value at both home and abroad. All concerned are benefited.

"I might go on further and point out to you many other activities that we have that are very seldom thought of, all of which must make a return. Still in addition to this, and it is an important consideration, through our growing production necessities we are required to make more and more of our own parts. We can usually make them better and at a lower cost. It is not, however, our policy to do this unless we can get such results, but where we can add value to our cars and make a reasonable return, we feel it is our duty to the responsibility we carry to do this. For all additional capital General Motors employs in extending its operations, other things being equal, there must be a proportionate increase in profits. Each dollar must make a showing for itself. If, for illustration, $100,000,000 were left in the business out of a given year's earnings, and a reasonable return on that $100,000,000 is 15 percent, then the next year's profits should increase, all things being equal, by $15,000,000 or we will not have done a constructive thing.

"A few words about our organization itself. We operate on the principle of what I might term a decentralized organization. I mean by that, each one of our operations is self-contained, is headed by an executive who has full authority and is responsible for his individual operation. We, naturally, think that this is the best scheme of organization or we would not adopt it. Our responsibilities are so great, the necessity of quick action and prompt decision is so great and contributes so much increased efficiency and effectiveness, that it is about the only way a business of the magnitude of General Motors could be conducted. It also, I think, has the very great advantage of developing executive ability and initiative on the part of a greater number of individuals. All the members of our organization appreciate what is absolutely true that they have a real function to perform and that upon their initiative, their industry, and the constructiveness of their decisions as a whole, depend the success of the institution as a whole. Coordination is effected through what we call InterDivisional Relations Committees, where those interested in the same functions of the important divisions meet together and discuss their own problems as well as the same problems from the standpoint of Corporation policy. For instance, our purchasing agents meet together in the form of a General Purchasing Committee, presided over by a vice-president of the Corporation. If it is found that one or more of the divisions can profit by purchasing as one unit, then we purchase as General Motors and all profit. If it is found that there is nothing gained, we do not do so. In that event the purchase is by the individual operations as their judgment may determine. In that way we get individual initiative, and at the same time we do not overlook anything from the Corporation standpoint.

"In addition to this, the Corporation maintains an organization in Detroit as an advisory service for the benefit of all. General Motors Research which you have visited, acts in a consulting capacity for the engineering departments of all the divisions s and IR addition to this, is constantly searching for new principles and ideas of a more fundamental and scientific character than would be possible for any of the individual engineering departments, which must be more concerned with immediate production problems. Legal and patent problems, as well as accounting and financial control, are handled in a similar manner. Sales research is also a very important activity. In no case, however, is the responsibility taken away from the head of each division. When differences of opinion arise, and differences of course do arise, they are discussed and considered from every standpoint. In every case that I can remember since I have been operating in the past three years, as a result of such discussion, everybody was agreed as to the proper course to pursue.

"Some months ago, I sent a message to our stockholders, entitled, 'How Members of the General Motors Family Are Made Partners in General Motors' In this I explained the attitude of General Motors toward its organization those who make it what it is, and what it will be in the future. I shall be glad to send any of you a copy of that message. After all, the tens of millions of dollars we may have in banks, the hundreds of millions of dollars invested in various ways, our plants and their equipment throughout the world, all are of comparatively little value without an intelligent and effective organization. It is easier to replace all the former than it is the latter. We have recognized that principle by developing plans whereby the organization itself profits through its own endeavors as partners in other words, receives something in addition to a daily wage. This applies to one and all. Time will not permit me to go into detail. I just want to say that I believe that this principle of the organization's participation in the result that they themselves accomplish is not only sound economically and equitably, but is the best kind of business from the stockholders' standpoint.

"A word about our policy of telling the facts about General Motors. We believe in frankly telling our stockholders and the public all the things that we can consistently. As I have already said, we report monthly our retail and wholesale sales so that everybody may know. We send quarterly, to our stockholders, detailed statements of the financial position of General Motors as well as its operating position. I make it a point to send messages several times a year to our stockholders, telling them of things that we are doing and why we are doing them so that they will get a complete understanding of our viewpoints. In our annual report we try to state all the facts the stockholders should know. We knowingly hold back no information that they are entitled to as partners in the business. I feel that this policy on the part of General Motors has contributed much to the feeling of good-will that exists toward the Corporation, not only throughout this country, but throughout the entire world.

"The story I told you about the position of General Motors in 1920 and my viewpoint of the greatest necessity in our dealer situation, leading up to the two words Proper Accounting might be expressed in still another way, this time in three words: Get the Facts.

"There is a very fundamental principle, the importance of which I am continually trying to impress upon our direct organization as well as our dealers; viz., Get the Facts.

"I would now like to tell you some of the things we do in General Motors to get the facts.

"Let me tell you about what we call our field trips. It may surprise you to know that I personally have visited, with many of my associates, practically every city in the United States from the Atlantic to the Pacific and from the Gulf of Mexico to the Canadian border. If any of you has done this, you realize what a big country it is. It has taken weeks and weeks of the hardest kind of work and continual travel to accomplish this. I wish that my duties were such that I could do more of it; and I am trying to arrange my affairs so that I can. In these trips I visit from five to ten dealers per day. I meet them in their own places of business, talk with them across their own desks and solicit from them suggestions and criticisms as to their relations with the Corporation; the character of the product; the Corporation's policies; the trend of the consumer demand; their viewpoint as to the future, and many other things that such a contact makes possible. I solicit criticism of anything and everything. I make careful notes of all the points that come up that are worthwhile, and when I get back home I study and develop those points and capitalize them so far as possible. The reason for all this is that irrespective of how efficient our contact through our regular organizations may be, our men in the field are charged with doing specific things and that takes all their time and effort. I go out from the standpoint of general policies and get the facts in a very personal way without the intermediary of an organization, which is apt to overlook the most important points and inject its own personal views on such points as it does get. I believe that this work that we have done has contributed much more than any of us appreciates to the progress that General Motors has made.

"You have gone through our Research Laboratories and have some idea of what it is all about. It is no different from our field trips, in principle. We are searching for the facts that we may know more about the fundamentals and be able to add value to the performance and effectiveness of our products, just the same as in the field work we are trying to learn more about the distribution of our products. We send representatives abroad to study foreign methods and foreign cars. We have an engineering office in London with representatives in other countries to keep us advised at all times as to what progress there may be along European lines that General Motors can capitalize. Again, we are seeking the facts.

"The Proving Ground here is also dedicated to the principle of getting the facts. As you see, we not only operate our own cars, but all competitive cars, both those made at home and abroad. We are seeking the facts about all of them to the end that General Motors cars are better cars. We are seeking in our sales activity here at the Proving Ground to impress upon our dealers the facts about our cars so that they may more intelligently present them to the consumer. "I wish I could get a definite measure of the relationship of the various things we have done, one to the other, and their contribution to the whole, but, naturally, a thing of that kind is impossible. I feel, however, that certain principles which we have developed are important factors in the progress we have made. One of these is that we have tried to treat everybody fairly and squarely. We have tried to recognize that our organization, which is contributing everything to the success of the Corporation, is entitled to participate in that success our partnership idea. We have also tried to recognize the position of our suppliers. Of our dealers I have already spoken. As to the public, you only have to compare any General Motors product of today and its price with the corresponding product of even two years ago and the price at that time: more beauty in design; more luxury in appointment; more comfort and convenience; increased performance at, in nearly all cases, a lower price. In other words, greatly increased values. We have worked conscientiously and intensively, always with the interests of our stockholders as our guide. Naturally, we are not perfect in carrying out these policies and ambitions of ours, but we do the very best that we can and feel that we are continually making improvement. Therefore, the first factor I believe should be trying to treat everybody fairly and squarely. Then we are trying, as I said before, to get the facts. My experience in business is that facts are too little considered in making decisions. It is difficult to get the facts to get all the facts, but it is worth every effort and we put forth that effort. Then, with the facts before us, we try to approach the decision with an open mind.

"We recognize that we must be conservative. Our responsibilities are too great to be otherwise. We cannot take chances, yet we must progress. We try to recognize that what we do today, whatever measure may be placed upon it, must be improved tomorrow; next week's performance must beat tomorrow's, and next month's and next year's must beat that of the previous month or year. Therefore, to sum up, we get these factors: a recognition of the equities of all concerned; getting the facts; analyzing the facts with an open mind. These, to my mind, are the principles which have contributed most to the present position of General Motors and to the progress that it has made during the past few years. This leads me to the last point the necessity of capitalizing the principles that we elected to guide us forward. This can only be accomplished in one way: hard work.

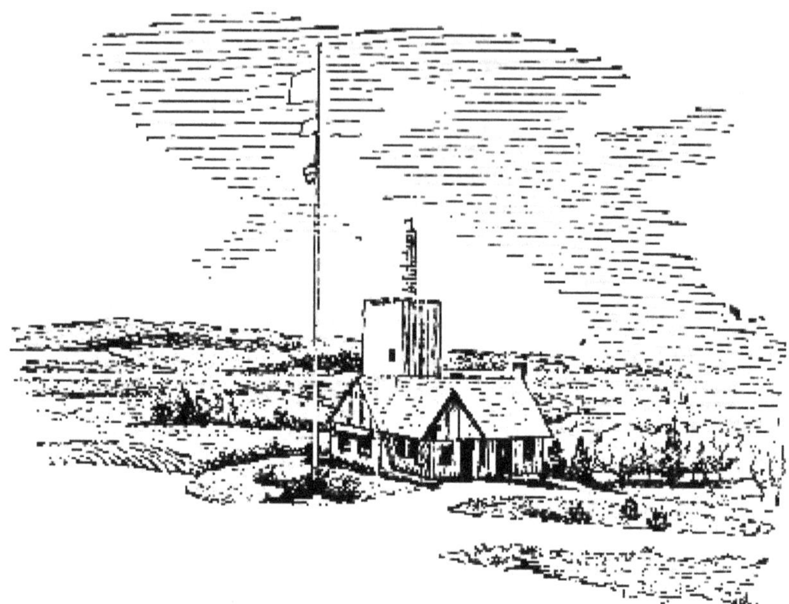

Weather Bureau Station, General Motors Proving Ground, Milford, Michigan

Entrance to General Motors Proving Ground, Milford, Michigan

"It seems sometimes each one of you has seen such cases that results are accomplished without hard work, but I think you will agree with me that in the long run it does not work out that way. In any event, I hope that General Motors will never attempt that sort of an experiment.

"Therefore, our principles completely expressed, as I see them, and they apply to every other business as much as they do to that of General Motors, are: Get the facts. Recognize the equities of all concerned. Realize the necessity of doing a better job every day. Keep an open mind and work hard. The last, gentlemen, is the most important of all. There is no short cut."

# Chapter XXV

## THE STOCKHOLDER INTEREST

IN DISCUSSING the purchases of interests in other companies, it was noted that in most cases General Motors followed the practice of buying all or part of the agreed shares in the open market, with the result that a number of purchases were completed without changing the stock structure of the Corporation even though the purchase agreement was stated in terms of stock.

After the General Motors Corporation succeeded the General Motors Company on October 13, 1916, only in the following cases was Common stock originally issued in the acquisition of properties:

$IOO PAR COMMON STOCK

| DATE | ADDITIONAL SHARES ISSUED | FOR STOCK OF |
|---|---|---|
| May 2, 1918 | 282,684 | Chevrolet Motor Co. of Delaware |
| Nov. 1, 1918 | 49,000 | Canadian Units |
| Dec. 31, 1918 | 99,401.7 | United Motors Corporation |
| Dec. 31, 1918 | 16,175 | Lancaster Steel Products Co. |
| Oct. 31, 1919 | 35,451 | Domestic Engineering Co. |
| Mar. 25, 1920 | 21,457 | Dayton Metal Products Co. |
| | **504,168.7** | |

## NO PAR COMMON STOCK

On March 25, 1920, 140,000 shares of no par stock were exchanged for all capital stock of the Chevrolet Motor Company of California.

As of June 30, 1926, 638,401 new no par shares of the Corporation were issued as part payment for the minority interest of Fisher Body Corporation.

In the foregoing, and other purchases, there were issued 390,708 shares of 6 percent $100 par Debenture stock and 60,000 shares of 7 percent $100 par Debenture stock. The last instance of this nature was in October, 1929, when 40,000 shares of 7 percent Preferred stock were issued as part payment in the acquisition of the North East Electric Company.

The era of prosperity began for the Corporation in 1923, when net sales reached a new high level, nearly $700,000,000 earnings on which left $65,121,584 available for dividends on Common stock, the largest earned for Common stockholders up to this time. Of this $24,772,026 was paid in dividends, and more than $40,000,000 reinvested in the business. The dividend on the Common no par stock had been raised to 30 cents a quarter on March 15, 1923, and was continued at that rate through the year.

This increase in dividends reflected both the growing prosperity of the country and its increasing acceptance of the automobile as a commonplace, a piece of property so widely owned that to be without one was to be out of fashion. All trades catered to the motorist. Dry-goods stores had added motoring accessories, sport clothes, and blankets. Hardware stores carried flashlights and tools. Automobile supply stores were growing in number and size a new development in merchandising. Hotels found a new and profitable business in the motor traveler. The farmer started wayside markets to reach the city dweller. A parts and accessory business had brought by 1923 a turnover in excess of $1,750,000,000 through retail trade, when clothing and other motorist supplies are included.

Also, an intense stimulus had been given to the search for substitutes, for new sources of raw materials and the development of by-products. A particularly noticeable instance was gasoline, which was originally the by-product of petroleum, while kerosene was the mainstay. The motor industry reversed their positions and built a far greater industry out of the formerly unimportant by-product. "Countless similar situations might be added by a detailed and careful analysis, all weaving closer and closer the vital and mutually advantageous relationship of the automobile and allied trades and industries." The motor car had become an essential of life for the American masses.

On June 16, 1924, the certificate of incorporation was amended to provide for a reduction in the number of no par Common shares, one share of new no par value Common stock being given for each four shares of no par Common stock then outstanding. This reduced the number of shares outstanding from 20,646,397 to 5,161,599. At the same time holders of the 6 percent Debenture and 6 percent Preferred stocks were given the right to exchange their stocks for a new issue of 7 percent Preferred stock, on payment of $10 per share in cash. The new Common stock after the exchange of one new share for four old shares, registered high and low prices of 66⅞ and 55¾ in 1924, but the next year saw it rise to a high of 149¾. This stock soared to 282½ in 1927 becoming the leader in the remarkable "bull" market of that hopeful year.

Both sales and earnings dropped somewhat in 1924. An initial dividend of $1.25 on the new Common stock, each share of which represented four shares of the old no par Common, was paid December 12, 1924. More than $19,000,000 was reinvested in the business. The Corporation's solid position, the greater market appeal of the new and higher-priced stock, and the flourishing condition of the country, all combined to focus attention on General Motors Common. When it was seen that the disposition of the directors was to pay in dividends roughly two thirds of earnings to Common stockholders, the market hesitation on the stock ceased. From 1924 through 1929 disbursements to Common stockholders ranged from 56.4 per-

New York headquarters of General Motors, 7775 Broadway, as seen from Central Park across Columbus Circle

cent of available earnings to 65.6 percent. With the decline of earnings in 1930, the ratio rose because of the Corporation's decision to maintain dividends as part of a policy to reinforce the general buying power of the public, a determination which increased the number of its stockholders all through that discouraging period.

Mr. Raskob, as chairman of the Finance Committee, exercised a great influence on the Corporation's dividend policy. A hopeful and sanguine man, he firmly believed that prompt and liberal distribution of earnings to stockholders not only tended to maintain general prosperity, but also that it made future markets for General Motors cars by increasing directly the buying power of those already well disposed to the Corporation's products by stock ownership. The various stock "split-ups" and stock dividends were motivated by the reasoning that the more persons financially interested in General Motors, the greater the number of potential customers, since, other things being equal, they would be disposed to prefer General Motors products to those of other manufacturers.

Sales for the year 1925 showed a decided increase, net income available for dividends exceeding $100,000,000 for the first time in General Motors history. Accordingly, extra dividends were paid on Common stock, $i on September 12, 1925, and $5 on January 7, 1926, and the regular quarterly dividend rose to $1.50. Notwithstanding these disbursements, the then record sum of $46,441,065 was reinvested in the business. The liberal investment policy being based on a financial position of increasing strength, the investing public pushed the price of the new no par Common stock from a low of 64⅝, to a high of 149 ¾ during the year.

A Common stock dividend was declared on August 12, 1926, at the rate of one half share of Common for each share outstanding. This change in the capital account raised the outstanding shares by 2,900,000 to a new high of 8,700,000 for the no par issue of 1924. Cash dividends for the year reached the then record of $103,930,993. Notwithstanding, almost $75,000,000 from earnings was reinvested in the business. It is

LAMMOT DU PONT
Chairman of the Board of Directors, General Motors Corporation, 1929

noteworthy that in 1926 General Motors paid in dividends a sum in excess of its net sales in 1915.

An important change in the Corporation's capital structure, approved by the directors on August 11, 1927, was adopted by the stockholders on September 12th, increasing the authorized Common stock from 10,000,000 no par shares to 30,000,000 shares of $25 par value. Two shares of the new $25 stock were issued for each of the no par shares, which resulted in increasing the number of outstanding Common shares from 8,700,000 to 17,400,000.

During 1926 the Corporation had put upwards of $100,000,000 into plant extensions and improvements, and in permanent investment in subsidiary companies. A portion of these expenditures was now permanently financed by the issue of 250,000 shares of 7 percent Preferred stock at $120, the marketing of the issue being taken over by J. P. Morgan & Company. Both the premium paid and the favorable rate at which the famous Morgan house undertook their distribution reflected the Corporation's favorable financial position.

From earnings of 1927 more than $90,000,000 was reinvested in the business. Cash dividends on the Common stock reached a new high at $134,836,081, $8 being paid on the old no par stock through the medium of three quarterly $2 dividends and an extra of $2 on July 5th. An initial quarterly dividend of $1.25 on the new $25 par Common was paid on December I2th with an extra of $2.50 on January 3, 1928. This was the equivalent of $15.50 on the old no par stock which closed its extraordinary market history on October 8, 1927, with a high record of 282%, after being quoted four years earlier at 55%. Since the number of shares had been doubled, the market price of new stock was reduced, but during the next three years ran from a low of 123½ to a high of 224¾

In 1928 when the American motor-car industry as a whole reached the new high of 4,601,141 units in production, General Motors sold 1,810,806 cars and trucks. Total sales were $1,459,762,906 and net income available for dividends $276,468,108, the highest in the Corporation's history. Its percentage of the total passenger car business was 41.4. Dividends reached an all-time high, with a disbursement of $165,300,002 on Common stock, and more than $100,000,000 the largest sum to date was reinvested in the business. The stock reflected these tremendous earnings and disbursements by reaching a new high of 224¾.

As prices in that elevated range were rather beyond the reach of small investors whom the Corporation desired to attract, stockholders were asked to approve an amendment to the certificate of incorporation effecting the exchange of their $25 par stock for a new issue with a par value of $10, thus increasing the number of outstanding shares two and one half times. This was done on December 10, 1928, the exchange of stock beginning on the January 7th following. On the completion of the exchange, the Common stock outstanding was 43,500,000 shares of $10 par value and no further change has been made. The highest quotation on these shares was 91¾ in 1929.

The year 1929, indeed, recorded new highs in many aspects of the Corporation's activities. In that year the Corporation made these records:

| | | |
|---|---|---|
| Unit Sales | 1,899,267 | passenger and commercial cars |
| Sold | $1,504,404,472 | worth of merchandise |
| Earned | $248,282,268 | available for dividends |
| Paid | $156,600,007 | in dividends on Common stock to 198,600 stockholders |
| Reinvested | $82,203,580 | in the business |
| Pay Roll | $389,517,783 | |
| Employees | 233,286 | |
| Bonuses | 167,378 | shares of Common stock |
| Its employees had | | |
| | $56,560,310 | balance in the Employes Savings Funds, and |
| | $38,762,678 | balance in the Employes' Investment Funds |

Its car divisions made unit sales to dealers as follows: Buick, 199,414; Cadillac, 15,416; La Salle, 21,498; Chevrolet, 988,191 passenger cars and 344,963 commercial cars; Pontiac (Oakland), 224,448; Oldsmobile, 99,435.

Including all cars and trucks, total unit sales were:

| | |
|---|---|
| Passenger cars | 1,554,304 |
| Commercial cars | 344,963 |
| **Total unit sales** | **1,899,267** |

From 1930 to 1933 the Corporation's record was one of meeting adverse circumstances as effectively as possible, maintaining a strong cash position in the face of shrinking volume and, through improvements in the design and workmanship of its products, striving to stimulate consumer interest. Declines, sharp when compared with the figures of 1922-30, appeared in sales and earnings. The latter years, however, do reflect a rise in the percentage of the total automotive trade enjoyed by the Corporation.

Also in gratifying contrast to declining sales, was the rise in the number of stockholders in the Corporation. The year 1929 witnessed a notable increase in stockholders, the number rising from 71,185 in the last quarter of 1928 to 198,600 in the last quarter of 1929. Since then the gains have been large. The figures for the last quarters of the past four years are:

| | |
|---|---|
| 1930 | 263,528 stockholders |
| 1931 | 313,117 stockholders |
| 1932 | 365,985 stockholders |
| 1933 | 351,761 stockholders |

The rise may be explained partly by the Corporation's strong cash position, the evident determination to maintain it, and the policy of paying out of surplus such dividends as did not threaten that position. Large blocks of stock became broken up under the financial pressure of the period and passed into small and scattered holdings.

By rigid economy the Corporation has escaped recording a loss from operations. Although margin of profit declined, cash and cash investments increased, reduction of inventories accounting largely for this increase in cash. Decisively the Corporation grappled with the problem of readjusting its idle real estate, plant and equipment account to salvage value, thereby eliminating the need for charging depreciation against this property. Consequently, the annual depreciation charge is reduced by some $7,000,000 a year.

The Corporation's forecasting experience stood it in good stead as was shown fairly early in the depression, when the necessary steps in retrenchment were taken and production was effectively controlled.

On May 26, 1930, the stockholders, to simplify its senior security structure, voted to replace outstanding senior securities by a new $5 series Preferred stock, exchangeable on the basis of 1.35 shares of the new $5 Preferred for each share of the 7 percent Preferred, 1.15 shares for each share of 6 percent Debenture stock, and i.io shares for each share of 6 percent Preferred stock. General Motors enjoys the unique record in the twenty-five years of its existence of never having passed a dividend on any of its senior securities, of which 1,875,366 shares of its $5 Preferred stock are all that are presently outstanding.

As the business skies began to clear in the spring of 1933, elements of strength at once appeared. Stocks of cars were so low that even a limited new demand caused factory activity. For the first time in the history of the Corporation, June sales exceeded May sales, and the period of brisk summer buying continued longer than usual. Moreover, by maintaining dividends to an ever increasing body of stockholders, without sacrifice of internal strength, the Corporation had commended itself to a large and important public. The essential soundness of its management and the solidity of its financial position had been demonstrated under the most severe financial strain ever experienced by modern industrialism.

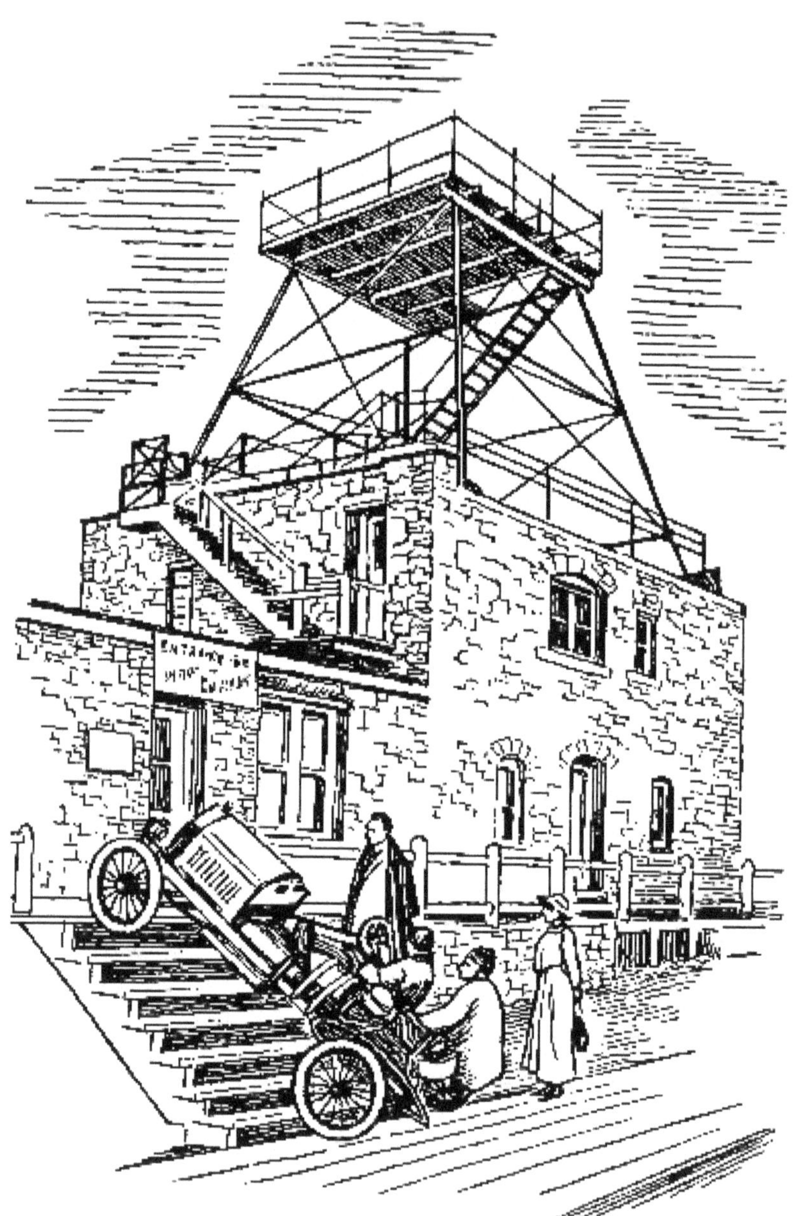

Buick on steps of Summit House, Pike's Peak, 1913. First car to climb the mountain entirely on its own power

# Chapter XXVI

## MARKETING THE MOTOR CAR

FROM its birth the automobile has been a public spectacle, a self-advertising piece of merchandise, performing in the gaze of all. Whenever a new car took the road in the old days, a crowd gathered to praise or criticize. Even yet a new model at the curb attracts passers-by. Thousands pay admission to automobile shows, and changes in design in well-known makes are accepted by press and public as outstanding news. But even so, automobiles have to be sold; the salesman has to get his order signed.

Old-timers say that for the first fifteen or twenty years of this century, automobiles sold themselves. That is true only in the sense that fewer cars were being built than the public was ready to buy. Nevertheless, there were sales problems. One was the early timidity of the public; an individual might want to buy and be quite able to buy, but his fears had to be overcome. Would the car run, and could he operate it? In such cases, salesmanship was largely a matter of demonstration. The salesman took his prospect for a ride, selecting a course on which excellencies would show and faults be overlooked. If the car was a good hill-climber, hills would be negotiated; otherwise the talk would be of speed on level stretches. With cars of substantial reputation, like Buick, demonstrating the car remained the central effort down to about 1918. Once the car had been demonstrated successfully, the prospect paid a deposit and the dealer entered his name on a waiting-list for future delivery. Many buyers were willing to pay premiums for early delivery.

But even in those years the manufacturer had his selling problems. First in the quantity field, R. E. Olds had to meet the problem of credit sales. Nearly everything hitherto manufactured for the public and sold through middlemen had been purchased on credit. Retailers did business partly on the capital of manufacturers and wholesalers, and in turn sold to many of their customers on open account. There were special reasons why this practice could not be followed with regard to automobiles. Manufacturers could not find capital enough to finance their trade on that basis. Each unit represented so much money that credit risks could hardly be taken. Cars depreciated rapidly in use, and at the start could hardly be considered as dependable, standardized merchandise. Consequently, Mr. Olds insisted upon shipping with sight draft attached to bill of lading, a cash transaction as far as the factory was concerned, since banks would finance goods in transit upon bona fide orders. That practice has been generally followed, though liberalized, as will be described later.

Under this system the dealer had to furnish all the capital and credit required to stock the car pending sale, and finance his deals. If a man of some means and substance, he could usually use local bank capital to pay his drafts and take over his cars from the railroad, but in many places banking facilities were meager, and bankers timid. So the manufacturer had to choose his dealers with an eye to their general credit as well as their selling ability. The seasonal character of the business made the dealer's ability to finance in advance of sale highly important to the manufacturer. March, April, May, and June were the best months for deliveries, but with limited facilities for manufacture, production had to proceed at a steadier pace. Unless the waiting-list was extra long, the factory could operate profitably only if dealers had credit enough to take cars in the winter months for storage in a reasonable quantity. In a vast marketing area like the United States, the inability of even 10 percent of the dealers to do this might be embarrassing. Therefore, although cash was the rule, credit qualifications had to be considered by the manufacturer. Various efforts were made to take short cuts to the market when not enough strong dealers could be found to give a manufacturer adequate representation in all sections. The factory branch store was one method which has been tried by nearly every leading manufacturer. At one time Chevrolet had over fifty of these stores, but found it advisable to abandon all but a few. The Buick branch in New York City outlived all its comrades. Also, distributorships were set up for large territories, a natural evolution in view of the time involved in rounding out a system of national representation on the basis of contracts with individual dealers. Some manufacturers operated sales zones on their own account; others allocated territories to independent individuals and firms. Some of the latter have been highly influential in the history of General Motors. An excellent example is Harry K. Noyes of Boston, Buick branch manager and distributor for New England for more than twenty-five years. In several other parts of the country continuity records of almost equal impressiveness have been made. Successful distributors often reinvested their profits with their dealer-customers, acquiring part ownership, and in some cases full ownership of important retail outlets in their territory.

In a trade of rapid growth and immense vitality, dealer arrangements are necessarily flexible, adjusting themselves to the needs of the territory and the business and financial capacity of the dealers. Hence sales policies which were too rigid to let the human element expand and flourish on the local side have usually come to grief. One of the early problems was to find and secure as dealers and distributors men who would keep on growing as the industry grew.

Despite the policy on the manufacturer's part to contract with substantial parties, the early dealers were usually men of more courage than capital. They had to be courageous, since they were staking everything on their faith in a new business geared to a scale of values beyond anything else ever taken to the market in quantity. In 1908 a good automobile cost as much as a small house; the difference was that banks, building and loan associations, and mortgage companies were solidly behind real estate, while the automobile dealer had to finance his operations on his own responsibility. At the start he usually operated "on a shoestring," until profits gave him a firmer base and longer line of credit. As a result, the survivors of these adventurous beginnings, now well along in years, still have the hearty air of men who in their youth grappled gayly with fortune, taking good luck with a grin and bad luck with a smile. They fought for public acceptance of the automobile ; upon their ability to convince all classes of buyers the new industry depended for the introduction of its products into areas as yet unprepared for it by advertising, good roads, or service facilities.

In New England Harry K. Noyes would break a trail over a mountain, sell his car, establish a dealer, and make his way back to the railroad in an ox cart. In Minnesota, Harry Pence would lead a string of Buicks westward through the mud and never think of returning until the last one had been sold. He might take wheat or potatoes in trade, but he sold the caravan and returned to Minneapolis for more. As one veteran said:

> We were business buccaneers on uncharted seas; we would gamble with anything except the good name and fame of the cars we were selling. I have risked my life more than once making demonstrations on hill roads and hairpin curves that I had never seen before. There wasn't the same uniformity in cars that there is now, and because one performed well that was no guarantee that the next one would. If it failed and we survived, we always had plenty of reasonable explanations on tap, made repairs at the nearest blacksmith shop and returned for another try. I made it a point of honor for years never to let go of a man who seemed to me to be the best dealer material in his community. Sometimes I had to wait years to get him, taking someone else in the meantime. But if the second choice didn't make good after giving him every chance and encouragement, back I would be after number one.
>
> We didn't know so much about General Motors then. We were Buick men or Cadillac men or Oldsmobile men, whatever the hook-up. That is still pretty much true in the field, but then we were treading on each other's toes all along the line, since less attention was paid then to keeping General Motors cars out of direct price competition with one another. That was just as well,

> for it kept us on our mettle at a time when our futures depended upon building the best dealer organizations possible, and our dealers likewise needed to reach and cultivate the best clientele in their neighborhoods. Later on, when our expenses of doing business were larger, and the market was less demanding, we welcomed the effort of headquarters to keep General Motors cars in price classifications where competition between them became marginal rather than general.

The first motor cars were offered to the public without top, windshield or lamps, and were serviced by blacksmiths or buggy builders. Dealers had to stock necessary equipment, create service organizations, and organize supply stations. A pioneer salesman invading new territory made it his business to talk country storekeepers into stocking gasoline, and to urge upon the livery-stable keeper the wisdom of providing accommodations for motor cars as well as for horses and carriages.

Fairs bulked large in the early sales efforts. The first motor cars were denied admittance to these gatherings, but that prohibition soon passed, and presently whole buildings came to be set aside for automobile displays. To Barnum & Bailey goes credit for breaking down the barrier against automobile showmanship in many backward parts of the country. During the entire 1896 season, Barnum & Bailey street parades were led by an automobile, and the public applauded its appearance in the ring twice daily.

Within a few years the automobile industry would be putting on its own shows or fairs in all the large centers. As the trade developed, this practice spread downward until the automobile show has become an annual feature in most county-seat towns. New York and Boston led off with automobile shows in 1900, following the precedents set by annual cycle shows. In showmanship, as well as in mechanics, the automobile trade owes a considerable part of its early momentum to the bicycle trade. By an easy transition firms making bicycles and bicycle parts began to make automobiles and parts. One of General Motors' chief accessory producers New Departure began its industrial career making doorbells and bicycle bells; then it passed to coaster brakes and latterly to ball bearings. Another started with bicycle gears. Hundreds of the most effective of the early automobile salesmen came directly into the new field from the bicycle trades; it was a commonplace to see an automobile in one show window of a store while the other window continued to display bicycles. And, of course, the "bicycle craze" of the early 'nineties brought the public around to the automobile as the next step in highway travel and rapid movement.

The bicycle craze was fast and furious while it lasted. It began with velocipedes and high-wheel boneshakers, and ran rapidly through to the "safety" and elaborate multiple cycles. Chief of these was the tandem, now remembered chiefly as "the bicycle built for two" of the famous "Daisy Bell" song. Those were the days of famous bicycle racers, of Tom Cooper, Eddie Bald, Harry Elkes, and Jimmie Michaels, paced by three-, four-, five-, and six-man machines. Probably the largest cycle ever built was a ten-seater upon which Boston belles used to take the air to the wonder and applause of pedestrians. To meet the demand for bicycles, some of America's most famous precision manufacturers the sewing machine and firearms makers turned part of their plants into production of cycles and parts, and so prepared themselves for similar attention to the coming automobile needs.

The swift rise and fall of the bicycle business caused many bankruptcies which prejudiced investors and bankers against the motor-car industry from the start. This prejudice remained active for years and had its influence on the early history of General Motors. When the then young Company needed capital in 1910 the old refrain was still being chanted, "See what happened to the 'bicycle craze.' 'The automobile business had to live down the difficulties of its predecessor.

Other established businesses which furnished large quotas of dealers and salesmen for the new automobile trade were the carriage and farm implement trades. The carriage trades, indeed, had developed the distributorship angle of vehicle selling before the automobile entered the American scene in quantity; in placing Buick distribution on that basis, Mr. Durant followed time-tested precedents at the outset, and of course improved on them, building by that means a staunch and comprehensive selling organization covering the whole country. The West, with its magnificent distances, especially required to be handled in this way, in great blocks rather than piecemeal, and Mr. Durant's early success there may be counted as chiefly responsible for his pushing Buick into first place in sales volume by 1908.

The "saturation point," one of the major delusions of the early period, kept conveniently receding as each year brought larger and larger sales. Prophets of doom were always talking about the mythical point where everyone would have all the motor cars he wanted. This opinion was based upon carriage and bicycle precedents and its disciples quite overlooked the fact that the automobile was an unprecedented development. Everyone wants a new motor car; production is limited only by buying power. Owners of bicycles and carriages were less keen for new bicycles and new carriages; and, in addition, persons who had never possessed earlier modes of transport came eagerly into the automobile market. So the "saturation point" remained in the realm of rhetoric for many years.

## AUTOMOBILE SHOWS

The big business of showing automobiles has developed new features of late. Dealers in the larger centers present what may be considered perpetual shows of all models in handsome showrooms. "Automobile Row" is usually the brightest spot in a city's mercantile section, though the tendency of dealers to herd together is perhaps a little less marked than it was. Nearly every sizable city now has its annual show in the late winter or early spring, with New York for years starting the schedule early in January. The New York show has grown steadily in color and popularity, drawing attendance from all parts of the country, and being always well patronized by a public willing to pay to view the new models.

In 1932, from April 2nd to 9th, General Motors held national exhibitions of its products simultaneously in the following fifty-five cities, located in thirty states and the District of Columbia:

| | |
|---|---|
| Albany, New York | Atlanta, Georgia |

Baltimore, Maryland
Billings, Montana
Birmingham, Alabama
Boston, Massachusetts
Brooklyn, New York
Buffalo, New York
Butte, Montana
Charleston, West Virginia
Charlotte, North Carolina
Chicago, Illinois
Cincinnati, Ohio
Cleveland, Ohio
Columbus, Ohio
Dallas, Texas
Davenport, Iowa
Dayton, Ohio
Denver, Colorado
Des Moines, Iowa
Detroit, Michigan
El Paso, Texas
Grand Rapids, Michigan
Houston, Texas
Indianapolis, Indiana
Jacksonville, Florida
Kansas City, Missouri
Los Angeles, California
Louisville, Kentucky
Memphis, Tennessee
Milwaukee, Wisconsin
Minneapolis, Minnesota
Nashville, Tennessee

Night Scene: General Motors Building at the Century of Progress Exposition, Chicago, 1933 New Haven, Connecticut

St. Louis, Missouri
New Orleans, Louisiana
St. Paul, Minnesota
New York, New York
San Antonio, Texas
Newark, New Jersey
San Francisco, California
Oakland, California
Seattle, Washington
Oklahoma City, Oklahoma
Spokane, Washington
Omaha, Nebraska
Springfield, Massachusetts
Philadelphia, Pennsylvania
Syracuse, New York
Pittsburgh, Pennsylvania
Toledo, Ohio
Portland, Oregon
Tulsa, Oklahoma
Rochester, New York
Washington, D. C.

There was also a special showing in connection with the General Motors exhibit at Atlantic City. This nation-wide selling effort received favorable publicity as the most determined effort made by any corporation to revive buying. The trade of approximately 2,000 of the largest and strongest General Motors dealers located in metropolitan areas was directly affected by the exhibits, and about 18,000 dealers benefited by the advertising and sales promotion campaigns. In Boston alone 240,000 persons, by actual count, visited the six-day Exposition with no other attraction than General Motors products.

General Motors products are displayed at all leading automobile shows in America and elsewhere, except in those foreign shows restricted to national products. In the case of the Chicago and New York shows, special General Motors exhibits are held during the period, the Hotel Astor in New York City and the Stevens Hotel in Chicago being the usual locations. In 1932 and 1933, the New York General Motors special showing was moved to the Waldorf-Astoria.

## PRICING AND SELLING

All the Corporation's price calculations come under review in a competitive market, which is the court of last resort, so that dealer welfare cannot be the only yardstick in fixing prices. The industry as a whole strains for reduced costs as the basis of appeal to the customer's desire and pocketbook. With the increase in manufacturing efficiency, General Motors has followed the plan of building more value into its cars year by year, adding to low-priced cars the features introduced on higher-priced cars. But there are evidently limits to this evolution, and the new Standard Chevrolet, with several features of the Master Chevrolet omitted, was a move toward a lower price level.

How much further manufacturers can go in this direction depends upon many factors, but obviously they cannot go on indefinitely without reaching such a point that cheapness fails to include the essentials of safety and dependability to which the public has become accustomed. In the meantime, the higher-priced cars tend to come down the price scale, partly through economies in production, partly through the introduction of smaller models bearing names which have achieved a reputation at high levels. Altogether the problem of price setting is one of many complications and cross-currents upon which the Corporation concentrates full attention, seeking to find each year a solution satisfactory alike to its customers, its dealers, and its stockholders. At best its decision must always be a compromise arrived at after considering all these interests.

The darker the night the brighter shines the light. There have been times in General Motors history when business rolled in with comparatively little sales and advertising effort, when the cry was for more cars in a seller's market, and the public took eagerly whatever was produced. These halcyon days had their satisfactions, of course, but they also had their disadvantages. Temptation reigned then to let well enough alone, to delay factory changes, and innovations in design. Under the stern challenge of depression General Motors has revised its processes in the interests of economy and efficiency, and its major products have been rigidly scrutinized in order to increase their appeal in a buyer's market, which demanded not only lower prices, but also higher quality and better workmanship.

In the United States the total number of automobile dealers has been decreasing steadily, due partly to decline in the number of manufacturers, and partly to the growth of combination dealerships, in which one firm handles two or more lines. The number of dealers declined from 50,868 in 1926 to 38,092 at the end of 1932. Exclusive dealership suffered heavily, while multiple-line dealerships increased 44 percent from 1926 to 1932.

One third of all dealers in 1932 handled more than one line of cars, while seven years earlier less than one fifth did so. In many localities General Motors contributed to this changing status by teaming Oldsmobile and Cadillac together, and Buick and Pontiac, a selling program facilitated by the temporary organization of the B-O-P Sales Company in 1932, (discontinued 1933). In 1932, Buick dealers handling only Buick cars decreased by 26 percent, while those handling one or more additional lines increased by 24 percent. Pontiac and Oldsmobile moved in a similar direction, but to a less degree. B-O-P in September, 1933, announced 6,883 direct dealer outlets, the largest number in five years divided as follows Buick, 2,105; Oldsmobile, 2,448; Pontiac, 2,330.

Of the 53,437 passenger car representations reported by Automotive Industries as in business at the close of 1932, General Motors passenger car divisions were credited with 16,000, or a little less than 30 percent of the total for the entire industry.

In the beginning automobile makers were financially assisted by their suppliers and their distributors. The former sold their goods on liberal datings, the latter made large deposits in advance. With these aids the manufacturer managed to produce a large volume on small working capital. Both the dealers and supply men were naturally consulted in planning a new model. Gradually, as engineering advanced and manufacturing companies came into large capital and controlled more of their supplies, the automobile makers advanced toward uniform control of distribution. Their contracts usually contained a thirty-day cancellation clause: in a "seller's market," with dealerships in demand, this clause was the "ace in the hole" through possession of which manufacturers could impress their will on dealers in the direction of adequate servicing, display, local advertising, and many other merchandising activities. Repair charges were fixed by the factory, and at one time dealer accounting was under direction.

In no retail trade is the dealer so closely controlled. Mr. W. C. Durant advertised one of the early Chevrolets to dealers in these words: U A little child can sell it." By systematizing retail efforts, and taking their message directly to the public through advertising, the manufacturers are sometimes accused of reducing automobile salesmen to the status of "order takers." This trend, especially

Chevrolet Assembly Line at Century of Progress Exposition, Chicago, 1933

strong in the sales programs of great quantity producers of the industry, reached its height in 1929. With the development of a "buyer's market" after 1930, selling became more difficult, and good salesmen were more appreciated. Consequently, the stronger dealers little by little have insisted on more leeway. Even though control aimed at, and usually resulted in, better merchandising and higher dealer profits, human nature even dealer human nature sometimes resents being "scheduled and programmed" too openly and too often. A good deal depends on the tactfulness with which the program is preserved. The new note in automobile salesmanship, and one which marks the eclipse of the high-pressure, mass-campaign methods, is emphasis on manufacturer-customer relations based upon consumer research, and a close study of markets.

Chevrolet since the end of 1928 has supplied more retail outlets than any other automobile manufacturer. Chevrolet has no distributors, as is the case with nearly all manufacturers, but deals directly with all its outlets, a situation making for coordinated action in the direction of a central program. Chevrolet's success in this regard is one of the reasons for its great strides, but changing conditions of the past three years have brought revisions even in Chevrolet's selling plans. With the discontinuance of B-O-P Sales Company, the direction of selling activities was returned to the divisions.

## USED CAR DISPOSAL FUND

No consideration of automobile marketing would be complete without reference to used cars. The year 1911 may be considered as marking the period in which used cars began to receive special consideration. In that year the Glidden-Buick Corporation was organized in New York City to sell used cars taken in exchange by the metropolitan branch, the conclusion being reached that in that area, at least, it was well to keep the merchandising of new and used cars in separate hands. Other areas operated on other plans.

In 1914, or thereabouts, used cars began to accumulate to the point of clogging trade. In that year the Curtis Publishing Company, after a searching investigation reported:

> Of all the problems that vex the automobile industry, probably the most serious and difficult of solution is that of the used cars. The second-hand car problem is likely to become more acute as the number of cars increases.

With the revival of prosperity the glut decreased somewhat, and little more was heard of it until the end of the war. Then with increased production of new cars and the strides taken in production methods, the used car problem began to assume serious proportions. The truth of certain shrewd observations which the Curtis company had made in its 1914 report began to be revealed. Many dealers proved to be better salesmen than they were buyers. The truth is that their buying experience had been somewhat limited, due to the fact that they took their cars at regular discounts from list prices fixed by the manufacturer. Also, hosts of dealers were weak in accounting, in common with most local merchants. Many of them thought that when they sold a used car for $100 less than they had allowed on a new one, they were making money, the commission on the new car being large enough to take up the slack. They overlooked the fact that in all probability they would have to furnish some service without charge, and the overhead cost was often completely disregarded. Dealers began to lose money without knowing quite why, until gradually their capital was either dissipated or frozen in the form of slow-moving merchandise on a used car lot. Two sales had to be made to produce one normal profit or, to be accurate, less than one normal profit, for interest and depreciation on cars carried over the winter cost the dealer heavily, and each spring he was likely to find himself with a carry-over of cars which had to be re-priced downward because they were a year older.

Destructive competition set in under the name of allowances, one dealer bidding more than the other. Instead of cutting the established prices on new cars, dealers effected the same result by allowing extravagant values on old cars, throwing in tires and other equipment, and guaranteeing extended free service. The used car surplus became the cause of price-cutting, direct or indirect.

Counsels of perfection, excellent advice to dealers, could make small headway against this condition. Various remedies were applied by the more progressive dealers; concentrating on the problem, they began to set up used car lots, which had the advantage of goodly space at small rentals. They organized and trained special staffs of salesmen and attendants to dispose of used cars. Manufacturers adopted various courses in assistance. They increased their advertising appropriations in an effort to bring the buyer to the dealer so convinced of a certain car's merit that he could not be moved away by competitors offering a little more for his old car. A determined effort developed to circulate among all dealers a fair price list on used cars, arrived at by expert calculation of depreciation a practice previously found effective by progressive distributors. General Motors entered upon a study of dealership costs, worked out comprehensive accounting systems for its dealers and installed them at reasonable expense through the General Motors Management Service, Inc. This activity has recently been discontinued, the reason for its existence having passed largely through the survival, under economic pressure, of those dealerships with sound accounting systems, competent staffs, and full reports from every part of their territories.

Notwithstanding all these efforts, used cars accumulated to the point at which manufacturers began considering the advisability of assisting dealers to clear the streets and lots of the more enduring and least valuable specimens.

Chevrolet was the first to act, setting up a Used Car Disposal Fund, financed by a specific levy against each new car manufactured. From this fund local dealers received credits of $25 each on certification that cars had been destroyed. Under this plan approximately 650,000 cars were junked from 1927 to 1930 inclusive. Later the National Automobile Chamber of Commerce issued a report advising action along this line, and several other companies established financial incentives for junking. Chevrolet's disbursements form a total sum not reached by any other manufacturer. The idea of the Used Car Disposal Fund originated with a Wyoming dealer who advanced it in a trade magazine.

Four out of every five sales of new cars now involve a trade-in. With used cars the percentage is lower, 40 to 50 percent. The percentage of trade-in deals has been rising in the case of both new cars and used cars for several years, but 1931 saw a slight drop in trade-ins of used cars. One of the chronic evils of the used car business is that to dispose of a used car taken in trade for a new car, the dealer in almost half his re-sales must take in another and less valuable used car as part payment for number one. Several deals may be necessary, each involving less reliable merchandise before the dealer can clear the transaction.

The partial paralysis of the market for cars in the past two years brought almost complete stability at a low level of production, as may be seen by comparing the number of cars scrapped and replaced with sales in the domestic market in 1931 and 1932. In 1913 three out of four sales represented additions to the whole number of motor cars in use, while the fourth took the place of the car previously in service. By 1924 this ratio had dropped from three-to-one, to two-to-one; by 1926 nearly half the cars produced went into replacements, and in 1927 replacements accounted for about three quarters of total production.

General Motors escaped almost entirely the pressure which this situation exerted on the industry in general in 1927. There were two reasons. One was the partial suspension of Ford activities coupled with the rapid rise of Chevrolet; the other was the favorable reception of the new Pontiac, 60,000 more Pontiacs having been sold in 1927 than in 1926. These two factors resulted in General Motors percentage of the industry's total business rising to 42.5 in 1927. Only once has this figure been exceeded for a full year's business: in 1931 the ratio was 43.3 percent.

In 1930 the American automobile industry approached a dead center, and in 1931 it passed that center, 755,000 fewer cars being produced than the number scrapped, replaced, or kept in storage by owners. In 1932 this situation was even worse due to the low new-car volume of that year. With practically no new buyers available, the strain on dealers naturally became acute, but for General Motors dealers it was somewhat relieved by the operations of the Used Car Disposal Fund, the combining of sales representations in many localities, and the Corporation's policy to prevent overloading dealers with cars for stock.

## DEALERS' STOCKS

Elsewhere has been related the growth of control over General Motors schedules in line with regular reports of dealers' stocks. Cars on hand, both with the divisions and dealers, were at peak in 1924. One must go back eleven years to 1922, to find as low a stock of cars as that of 1933. This reduction measures both the effectiveness of the Corporation's control program and the desire of the Corporation to avoid pushing its dealers into difficulties by selling them too many cars.

Along with a detailed program of sales promotion went a continuing effort to improve the financial status of dealers wherever weaknesses appeared, the inquiry being pursued tactfully, and remedies being proposed in a spirit of cooperation. Of course, where dealers who consistently fell below "bogey" the national average for dealers in the appropriate price class failed to avail themselves of the advice and assistance offered, changes in representation were made.

Another reform instituted was distinctly in the dealer's interest. It had been the habit, in the early days of the industry, for dealers to contract with the factory for a year's requirements, taking cars whenever the factory shipped them, a fact which contributed to the dealers' overstock of 1924. The new plan, which has been used since, called for firm orders from dealers for one month in advance and tentative orders for three months in advance, efforts being continued to turn the tentative orders progressively into firm orders. This has had the effect of relieving dealers of uncertainty, and in practice goes far to offset the statements often heard to the effect that factory domination has decreased dealer initiative. In this vital matter of control of shipments, dealers are freer and safer than they were ten years ago.

Until fairly recently nearly all manufacturers, with an eye to production totals, were disposed to crowd cars upon dealers in quantities likely to prove embarrassing in a falling market. In the early days of the industry this was necessary, because manufacturing and storage facilities and financial capacities were insufficient to meet the seasonal demands of spring and early summer. A beneficial change in this regard came with the rise of closed cars to favor. At present the market takes approximately 44 percent of its automobiles in March, April, May, and June, and the first six months of the year develop 60 percent of the annual sales. The business is still seasonal, but slightly less so than it used to be. However, with mass production methods to the fore, the temptation arose to force cars on dealers faster than they could sell them, using a "floor plan" loan to finance all or part of the transaction. A turn of the economic pendulum soon revealed the danger. Overloaded with new cars which must be sold soon after receipt, a dealer becomes vulnerable to the point where he must accept at uneconomic valuations more used cars than he can dispose of quickly and profitably. After one experience of this nature General Motors has set its face against overloading dealers, in the confident belief that the Corporation should not seek to improve its position at the expense of, or to the embarrassment of, its dealers, upon whose continued solvency and aggressiveness so much depends.

## GENERAL MOTORS HOLDING CORPORATION

Another effort of the Corporation to assist dealers goes forward under the General Motors Holding Corporation, formed on June 22, 1929, to invest in dealerships where more capital seemed to be required. The Holding Corporation maintains six regional offices: in New York City, Chicago, San Francisco, Dallas, Atlanta, and Detroit. After four years of operations it has invested Corporation funds in dealership companies, in circumstances and under conditions which seem at once to safeguard the investment and promote effective representation.

Applications for the Holding Corporation's cooperation come through the sales offices of the car divisions which base their recommendations upon a careful investigation of the dealer's situation, character, and prospects. A check-up by the Holding Corporation is then made. Usually the Holding Corporation's capital is used either to retire silent partner interests and/or bank loans, or to increase facilities where they seem inadequate to the potential market. Holding Corporation investments are not regarded as permanent, the expectation being that the dealer's share of the profits will retire the Holding Corporation's investment. The Holding Corporation is regarded as operating a revolving fund to be used capital-wise to improve sales representation which, except for under-capitalization, is healthy and well-managed. While General Motors' interest remains in the investment, the Holding Corporation exercises a considerable measure of control, and endeavors to reduce expenses and increase profits.

## GENERAL MOTORS FLEET SALES CORPORATION

The special problem of fleet sales to large users resulted in the incorporation of General Motors Fleet Sales Corporation in 1930. Deliveries were handled through dealers, under a special contract, but Fleet Sales Corporation dealt directly with national and state governments. The fleet users' agreement applied only to operators of at least 100 cars who agreed to buy a minimum of $20,000 worth of General Motors cars and trucks annually.

In 1933 the minimum Fleet sales contract was reduced to $15,000; also dealers began to deal directly with national and state governments.

With the formation of this corporation, the entire fleet activities of individual General Motors divisions were taken over by the General Motors Fleet Sales Corporation. This consolidated plan of selling resulted in better sales coverage of large national fleet users at a lower sales cost, due to a reduced total personnel. In addition, it made it possible for one General Motors organization, specializing in a particular field, to sell the entire line of cars and trucks to the large fleet user, and thus better fit General Motors products to each of the transportation needs of the user.

This corporation also handles the sale of General Motors cars and trucks to the United States government, the various states, and the District of Columbia. During 1932, 2,101 General Motors cars and trucks were sold to the national government, for $1,404,330. In the first seven months of 1933, government awards to General Motors totaled $5,730,634, representing 11,917 cars and trucks. This large increase is principally due to the Reforestation program, and the motorizing of certain sections of the National Guard. Awarding of government orders proceeds under strict competition both as to price and quality, yet General Motors secured 78 percent of this business on the basis of units, and 75 percent on the basis of dollar volume. The outstanding item in 1933 government awards was Chevrolet.

Dr. Carlos C. Booth, first American physician to use an automobile in his practice. Car designed and assembled by himself, 1896

Even in the automotive industry itself, relatively few persons comprehend how the fleet business has expanded. Within five years General Motors divisions have sold 64,339 units to 232 companies. Each of 84 companies has purchased an average of 567 Chevrolets during this period, the heaviest buyer taking 3,665 units. In the same period, 3,905 General Motors trucks have gone to 41 fleet users, the average buyer taking 95. Truck business, including both Chevrolet and General Motors trucks, runs at about 25 percent of total fleet sales, passenger cars for business purposes, 75 percent.

Through its studies of mass transportation problems and the special needs of large users, General Motors Fleet Sales Corporation has brought elements of strength into dealer relations with the larger purchasers of automobiles.

It has been said that the first automobile was bought, not sold. That was probably true, also, of the 10,000th automobile; a demonstration sufficing to convince the customer, but the car had to be sold to the dealer. Today, with dealer organizations spanning the world, automobiles have so increased in quality and numbers that there must be a closer relation between output and potential sales, between manufacturer and consumer, between dealer welfare and producer's prosperity. In the enlightened merchandising of the future, the indirect approach based upon consumer research and close knowledge of market needs is likely to replace the former emphasis on high-pressure salesmanship.

Some of the glamor of the pioneer days has vanished as the automobile trade has become established as a huge and stable business, operated on scientific lines with all the safeguards which can reasonably be set up for the protection of its representatives in a freely competitive market. But the marketing of automobiles still attracts large numbers of keen, driving men, full of confidence that the automobile will reward their efforts amply as the years roll on.

# Chapter XXVII FINANCING AND INSURING THE BUYER

AUTOMOBILES, as we have seen, went from factory to dealer on a cash basis in the early days of the industry, and were generally sold by the dealers for cash. Dealers soon discovered, however, that some excellent sales required credit, and they began to accommodate purchasers with good credit or satisfactory records under special circumstances. As they found more sales could be made where credit was available, the more daring of them began to work out plans for instalment selling, getting a substantial down payment, and holding title to the car until it had been completely paid for. It is said that the first systematic use of instalment selling in the automobile field occurred in San Francisco in 1913, but this priority is disputed by those aware of similar early arrangements elsewhere. A trade growing vigorously, with so many keen, driving men at work over so wide an area, was certain sooner or later to find ways of using credit in large sums.

The basis of these instalment transactions was, of course, the dealer's knowledge and vital interest in the collectibility of his account; this element is continued under the recourse or full endorsement plan wherein the dealer's responsibility remains behind the obligation until it is paid in full.

There grew up, naturally, finance companies willing to purchase automobile paper from dealers at a discount on well-sold and well-insured cars. Bankers might not feel able to take automobile paper in quantity, but they were quite willing to lend money to sound finance companies so that the latter could do a large volume of business on comparatively small capital. In general, these specialized finance companies prospered, because their differentials or charges (the difference between the cash and the time or instalment purchase price) were relatively high, and the risks comparatively low. As experience accumulated, it appeared that a good, insured automobile in the possession of an honest man or woman, and under the complete legal protection with which all states surround property held under lien, was a fairly safe investment, provided that down payment and schedule of payments anticipated the depreciation of the security in use.

The finance companies developed another line of automobile investment that of financing dealers' stocks. The day had passed in which one sample car represented the only large investment the dealer had to carry. As the factories enlarged their lines, the dealer could provide adequate representation only by showing each model in its various body styles. His own capital and earnings were usually absorbed in his growing business, providing more floor space for showing and stocking cars, a larger supply of parts, better service facilities, and of course used cars taken in trade. Arrangements were made for stocking new cars under a variety of agreements which were generally known as "floor plans." When these cars were sold for cash, the liens on them were quickly released. At first it was necessary for the dealer to pay and secure title to the car before he sold it at retail on instalments, but as the business developed many transactions consisted of merely accepting the retail contract in part or full payment of the wholesale obligations. The dealer, of course, remained liable for all unpaid balances, so that the finance company always had as security back of the transaction the merchandise itself adequately insured, the credit of the purchaser, and the general credit of the dealer.

The keen financial mind of John J. Raskob began to work on this financing problem shortly after he entered General Motors. Need for a finance service for the dealer as well as the retail purchaser at moderate time-price differentials, and functioning uniformly as to policies and rules, was most apparent and was urged upon the Corporation from many quarters. If General Motors could set up the machinery to provide such service at reasonable profit, the result would be increased sales, improved dealer earnings, and greater consumer satisfaction. Accordingly, on January 29, 1919, after a thorough study of the situation, and weighing both the advantages and hazards of the innovation, General Motors Acceptance Corporation was incorporated under the investment section of the banking law of the State of New York, with 20,000 shares of Common stock, $100 par. General Motors Corporation bought this stock at $125 per share, enabling the new subsidiary to start life with a capital of $2,000,000, and a surplus of $500,000. The Corporation's subsequent purchases of GMAC stock have been at that average price, and the sums involved were distributed between capital and surplus in that proportion.

GMAC'S purpose has been condensed to this statement:

> Briefly the aims and purposes of the Corporation are to assist, through the proper application of the credit function, in the orderly distribution and sale of the products of the General Motors Corporation in such a manner as will make for a sound and healthy manufacturing and merchandising condition.
>
> It being foreseen that credit would inevitably play a most important part in the development of the automobile market, it was the aim of General Motors Corporation to keep within its control, to the greatest extent possible, every factor of importance in its development. At the same time it was understood that the Corporation (GMAC) was to function quite independently and be free from any pressure of expediency which might impair its character and purpose, and so defeat the very aims which brought it into existence.

In order to fulfill these functions the organizers of GMAC developed complete plans for financing sales of General Motors cars and products. It set up also a Financial Sales Department, through which the obligations of the Acceptance Corporation were offered independently to banks and investors throughout the country, thus creating a stable source of accommodation through the use of which additional business would be possible beyond the extent of GMAC'S own capital. In this department are centered all operations through which the Corporation

(GMAC) finances itself, including not only its sales activities, which parallel somewhat the lines of operation employed by commercial paper brokers, but all its bank relations and day-to-day borrowing."

Under the GMAC wholesale plan, General Motors distributors and dealers, after credit has been established, may purchase new passenger cars, commercial vehicles, and other products of the Corporation directly from the sales Companies of General Motors by paying a small amount in cash; the balance as the cars or products are released from trust, or at an agreed date after shipment. Merchandise so financed may be stored in the dealer's showroom for display, in warehouses under the control of the dealer, or in public licensed warehouses under pledge of warehouse receipt.

General Motors Acceptance Corporation retains title to the products financed until full payment is made. The plan provides that a dealer may pay the amount due on a car, and secure immediate release. After release the dealer has full title to the car.

The GMAC wholesale charge includes insurance protection for the dealer for the full laid-down price of the car against loss or damage arising through fire or total theft.

Under the GMAC retail plan, General Motors dealers are urged to sell the products of the Corporation to customers in good credit standing upon terms properly suited to the purchaser's income. The buyer pays a portion in cash, and/or trade-in (usually from 30 to 50 percent, depending upon his circumstances) and gives an obligation for the remainder payable in equal instalments adjusted to his income. If the transaction is in accordance with sound credit merchandising policy, and the resultant obligation is suitable, GMAC purchases the time sales contract from the dealer, and carries it to maturity.

General Motors Acceptance Corporation has recourse to the dealer on all obligations of purchasers which are bought by the Corporation. The deferred payments are made by the purchaser direct to the Acceptance Corporation, relieving the dealer of routine collection details. Adequate reserves are set up to protect the dealer as well as GMAC against loss.

The GMAC retail plan provides fire and theft insurance supplied by the General Exchange Insurance Corporation, protecting the purchaser and dealer as their interests may appear. A policy is issued to each purchaser so that he may be properly informed of the exact nature of his protection. The purchaser may secure, at his option, collision, or all property damage insurance.

The total retail GMAC financing differential varies according to territory for the reason that insurance protection is included as part of the differential, and insurance premiums vary according to territory and type of car. Low differentials are made possible by large volume and efficient and economical operation.

In contrast with other forms of instalment sales, the amount of the differential in the time price paid by the purchaser depends upon the term and amount of his obligation. He pays only in proportion to the accommodation he receives. The larger the down-payment, and the fewer the number of monthly payments, the less the differential between the cash and time prices.

GMAC recognizes no fixed terms GMAC believes that the sale of cars on deferred payments as well as for cash should be based primarily upon the quality of the car at the price, and not on the lure of so-called "easy terms, " terms being regulated to the income and circumstances of the purchaser, and its financing operations are carried on in accordance with this principle.

In The Economics of Instalment Selling[1] Professor E. R. A. Seligman describes the recourse, the non-recourse and the repurchase system of adjusting the difficulties arising from delinquencies:

> The recourse system is so called because the customer's notes, which are handed over to the finance company by the dealer, are endorsed by the dealer so that the finance company, in case of the purchaser's failure to pay, will have recourse against the

---

[1] Harper & Brothers, New York City, 1927

> dealer. The dealer, therefore, in order to discharge his liability to the finance company, is compelled to take steps looking to the repossession of the car (unless, indeed, as is sometimes done, the finance company relieves him of this responsibility and repossesses the car for him) which he may then sell as a used car to meet his liability to the finance company. The dealer thus assumes the responsibility involved. If he sells the used car for more than the balance of unpaid instalments, he makes a profit; if he sells it for less than this balance, he undergoes a loss. The important factor, therefore, is the difference between the actual value and the amount of unpaid instalments.
>
> Opposed to the recourse system is the non-recourse system, or in other words, the system whereby the finance company has no recourse against the dealer. The finance company here assumes all of the responsibility. It not only makes the original advance to the dealer and receives the instalments as they become due but, in case of default, institutes its own methods for securing repossession of the car; and subsequently, after salvaging and reconditioning the car, the finance company itself disposes of the used car in the market. Under this system the dealer becomes to all intents and purposes the agent of the finance company for selling a new car. The decision, and frequently the investigation, as to whether the individual purchaser is a good risk is relegated to the finance company; and virtually all that remains for the dealer to do is to display the car, and to arrange with the purchaser as to terms, which in most cases are the standard terms fixed by the finance company.
>
> The third or intermediate plan is the repurchase method. Under this plan the finance company has no recourse as such against the dealer, but enters into a contract with him whereby he agrees, in the case of repossession by the finance company, to repurchase the car. In this way the finance company is saved the necessity of attempting to dispose of the used car, and limits itself, in a large measure at least, to its purely financial function.

In principle, GMAC has held firmly to the recourse system, continuing the dealer's responsibility in the transaction until it is finally closed. This is considered highly important, since it means that control of credit sales is thus primarily in the hands of the dealer, his judgment being supplemented by the broad experience of a highly trained credit organization. It follows that losses therefore will be kept to a minimum, because the dealer will exercise care in selecting his instalment customers. This fundamental difference between recourse and non-recourse companies accounts for the higher differentials generally charged by the latter, since they are not in a position, as the recourse companies are, to reduce risk by keeping the dealer back of the paper originated by his sale.

Under a strict recourse plan, no losses could theoretically be sustained by the finance company except through dealer insolvency. From inception until August i, 1925, under the GMAC plan the dealer had to meet all losses up to the limit of his ability on certain risks against which he could not well protect himself cheaply and efficiently. These risks of conversion, confiscation, and collision being insurable, GMAC covered the risks by insurance and relieved the dealer.

From 1919 to 1932 inclusive, GMAC purchased in the United States and Canada instalment contracts on 3,775,777 new General Motors cars, or 32.6 percent of all General Motors dealers' sales to users, and 3,564,468 used cars. Under the wholesale plan it "floor planned" for dealers 3,777,703 new cars or 32.7 percent of General Motors sales to dealers.

The record reveals a sudden rise in GMAC contract purchases from 1925 to 1926. Before 1925, approximately 20 percent of new General Motors cars were sold in the United States and Canada on the retail GMAC plan. Since 1925 the average has been about 40 percent. General Motors production turned sharply upward in 1926. Furthermore, GMAC'S service had been augmented in 1925 by the creation of an insurance company; and by the inclusion in its financing plan of protection for the dealer against the hazards of conversion, confiscation, and collision. As time sales became more popular the percentage of a dealer's business represented by instalment sales increased and, likewise, the repossession risks in relation to his resources. Therefore GMAC instituted a reserve for the dealer to protect him against the costs and losses involved in handling repossessions; this dealer-reserve probably was a big factor in the increase in GMAC'S volume after 1925.

Since, in addition to the moral factor involved, the security and liquidity of any instalment contract is definitely influenced by the size of the down payment and the schedule of instalments, the effort has been to avoid long commitments and small down payments. The most inclusive category is that which has one third to one half of cash value paid down with the balance spread over twelve months.

It must not be inferred that the comparatively small losses result altogether from the relatively safe character of the business; on the contrary, they are kept in control by good management exercised at many points in the transaction. By no means is every car paid for on schedule, and when payments fall behind it is a nice problem whether to extend or renew the paper or to repossess the product. In all such cases the GMAC policy is to permit the product to remain with the purchaser whenever it is reasonably safe to do so, and whenever it appears that with considerate and helpful service the purchaser can and will pay. The result is that retail renewals, in consideration of unusual circumstances, became necessary at times. The Corporation's policy in this regard is regulated by conditions, with due thought given to both the interests of the dealer and the purchaser.

To finance the growing volume of business, GMAC sold $50,000,000 of 5 percent serial gold notes, dated March 1, 1926, and maturing at the rate of $5,000,000 annually, through a banking syndicate composed of J. P. Morgan & Company, the First National Bank, the Bankers Trust Company, and The National City Company, all of New York. Before this sale GMAC had been depositing its domestic receivables under a trust deed, against which security collateral gold notes were issued. From 1919 to August, 1925, the trustee holding the documents as security for GMAC collateral gold notes was the Irving Bank Columbia Trust Company. A personal trusteeship was arranged in 1925 as a matter of economy. The $50,000,000 long-term loan of March, 1926, was not secured by collateral in trust, but an indenture in behalf of the holders of the notes was executed to the Bankers Trust Company. Since April 1, 1926, GMAC has not pledged its domestic and Canadian collateral. The previous trustee then became comptroller; receivables were segregated under his control. The comptroller certifies to the possession of live receivables and/or cash equal to GMAC'S obligations.

Short-term borrowings were further funded on February I, 1927 when through the same banking syndicate $50,000,000 of ten-year sinking fund 6 percent gold debentures were sold to the public.

The Financial Sales Department, handling domestic borrowing operations, has established a national market for its short-term paper through banking institutions in every state in the Union. More than 10,000 banks have purchased these notes, this clientele standing at 7,200 on December 31, 1932. Other customers are insurance companies, corporations, and individual investors. Notes are offered in denominations of $500 to $1,000,000 at prevailing discount rates for commercial paper and are payable in many of the principal cities of the country.

Canadian business is financed, primarily, by bank credit obtained in Canada. Financing of overseas branches is handled through the Overseas Bank Relations Department, the funds coming from both domestic and overseas banking institutions.

In addition to automobile paper, which forms the bulk of its business, GMAC has retail sales plans for the purchase of paper arising out of sales of other General Motors products, including:

> Delco farm-lighting and power plants, oil burners, radios, Delcogas machines all manufactured by Delco Appliance Corporation of Rochester, New York; Frigidaire household electric refrigerators, electrical refrigeration for commercial establishments, water coolers, ice cream cabinets, bottle coolers, and air conditioning equipment manufactured by the Frigidaire Corporation, Dayton, Ohio.

## INSURANCE

The financing of motor-car purchases has always involved the question of adequate insurance protection. Serious damage to the automobile, from any cause whatsoever, affects in many cases the willingness and even the ability of the buyer to pay out what is still due on the contract. Prior to August, 1925, GMAC did not require insurance, but left the matter of proper insurance entirely up to the buyer and the dealer. General Motors, through General Exchange Corporation, maintained a service which assisted and advised dealers in their insurance problems, counseled with the manufacturing divisions on matters of construction, and assisted buyers and dealers to secure adequate insurance protection through reliable sources. It reached a volume of $1,200,000 in annual premiums in 1924, writing policies on 250,000 cars. Its experience resulted in the extension of such coverages as confiscation, conversion, and single-interest collision, and other forms of protection to meet the needs of the dealer in connection with his time sales. Analysis of many claims resulted in recommending changes in construction and led to the improvement of antitheft locks, better wiring, and other improvements built into each General Motors car.

Until succeeded by General Exchange Insurance Corporation, GEC built up a body of experienced insurance personnel and statistical knowledge which was of great value to its successor. Its general manager was W. A. Edgar from 1920 to 1922, when he was succeeded by Livingston L. Short, who later became president of the General Exchange Insurance Corporation. Curtis C. Cooper was president of GEC and chairman of GEIC'S board during his service as president of GMAC.

Accumulating experience left no doubt that insurance was urgently needed to protect the various interests of the buyer, the dealer, and the purchaser of the contract in all instalment sales of automobiles. Only 57 percent of buyers were voluntarily insuring their cars for their own protection. Thousands of instalment buyers suffered heavy losses without recovery, and of these many were financially unable to replace or repair their cars. As a result General Motors and its dealers lost many customers, and both the dealer and GMAC were in a position to lose directly, since sums payable on seriously damaged, stolen, or destroyed cars were frequently left unpaid.

Once the situation had been thoroughly canvassed, it appeared evident that the buyers of General Motors products, the dealers and General Motors itself had so much at stake in the provision of adequate automobile insurance that it became necessary to go beyond the service offered by General Exchange Corporation. Accordingly, the formation of General Exchange Insurance Corporation was authorized on June n, 1925, with a capital of $500,000, and surplus of $1,000,000. GEIC began to function in August, 1925, being chartered under the insurance laws of the State of New York. Within a few months it was admitted to do business in all the states, the territories of Alaska and Hawaii and the Dominion of Canada, offering insurance coverage in each policy protecting the interests of the buyer, the dealer, and GMAC. The success of this enterprise was immediate. On June 30, 1927, after little more than two years of operation, the net worth of the company had grown to $2,176,000 from the original investment of $1,500,000. The capital of GEIC has since been increased to $1,000,000, and the surplus to $1,500,000.

The insurance company has gone on expanding. In practically every year since it began operations it has written more automobile insurance than any other company. In 1929 there were 495 stock insurance companies in business writing all lines of fire insurance. During 1928 the General Exchange Insurance Corporation stood tenth as to dividends paid, twenty-second as to volume of premiums written, thirty-ninth as to assets, and forty-ninth as to capital funds, although many of the other companies insure classes of merchandise which are not included in GEIC coverages. Furthermore, GEIC solicits neither renewals on fully paid for cars nor policies on cars sold for cash, although a small volume of this class of business is written which comes to it unsolicited.

Field experience showed the necessity for prompt settlements. Coverage had to be dependable and loss-settlements very prompt, in order that the buyer of the car could have his car repaired as soon as possible after a loss occurred. Ill-will quickly develops in the case of delayed settlements, the owner's impatience turning into criticism of the car itself, the dealer, and the Corporation which stands behind both. Slow settlement of claims is a blow to the prestige of an insurance company, for delay in securing service or needed parts keeps the buyer out of his car, on which his pleasure, and frequently his business, depend. Obviously, these considerations require special treatment, and are largely responsible for the policy which GEIC follows in settling claims; namely, to give the insured exactly that to which he is entitled under his insurance policy, and to make it available as quickly as possible.

The speed of settlement is quite remarkable. More than 7,000 claims are settled in an average month by GEIC'S own personnel in the field. Although settlements are made in distant areas and with motorists away from their place of residence, only 15 percent of claims against GEIC (other than total thefts) remain unsettled more than thirty days after date of loss. This includes suspicious and controversial claims. Total thefts, however, are subject to the usual sixty-day waiting period, a procedure followed by all insurance companies in order to let the police act toward recovery of the stolen car. Unless the car is recovered within sixty days, the claim is paid on a total loss basis, the purchaser receiving a sum equal to the value of his car at the time of loss. GEIC acts so promptly in these cases that within five days after the waiting period 95 percent of the owners receive checks covering their losses.

In settling 85,000 insurance claims a year for a gross sum of $5,000,000 GEIC comes in contact with many purchasers under conditions making directly for building of good-will. The staff adjuster, who is a GEIC employee, visits the assured in his home or office promptly, and arranges a fair settlement without delay. Where repairs are necessary, the work is done wherever possible by the dealer who sold the car, assuring the purchaser in this way the use of genuine parts, and authorized General

Motors service. Upon completion of the work, payment for the repairs is usually made by GEIC directly to the dealer. A satisfactory claim settlement and satisfactory repairs assist the dealer in cementing his relationship, and that of General Motors, to the buyer.

Among the coverages which were added to the GEIC'S service in 1930 with the approval of both dealers and buyers is the Accidental Physical Damage (A.P.D.) policy covering damage to automobiles from all the usual causes, and some most unusual ones, including floods, tornadoes, riot, civil disturbances., and airplanes. In the A.P.D. policy collision insurance is included. The advantage to the buyer of this type of policy is apparent, since it results in the almost automatic recovery of his means of transportation with the least possible delay. The advantage to the dealer is also evident, since this insurance provides cash for the payment of repairs. The Accidental Physical Damage policy is regularly offered by GMAC as an integral part of its plan. Buyers may waive the collision feature if they so desire, but more and more are availing themselves of it.

When General Motors entered the insurance business, the wisdom of the move was questioned by many insurance men, even though the Corporation made clear its intention to operate only in one highly specialized field. It was said that GEIC'S business would be at the expense of other companies. This has not been the case. Through the united efforts of GMAC, GEC, and GEIC, the education of the motoring public to the benefits of automobile insurance has been pushed decisively, with the result that the business of other companies has increased along with that of GEIC. Other companies have benefited by writing policies on renewals after GMAC has been paid out, and on cars sold for cash to persons initiated into the merits of automobile insurance by GMAC requirements on a former purchase.

The growth of the General Motors Acceptance Corporation since 1919 follows the upward curve of the industry and the Corporation's business in an accelerated tempo as the popularity of instalment buying has increased. Due to purchases of its stock by the General Motors Corporation year by year until 1929, plus a transfer of $13,750,000 from undivided profits to capital and surplus, GMAC capital and surplus rose from the original figure of $2,500,000 in 1919 to $70,000,000 in 1929. In the same period time sales transactions, both wholesale and retail, originating with General Motors dealers, and financed by GMAC, increased from $20,881,000 in 1919 to $1,133,117,000 in 1929, the peak figure of GMAC history.

Starting with six branch offices in 1919, and 336 employees, it grew to 107 branch offices in 1930, with seventy-three in the United States, nine in Canada, and twenty-five overseas. The maximum number of employees was reached in 1929, with 5,532. During 1931 and 1932, twenty-one branches were closed, and the personnel stood at 3,128 on December 31, 1932.

Under the most adverse conditions of the depression, loss ratios on automobiles have been kept down to a point at which the discounting of an automobile sale is demonstrated to be one of the safest forms of business developed. The advantages to General Motors resulting from this large financial operation, and the insurance business of its subsidiary, are not confined to its balance sheet. By offering to dealers plans which provide reasonable accommodation and adequate protection to themselves and to the buying public, GMAC has assisted the parent corporation to reach new high levels of production, built goodwill for the Corporation's products, and strengthened the morale of its selling force the world over.

General Motors Acceptance Corporation and General Exchange Insurance Corporation have both prospered. GMAC'S rate of earnings depends, of course, upon the general volume of trade and the amount of money it can use. Since trade declined in 1930, GMAC borrowings from banks have declined simply because it could find no use for the money freely offered to it at low rates. The Corporation called for redemption on August 1, 1932, $5,000,000 of its 6 percent Debentures, and on February 1, 1933, it redeemed all of the remaining $30,000,000 of its 6 percent Debentures.

The present chairman of the board of General Motors Acceptance Corporation is Mr. Alfred H. Swayne, elected in March, 1921. The president of GMAC from its incorporation in 1919 until March, 1921, was J. Amory Haskell, C. C. Cooper then being general counsel. The latter succeeded Mr. Haskell in the presidency and remained there until October, 1929, when he was in turn succeeded by John J. Schumann, Jr.

# Chapter XXVIII COOPERATIVE PLANS

BEHIND all cooperative plans is the desire to arouse, for mutual benefit, the individual interest of all who share in the work. General Motors has fostered several plans which, after close study, promised to be beneficial both to the Corporation and to its employees. The extra rewards of General Motors have been based on a desire to meet, as far as could be foreseen, and within the frame of other primary corporate interests, the needs of each of several well-defined groups.

## BONUS PLAN

In 1918 the Bonus Plan was adopted. This plan provided for annual awards of General Motors Common stock, or its equivalent, to employees who had contributed to the success and prosperity of the Corporation in some special degree by reason of their ability, industry, and loyalty. During the first four years of this plan, awards were made in two divisions (1) Senior Awards to employees earning $2,400 and over, and (2) Junior Awards to all other employees. The plan at this time provided for a bonus fund equal to 10 percent of the net profits of the Corporation after deducting 6 percent on the net capital employed. Beginning in 1922 the amount set aside was 10 percent after deducting 7 percent on the net capital employed, and employees with salaries of $5,000, and over, were eligible for bonus awards. Although other features have been changed, the 10 percent deduction of net profits after capital service has been maintained for bonus purposes. In 1931, employees earning $4,200, and over, were eligible for bonus awards. Beginning in 1923, at which time the Managers Securities Company was organized, the amount set aside for the bonus plan was 5 percent of net earnings, after 7 percent on net capital employed, which amount was distributed in the form of Common stock through the Bonus Plan, the other 5 percent being paid under contract to the Managers Securities Company. Beginning in 1930, the contract with the Managers Securities Company having been terminated, the full amount of the Corporation's bonus provision of 10 percent of the net profits of the Corporation, after the deduction of 7 percent on the net capital employed, was paid to the General Motors Management Corporation. One half of this payment takes the form of a subscription by General Motors Corporation for Class A stock of the General Motors Management Corporation. The Class A stock so acquired is distributed to employees in accordance with the Bonus Plan. Class A stock received by employees is equivalent share for share to General Motors Common stock, and is convertible into the Common stock of General Motors Corporation at the option of the recipient. The General Motors Common stock so used for bonus purposes is purchased in the open market.

Since the inauguration of the Bonus Plan, it has received the commendation of economists and other authorities as being one of the best plans as yet developed. Perfection, however, has not been attained; changes have been made, and more will be made. It is worthy of note, however, that the bonus plan is entirely based on rewarding employees with stock, not cash, thereby setting the stage for a continuing property interest in the Corporation, and also that a real effort is made to discover meritorious performance.

Bonus stock is delivered to employees in four lots, one fourth at the time of the award, and the remaining three fourths is held in trust to be delivered in equal instalments at the end of each of the three following years. As 7 percent

Cooperative Plans

was not earned on the capital employed in 1932, there were no bonus payments in that year. A record of the awards follows:

| Year | Number of Bonus Awards | Shares of Common Stock (b) | Year | Number of Bonus Awards | Shares of Common Stock (b) |
|---|---|---|---|---|---|
| 1918 | 3,884 | 490,238 | 1926 | 1,513 | 428,170 |
| 1919 | 6,453 | 402,485(c) | 1927 | 1,998 | 272,798 |
| 1920 | 6,578 | 159,312(c) | 1928 | 2,513 | 195,570 |
| 1921 | (a) | (a) | 1929 | 2,840 | 167,378 |
| 1922 | 550 | 179,732 | 1930 | 1,929 | 117,624(d) |
| 1923 | 647 | 226,278 | 1931 | 1,378 | 65,954(d) |
| 1924 | 676 | 115,272 | 1932 | (a) | (a) |
| 1925 | 943 | 345320 | TOTAL | **31,902** | **3,166,131** |

(a) No bonus was available for the years 1921 and 1932.

(b) Equivalent number of shares on basis of $10 par value Common stock.

(c) In addition to the Common stock awarded in 1919 and 1920, 18,934 shares of 7 percent Preferred stock were awarded, of which 14,191 shares applied to the 1919 awards, and 4,743 shares to the 1920 awards.

(d) Awards in 1930 and 1931 in Class A stock of General Motors Management Corporation equivalent share for share to General Motors Common stock.

## MANAGERS SECURITIES COMPANY

The Managers Securities Company, as the name implies, was an effort to reward, consistent with their responsibilities, those holding the more important managerial positions.

Decentralized organization, the term applied to General Motors system of management, presents a problem of effective control as between the several operating units and the Corporation as a whole. In the decentralized system the manager of a unit is the chief executive, the equivalent of the president of a separate corporation. Conceivably he could conduct the affairs of his unit with little regard for the other units of the group, if he thought that in this way his own unit would most benefit. The Corporation on the other hand cannot well have one unit reap advantage at the expense of another; also it is desirous that the benefits obtained by one unit shall be made available to the others, so that each one may be advancing not only its own but the common good of all, which, of course, is the Corporation itself. These benefits of counsel, cooperation, and teamwork are necessary for effective control, and it was apparent that they could be secured best by enlisting individual self-interest.

The tendency for management to become divorced from stock ownership in large enterprises is a commonplace of industrial evolution. Both advantages and disadvantages of this trend have been noted. Among the advantages frequently cited are these: creation of broader public interest in the Corporation through wide distribution of its stock, of especial moment to a manufacturer whose goods are in everyday use by millions of persons; the greater freedom which management enjoys, in its dealings with the market and its employees, when personal participation in profit by management is relatively small compared to the whole stockholder interest. The point has been made that a corporate management so circumstanced can develop a balanced policy based on long-range views toward the well-being of each of the three main interests toward which it is responsible: toward the market into which it brings goods and services to the public; toward its employees to whom it supplies the means of life and improving status; toward the stockholders for whom it earns dividends proportionate to their investments in the property.

On the other hand, it has been noted that too complete a separation of ownership from management may result in loss of interest and wasteful administration, since one of the direct incentives to efficiency is lacking when management's share in profits is too small to stimulate to utmost efforts. This feature received unusual attention in the autumn of 1923. Mr. Pierre S. du Pont had retired from the presidency of General Motors with E. I. du Pont de Nemours & Company owning directly, or indirectly through its affiliate General Motors Securities Company 7,500,000 shares of General Motors Common stock, or more than one third of the total issue of 20,646,397 shares then outstanding. This represented a higher percentage of the Common stock than the members of the great Delaware family and their various corporations have held since. The block of 7,500,000 shares represented not only what the du Pont interests had bought as a matter of settled policy, but also their purchase in the 1920 emergency.

In 1923 the time seemed propitious to reduce this great block by permitting senior executives of General Motors to secure holdings, thereby insuring keen personal initiative, and continuity of management. Accordingly General Motors organized the Managers Securities Company of Delaware with an authorized capital stock of $33,800,000, divided into three classes: $28,800,000 7 percent cumulative nonvoting convertible Preferred stock; $4,000,000 Class A stock; and $1,000,000 Class B stock. The du Pont Company agreed to sell through its affiliate, the General Motors Securities Company, which held as its sole asset 7,500,000 shares of the Common stock of General Motors Corporation, 30 percent of its holdings to Managers Securities Company. Therefore 2,250,000 shares of General Motors Common stock were sold to the Managers Securities Company, on October 15, 1923. Payment for this stock was made with $28,800,000 7 percent Preferred stock of the Managers Securities Company, and $4,950,000 in cash, making a total of $33,750,000, or $15 per share.

General Motors subscribed for all the Class A and Class B stock of the Managers Securities Company, paying $5,000,000 in cash, and entered into a contract with the Managers Securities Company, agreeing to pay to the latter each year from 1923 to 1930, inclusive, 5 percent of its earnings in excess of 7 percent on the net capital employed.

Under the trust indenture the Managers Securities Company agreed to pay to the trustee annually a minimum amount of $1,750,000, and such portions of the remaining amounts earned under its contract as were in excess of normal income taxes, dividends on outstanding 7 percent Preferred stock, and dividends on the Class A and B stocks limited to 7 percent as long as there were outstanding 7 percent bonds or Preferred stock. General Motors Corporation thereupon sold to certain managers of the General Motors Corporation, and subsidiaries, such amounts of Class A and Class B stock as were determined by a special committee of the board of directors of General Motors Corporation.

The plan was a significant example of Corporation thought and practice. It provided for all the contingencies which could be foreseen, such as changes in personnel through resignation, dismissals and death, with provision that the determination of the Finance Committee should prevail in all matters of discretion.

In all, eighty men were selected, to whom Class A and Class B stocks were sold in varying amounts. A unit of 200 shares of Class A stock, and 200 shares of Class B stock cost $25,000. The number of units allotted to an individual depended upon his significance to the Corporation.

Cash dividends approximating $350,000 were paid to each unit of Class A and Class B stock during the seven years of the company's contractual life. At the start, the holder of one unit received dividends of only $600 a quarter, the balance of earnings going to debt service. As the debt was reduced, the quarterly dividends paid to the managers rose. The sums devoted to debt service reduced the company's obligation to such a point that in 1926 the balance of the purchase price was financed at a lower interest rate. The original indebtedness of $28,800,000 was fully paid by 1927.

The rapid rise in earning power of General Motors Corporation during these years brought a phenomenal increase in the value of these units. Each beneficiary of the plan receives eventually, for each original unit bought, 22,545 shares of $10 par value Common stock; and even at the low price prevailing in 1933 the value of an original unit was more than ten times its original cost.

It has been said that General Motors made many men independently rich through its Managers Securities plan. Of course, heaping up wealth for them so abundantly and swiftly was not contemplated when the proposal was made and the plan drawn. At that time the prospect was that the managers, with good luck and the best of team-work, would reap perhaps a fraction of what they actually received. No one could anticipate the rosy future which was destined for the Corporation and the country at large during the six years that followed. To what extent the stimulation of managerial interest added to profits cannot be determined accurately, but this was probably a very considerable factor in the strides which the Corporation took. To some extent, the managers rewarded had created their own fortunes.

The Corporation's contract with the Managers Securities Company was to have expired on December 31, 1930, but was terminated by agreement on December 31, 1929, in order to organize the General Motors Management Corporation along somewhat the same lines as the Managers Securities Company.

## GENERAL MOTORS MANAGEMENT CORPORATION

With the termination of the contractual relationship with the Managers Securities Company as of December 31, 1929, the advantages of the plan were so manifest that the General Motors Management Corporation, organized in March, 1930, consolidated the Corporation's two previously mentioned profit-sharing plans.

The General Motors Management Corporation was formed in Delaware with an authorized capital of $10,500,000, consisting of 50,000 shares of Common stock, 500,000 shares of Class A stock, 500,000 shares of Class B stock, all of $10 par value.

This capital was authorized to cover the functions previously carried on under the Managers Securities Company and the Bonus Plan, to carry both profit-sharing arrangements forward for seven years, to March, 1937.

General Motors Corporation sold to the Management Corporation 1,375,000 shares of General Motors Common stock at $40 per share. This stock had been purchased from time to time in the open market by General Motors Corporation in anticipation of a profit-sharing plan to replace the Managers Securities Company. The Management Corporation financed this purchase by the sale of 50,000 shares of Common stock at $100 per share, and by the issuance of $50,000,000 of seven-year 6 percent serial Debenture bonds. General Motors Corporation subscribed to the 50,000 shares of Common stock, and in turn sold the bulk of the shares to some 220 of its executives at $100 a share.

General Motors entered into a contract with the General Motors Management Corporation, agreeing to pay the latter yearly on or before March loth of the succeeding year for a period of seven years ending December 31, 1936, a sum equal to 5 percent of the net earnings of General Motors during the preceding calendar year after deducting 7 percent on the capital employed. In addition, General Motors agreed to subscribe an amount equal to an additional 5 percent of its earnings to the Class A stock of the Management Corporation, at the book value thereof, based upon the cost of the General Motors Common stock which the Management Corporation was to purchase from time to time in the open market.

The payments on account of earnings, under the contract, after making provision for income taxes, accrue exclusively to the benefit of the Common stock. These net earnings are capitalized and paid to the Common stockholders as a dividend in Class B stock. There is allocated to each share of Class B stock one share of General Motors Common stock.

## EMPLOYES' SAVINGS AND INVESTMENT PLAN

One of the most significant, comprehensive, and interesting plans in the history of industrial relations is the Employes' Savings and Investment Plan, established by the Corporation in 1919. As the Corporation grew, and the number of employees increased, many plans were considered for providing for them in their later years when they were retired or to tide them over times of need. Instead of following a more conventional corporate plan, John J. Raskob conceived the idea of enlisting the interest of employees throughout the organization (in 1919 numbering 85,000) in a plan of systematic saving and the building up of individual estates through these savings, aided by proportionate contributions on the part of the Corporation these contributions being made in such a manner as would give employees an opportunity of sharing in the development of the business. It was felt that, by helping those who helped themselves, they could attain positions of individual independence that would enable them to care for themselves and their families in later years. Since the inauguration of the plan, experience has brought out the values of various features in it, additional to those with which it was originally and chiefly concerned.

The plan provided for the establishment of two funds, a Savings Fund and an Investment Fund, a new class to be started at the beginning of each year, and to mature in five years. As indicated previously, the plan originated with the class of 1919. Into the Savings Fund of each class, as it was formed, any employee who had been in the service of the Corporation for three months or more was allowed to pay 10 percent of his earnings, not to exceed $300 for the year. The limit in 1927 was increased to 20 percent but, beginning with the current class of 1933, it has been revised to 10 percent, as it was originally. Payments into the fund were facilitated by having the employee designate the amounts he wanted to save periodically, and then deducting such amounts from his regular wage or salary. For each dollar thus paid into the Savings Fund by the employee, and remaining to his credit at the end of the year, the Corporation contributed a given amount to the Investment Fund for the account of such employee.

The Corporation's contribution, under the plan operative .for the classes of 1919, 1920, and 1921, ran as high as dollar for dollar; for the classes, 1922 to 1932, inclusive, the contribution was fifty cents for each dollar; and, under the new plan beginning with the 1933 class, twenty-five cents for each dollar will be paid. The monies thus paid in by the Corporation are invested in Common stock of General Motors Corporation, as is also the income therefrom during the life of the class, and accrue to the benefit of the employee. On payments into the Savings Fund, the employee is credited with interest at the rate of 6 percent per annum, compounded semi-annually, except for the last four classes, in which interest was compounded annually. Beginning with the current class of 1933, in view of the lower prevailing interest rates, the rate in Savings Fund payments has been changed to 5 percent, compounded annually.

As a new class is started at the beginning of each year, it is possible for each employee to have at one time 10 percent of his annual wage, up to $300, invested in each of six consecutive savings classes, or a total maximum investment of $1,800. Means are provided whereby the employee may withdraw his funds with accrued interest any time he wishes. In the event of his leaving the Corporation, he must withdraw his funds except in special discretionary cases. If he makes such withdrawals, he receives back not only his original savings, with accrued interest, but a certain portion of the Investment Fund which has been credited to his account, in relation to the length of time his savings payments remained in the fund. Under the plan governing the first three classes, 1919, 1920, and 1921, the unaccrued portion of the Investment Fund of the withdrawing employee was distributed to the surviving participants at maturity, but this tontine feature was eliminated beginning with the class of 1922 the forfeited Investment Fund portion subsequently reverting to the Corporation.

Nine classes have matured since the plan was inaugurated, and receipts by employees in the various maturities have ranged from a return of more than two-to-one on their original investment to more than nine-to-one, the average for the nine classes being better than five-to-one. An employee who saved in these nine classes, $300 in each class, $2,700 in all, has received $4,204.80 in cash and Common stock equivalent to 367½ shares of the present $10 par value stock.

In connection with the plan, the Corporation permitted an employee to have his payments in the Savings Fund applied on the purchase of a home and, at the same time, receive the full benefits from the investment fund at maturity. Over 43,000 employees have been assisted in buying homes through this arrangement. In 1929 the participation in the Employes' Savings and Investment Plan reached the highest point, with 185,000 employees participating in one or more classes.

At the beginning of the industrial depression, the participating employees of the Corporation entered 1930 with a reserve of approximately $75,000,000. In addition to this, there was an equity of approximately $15,000,000 which had been diverted to the purchase of homes or, in other words, an accumulated fund of approximately $90,000,000 was available as a result of the previous five years payments and of the contributions to the Employes' Savings and Investment funds.

There were further payments made into the funds by both the employees and the Corporation, from January 1, 1930, to April 30, 1932, at which time the plan was suspended. During the three-year period, 1930 to 1932, inclusive, there was disbursed a total of $78,000,000. At the end of 1932 there was still available, in both funds, approximately $47,000,000, to which should be added $13,000,000, remaining equity in the purchase of homes, or a total of $60,000,000.

The largest disbursement in any one year took place in 1932 when approximately $44,000,000 was distributed. This total distribution represented the amount received by participants in the matured class of 1926, and the amount withdrawn from the fund by employees whose services with General Motors were terminated, or who required money to meet their various needs. Of the total disbursement, $30,719,705 constituted the employees' own original savings, and the balance of $13,421,913 interest and Investment Fund benefits (cash and securities) paid by the Corporation.

As can be readily appreciated, the plan has proven of incalculable benefit, particularly during periods of stress. It has given a large measure of security to employees who were laid off or put on part time, as well as to those who, though working, have had to meet various emergencies. It has enabled large numbers to go back to farming; has provided money to many to engage in small business enterprises of their own; and has assisted thousands to make adjustments to new conditions. In many communities, through the distribution of these funds accumulated during better times, it has relieved public and private welfare agencies of a tremendous burden.

During the fourteen years the plan has been in operation, up to January 1, 1933, cash in the sum of $105,439,122 was paid in interim settlements, that is, settlements made prior to maturity. In addition, there was paid, through the nine matured classes, $51,737,237, representing $30,406,661 in cash and 1,624,414 shares of General Motors Common stock; valued at the market prices prevailing at the different maturity dates, this figure would be $75,912,990. If the market value of the securities at the time they were distributed is taken into account, the total value to employees was $232,455,351.

This large figure indicates to some extent the size and scope of the Employes' Savings and Investment Plan. Further payments into the fund were temporarily suspended after April 30, 1932, because of the economic situation. On August 1, 1933, in view of improved conditions, the Savings and Investment Plan was resumed with certain modifications, as previously explained, to meet the new situation.

## EMPLOYES' PREFERRED STOCK SUBSCRIPTION PLAN

Related to the Employes' Savings and Investment Plan, the Corporation had another plan, the Employes' Preferred Stock Subscription Plan, which was in operation from 1924 to 1930, inclusive.

Any employee who wished to invest could buy proportionately to his salary 7 percent Preferred stock of General Motors in amounts from one to ten shares, each year, at a price fixed annually. The cash proceeds in the Savings and Investment classes, as they matured, could be applied to the payment for the stock or payment could be made out of salary or wages over a period of a year.

To make the plan more attractive, there was made each year, for five years (provided the employee remained with the Corporation) an extra payment of $2 a share, in addition to the regular $7.00 a share dividend. This plan was discontinued at the close of 1930. No further Preferred stock offering has been made to employees since, but the $2 extra payment was continued during the life of present classes.

## GROUP INSURANCE PLAN

In order to encourage employees to protect their dependents, General Motors Corporation arranged with the Metropolitan Life Insurance Company for the issuance of an employees' cooperative group insurance policy, effective December 1, 1926. Under this plan all applying employees of

General Motors, its subsidiaries, and affiliates may be insured without medical examination, provided they had been employed for three months or more. The cost is shared by the employees insured and the General Motors Corporation. In addition to regular life insurance, employees who become totally and permanently disabled before the age of sixty receive the face value of their policy in twenty equal instalments, the first monthly payment being made three months after proof of the disability. If death occurs during such disability, the remaining value of any instalments unpaid will be paid in full to the beneficiary. Any insured employee leaving General Motors may obtain from the insurance company within thirty days, without medical examination, an equivalent amount of life insurance at rates applicable to his age and class of risk.

In 1928 the insurance plan was enlarged to include increased death and total disability benefits, and in addition, health and non-occupational benefits were added at a small increase in the cost. Under the new plan an employee could receive sick benefits for several different disabilities during the same year, each benefit covering a period of not more than thirteen weeks. Furthermore, in the case of permanent and total disability an employee could receive temporary sick benefits for thirteen weeks, and then be eligible for total and permanent disability benefits for a period of forty months. Thus the combined benefits for a totally and permanently disabled employee might cover a period of three years and seven months. In 1931 the total and permanent disability benefits were restricted to those individuals who had been continuously employed by the Corporation for two full years. At the beginning of 1933, it was deemed advisable to eliminate the total and permanent disability benefit because the experience of the insurance company, along with that of all other companies, had been such that this benefit could not be continued without an appreciable increase in contributions by the employees. Under the new arrangement a group policy provides that if an employee becomes permanently disabled after he leaves employment, and dies prior to his sixty-fifth birthday, and within twelve months after leaving, the amount for which he is insured will be paid in full to his beneficiary. Under the old plan an employee must have been insured for two years before becoming eligible for the benefit under the total and permanent disability clause. The provisions in regard to regular death benefits remain the same, while temporary disability benefits have been reduced slightly.

During 1932 the Corporation lost 1,048 of its employees through death or permanent disability, on account of which, payments totaling $2,273,306 were approved for the benefit of such employees or their dependents, and, in addition, there were 9,834 employees who received benefits amounting to $827,739 on account of temporary disability resulting from sickness or non-industrial accidents. Under the group insurance plan approved claims for the benefit of the Corporation's employees and their families have totaled $16,491,080 from the inception of the plan on December 1, 1926, up to and including December 31, 1932.

At the end of 1932 over 99 percent of eligible employees were participating in this plan.

## EMPLOYEE EDUCATION AND TRAINING

Any list of activities of mutual benefit to employees and the Corporation should include the pioneer work that has been done by the Corporation in the field of employee education and training, which is now largely centralized through the General Motors Institute at Flint. The General Motors Institute itself developed out of a plan suggested by Mr. J. Dallas Dort of the Durant-Dort Carriage Company about twenty-five years ago, to secure cooperation between management and wage earner. Flint's first effort in this direction was the Flint Vehicle Workers' Mutual Benefit Association, which began as a mutual insurance society and gradually developed social, athletic, and entertainment features. This grew into the Industrial Mutual Association, huge in membership, occupying large clubrooms, and sponsoring a broad program of recreation and education in addition to maintaining its insurance and welfare features, managed by directors elected by the members of the Association, most of whom are General Motors employees. The I.M.A. auditorium in Flint is the largest assembly hall there, and the scene of the city's popular concerts, athletic contests, and mass meetings. As an employee managed enterprise, I.M.A. has been remarkable not only for its size and success, but also for its public spirit, humanitarian leanings, and educational enterprise.

The educational program originated by the association was based upon the principle that the best way to help an individual is to help him help himself. It began as an evening school, with convenient short-unit courses closely related to the requirements of the automotive industry, and of such a nature that they could be arranged in flexible sequences to meet the needs of individual employees from the plants for further training.

As its courses of instruction increased both in popularity and number, particularly in technical lines, a closer union with the factories seemed advisable, since students were preparing themselves for increased usefulness in the automotive trades. Thus the educational activities of I.M.A. came to be organized into the Flint Institute of Technology. Mr. Harry H. Bassett, then president of Buick and since deceased, took an active interest in the school, and in 1926 enlisted the support of General Motors, after which it became General Motors Institute of Technology. Suitable buildings and grounds were secured for the school to meet the requirements of divisions and subsidiary companies of General Motors. There are facilities for 2,800 day and evening school students; the faculty has numbered forty-two members giving full time, besides sixty part-time instructors.

To this institution then fell the task of providing a means through which the effort and experience of the Corporation might be pooled in the development of a sound training program for the various divisions and their employees. It also provided an agency for types of training which could best be conducted for the Corporation as a whole through a central program.

Harry H. Bassett (1875-1926}, President and General Manager, Buick, 1919-26

The training program, as it developed, divided itself into two main branches: first, that designed to contribute to the development of employees of the existing organization and, second, foundation or apprenticeship training of young men for beginning connections with the company. The first is conducted through spare time and extension courses, the second through full-time and cooperative training courses.

In the extension program, a major emphasis has been placed upon executive training for foremen and prospective executives. Here the wide pioneer experience of the entire Corporation from the very inception of the movement was pooled in the development of the executive training program. There was thus provided a uniform program representing the best experience and executive policies of the Corporation in form available to all divisions in Canada and overseas, as well as in the United States. During the years of 1928 and 1929, this program reached as many as 5,000 employees of the Corporation, representing practically all of the divisions.

In the instruction of young beginning employees, the major emphasis has been placed upon cooperative training the General Motors Cooperative Training Program for manufacturing units, and a corresponding Dealer Cooperative Service Training Program for the service field. These programs combine with a practical experience in the plant or service station an intensive, related program of technical instruction given at the Institute through the cooperative plan of alternate periods of work and study. The program thus gives the young men both the practical and technical equipment needed to meet the requirements of the field for which they are being trained.

The Institute, because of its central position, thus becomes practically a research laboratory in training. It conducts certain types of training through a central program for the entire Corporation, and in addition, in cooperation with the divisions of the Corporation promoting an interest in personnel development, it develops methods and techniques, collecting and correlating the experience of the entire Corporation, and bringing to each division the advantage of the knowledge thus gathered. The single objective is that of developing more effective means of personnel training for the Corporation.

During the later stages of the depression, in common with most educational efforts, the work of "General Motors Tech" has been somewhat curtailed, and its program revised to correlate more closely with the changing requirements of the industry. The Institute, and the program which is centralized through it, have been recognized by educators and industrial executives as one of the best managed industrial educational training programs in the country.

*Grand Motors Institute, Fleet, Michigan*

## HOUSING

Reference has been made to General Motors' extensive house-building programs in a number of the cities where its large manufacturing plants are located. Back of all these construction projects was a definite desire to provide better housing for employees in swiftly growing cities where private building operations lagged behind the needs of the wage-earning population. Large subdivisions were plotted, streets opened, sewer systems installed, and connections made with public utilities. In some cases community halls were erected for public and social gatherings. The projects were carefully planned from the standpoint of attractive street layouts, open spaces, and landscaping. The best possible house was built for the money, with all benefits of large-scale buying inuring to the purchasing employee. Complete in all respects, the houses were sold to employees on reasonable deferred payments which might be deducted from wages. Through the operations of the Savings and Investment Fund, from which deductions could be made on house-buying account without diminution of the employee interest, the Corporation considerably eased the lot of thrifty employees in their progress toward house ownership.

## COOPERATIVE GARDENING

In the past few years some of the units of the Corporation have fostered cooperative gardening for the benefit of employees, both those on the pay roll and those temporarily laid off. As a general practice, the unit provides the land, renting it if necessary; attends to the plowing; provides the seeds and plants, and competent direction. As instances along this line may be cited the Harrison Radiator Corporation, Lockport, New York, and the particularly successful experiment of the McKinnon Industries, Ltd., St. Catharines, Ontario.

## WELFARE WORK

Nearly all General Motors plants maintain various services for their employees. These are under the direction of plant managers, and are of so many sorts that no adequate description can be given of them here. Historically, Chevrolet seems to have been the leader of General Motors in developing work of this nature, and many of its pioneering welfare activities have been widely copied both inside and outside of the Corporation.

In the Corporation's annual reports various of these plans and programs have been discussed in great detail, with emphasis on the fact that they have been inaugurated "for the purpose of promoting the effectiveness and well-being of the Corporation's operating personnel." The 1932 report says:

> It has been stated that the fundamental objective in all these plans has been to help the individual to help himself ; to make him a better citizen; to give him the opportunity to become independent.

In the main this objective has been gained, and in those particulars where the greatest measure of successful cooperation has been achieved, it is to be expected that the Corporation will go considerably farther in the direction of employee-benefits.

The oldest and most important of General Motors cooperative plans have been revised freely under the influence of changing circumstances.

# Chapter XXIX PUBLIC RELATIONS

PUBLIC RELATIONS arise out of a corporation's policies in its dealings with actual and potential customers. Good public relations make for confidence in an institution. Yet mere goodwill that leads nowhere is not enough. What a corporation wants is the expression of a belief in its integrity through the purchase of its products or services; and the confidence of those whose capital it uses and those with whom it transacts business while producing those goods and services.

Good products and good services are a fundamental objective of good management. So also are satisfactory public relations, which have a direct bearing on earnings and employee morale. General Motors has defined the scope of its activities in this field as follows: It is the task of Public Relations to "represent and interpret General Motors to the public in such ways as will favorably dispose the public toward the purchase of its products."

As a corporation grows in size its operations inevitably take on more social significance. Any vast industrial organization in which large numbers of persons use the capital of the public in the production of goods becomes a factor of importance in the life of the nation. General Motors operations in a vital way touch, directly or indirectly, people in all parts of the world. Its products bear intimately on the daily lives of all members of the family in countless homes and in every walk of life. Historically, the automobile industry evolved swiftly toward a manufacturer-consumer relationship. The nature of the product a machine whose effective operation depended upon skilled servicing required that the manufacturer stand behind his wares beyond the general practice of the time. His goods were style goods, not staple goods; as such they had to be advertised and sold, not merely as automobiles, but as Buicks, Cadillacs, Oldsmobiles, and the like, since their styled identity persisted through the lifetime of each car. This had been true also of the carriages, and agricultural implements that preceded the automobile, but the automobile manufacturer had to carry consumer service to an unprecedented point. His products were more complicated and more mobile. The ease with which the public could secure repair parts and skilled service almost from the beginning was a vital element in maintaining popular favor, and had definite good-will value for the manufacturer. No other industry of equal magnitude stands in quite the same position with respect to the public.

Organized as a holding company, the pressure of events converted General Motors naturally into an operating company. The identity of manufacturing groups was maintained, each with its selling organizations, so that General Motors products were, and are still, sold, in the main, under division names. But since General Motors had made itself responsible for divisional operations, the old law of the industry, early recognized and never for long disregarded, that the maker must stand behind his wares morally as well as financially, manifested itself anew. General Motors found it expedient to show, through institutional advertising, that it stood behind divisional products, because the man in the market-place had been taught to look, not only at the merchandise offered, but also at the character, resources, and policies of its markers. The Corporation found that an important part of all the influences which cause people to buy General Motors cars consists in what they believe about the Corporation as a whole rather than what they believe about each car in itself. From the beginning the Corporation's concern for its public relations has been a logical growth, in the course of which its position has been clarified by specific events and expressed many times in both word and deed.

In the twenty-fourth annual report for the year 1932, this statement appeared:

> It is recognized that the Corporation's most vital relationship is with the public. Its success depends on a correct interpretation of the public's needs and viewpoints as well as on the public's understanding of the motives that actuate the Corporation in everything it does.
>
> In order to formulate its policies in harmony with this basic principle, no effort is being spared to analyze and evaluate the public forming the Corporation's actual and potential customers, in its thinking with respect to all things in which the Corporation plays a part.
>
> This represents, however, but one phase of the Corporation's public relations policy, for, while it is essential that the Corporation understand the public, it is equally essential that the public understand the Corporation. Good-will is established and maintained not alone by excellence of product and service, but by a combination of this with public knowledge and acceptance of the policies of the Corporation.
>
> The correct interpretation of such policies to the public is regarded as a primary function of management. Progressive industry today places the formulation of sound public relations policies on a parity with the formulation of other major policies. In fact, it goes further and recognizes that every forward step in procedure should be subjected, whenever possible, to advance appraisal from the standpoint of the public interest in order to ensure any contemplated action meeting with public acceptance.
>
> The Corporation is keenly alive to the importance of this responsibility to the public.

One way in which this responsibility has been met fully has been through reports to stockholders. General Motors was one of the first American corporations to issue complete quarterly statements. Though addressed particularly to stockholders, these and other statements reach the public through the press. On a scale which draws favorable comment from serious students of corporate affairs, the Corporation seizes upon the opportunity offered by its annual reports to stockholders to make a detailed accounting of its stewardship. In these reports will be found not only statistics revealing the Corporation's financial position, but also explanations of procedure and policy. These reports go also to banks and trust companies; the information they contain becomes part of the back-log of confidence upon which rest the Corporation's relations with thousands of suppliers and a multitude of financial concerns through which General Motors transacts its world-wide business. The stability of the Corporation's security structure is in part due to the frankness and completeness of its published statements.

As an example, let us consider the 1932 annual report, a booklet of thirty-six pages distributed to 372,000 stockholders, many financial institutions, and the press.

> The statistical body of the report includes, in addition to the annual statement with appropriate comparisons for the various items, a complete record of net sales, earnings, and dividends from 1909, an analysis of unit sales for eight years, statistics and charts covering total automobile registrations, overseas sales, Employes' Savings and Investment Funds, pay rolls, and number of employees, bonus awards, number of stockholders, and a roster of the Corporation's divisions, subsidiaries, and affiliates.

In addition the report carries the following explanatory features:

A Financial Review of four pages on earnings; dividends; net working capital; real estate, plant and equipment account; and investments.

An Operating Review of two pages, not only giving facts but elucidating them.

A Report on The Industrial Depression: Its Economic and Operating Influences, five pages.

Operating Reports on General Motors Overseas, and General Motors Acceptance Corporation, one page.

Cooperative plans, including General Motors Management Corporation, Bonus Plan, Employes' Savings and Investment Plan,

Group Insurance Plan, Housing for Employees, two pages.

Good-will and Patents, Public Relations, one page.

This consistent policy of frankness has won both public confidence, and expert approval. In the April, 1933, issue of Scribner's Magazine, Anderson F. Farr has a comprehensive article on the public relations of large corporations entitled "Give the Stockholders the Truth." It deals with "restoring confidence in American finance," and makes the following significant statements about General Motors' annual report to its stockholders:

> Then there is the outstanding example of clearness, sincerity, and reliableness as reflected in the twenty-third annual report of the General Motors Corporation covering the year 1931. The General Motors Corporation is one of the very great business enterprises of the world, with 313,117 stockholders. Its importance is tremendous in many lines of business activity. Its capital and surplus amount to $926,000,000, and its employees are measured by the tens of thousands. It behooves a concern of such power, and with such ramifications, to set an example of positive integrity and rugged simplicity in giving an accounting to its stockholders so that when a report gives a figure for profits there is no chance whatsoever of misinterpretation. Such an example has been set. The report for 1931 is one of the most complete reports issued to stockholders, containing a wealth of interesting facts and information, put together in a manner to indicate that the officials recognize a duty to stockholders, and a high degree of conscious moral and financial responsibility.
>
> The consolidated income account is given in comparison with the income account for the preceding year, and every item representing a loss is charged to income fully as much as any of the usual charges of expense. There is even an extraordinary and nonrecurring charge of $20,574,514, largely representing revaluation of net working capital abroad to a dollar-value basis, and including a revaluation of security investments to market value. The account is clear from beginning to end, edited as far as possible, which is far at that, in non-technical language. Moreover, no part of the accounting exhibits is at variance with figures or implications contained in the message to stockholders. That also is as it should be.

Communications to stockholders are not confined to printed documents. The president writes a letter to each incoming stockholder, and also a letter to each outgoing stockholder. These bring many replies, each of which is answered. This correspondence gives evidence that contacts so established are influential in encouraging new stockholders to regard their General Motors investments as permanent, and in influencing them to become actively interested in the progress of the Corporation.

Through the newspapers General Motors acquaints the public with important new developments as they occur, and provides the nation with a business barometer on the eighth day of each month, when it publishes the unit sales and deliveries of its car divisions for the foregoing month. These figures are accepted as one of the more dependable indications of the business trend, comparable to the figures on tonnage shipped put out for so many years by the United States Steel Corporation on the tenth of each month. They are published on the earliest possible date consonant with careful compiling without waiting for a circulation of the figures first within the organization, on the theory that the public is entitled to news which by common consent is deemed to possess a peculiar barometric value in the interpretation of economic events.

The Corporation maintains a public relations staff in its New York and Detroit offices, to make effective the broad policies of the Corporation in its widely ramifying relations with the public. It answers inquiries concerning the Corporation and its activities which pour in from all parts of the country, the schools being especially prolific sources of correspondence. However, while the Public Relations department acts mainly in a staff capacity on matters affecting the Corporation's public relations, General Motors stresses the fact that the duty of establishing and maintaining good relations with the public rests upon every division and department and every employee and dealer.

President Sloan expanded this theme forcefully in a circular letter to all the Corporation's executives and general staff officers on July 6, 1933, as follows:

> I call attention to the fact that General Motors, while a private Corporation, nevertheless is so broad in its ramifications that both our thinking and our action become a matter of more or less public concern. This is particularly true because we are different in the sense that we deal directly with the public, and the public is, therefore, concerned with our operations on account of their direct use of our products. The greatest asset that any organiza-

> tion like ours can have, is public good-will ... every individual, however unreasonable he may be, has a certain sphere of influence, and ... we must remember that such influence is tremendously effective because it is looked upon as unprejudiced even as against facts that we might present contrariwise...
>
> With the above introduction to the subject, I urge upon every executive to keep this matter prominently in mind, and would appreciate every executive transmitting to those under his jurisdiction, such observations as he might see fit to make on the subject that I am now presenting, for their information and guidance. Let us give particular attention to that phase of the general question which finds expression in prompt and courteous replies to suggestions, criticisms, and everything of similar nature, whether directly or indirectly concerned with our immediate problems, which arise in the voluminous correspondence that our executives must necessarily carry on with individuals, both without and within the Corporation's activities.
>
> I appreciate that it takes a little time to do these things. I appreciate that when we are worried and have so much to do, it seems foolish to divert our energies in this direction. I only want to point out that in my many years' operating experience I have found that it is more than worth while.

The common sense of this policy will scarcely be disputed as a practical matter. In fact practicality, instead of a blind groping after prestige, dominates General Motors' work in this field. For itself, General Motors has defined its Public Relations purposes in the frankest possible terms as quoted at the beginning of this chapter. Even though this definition is eminently businesslike, it is nevertheless so broad that not all the various ways in which Public Relations work goes forward can be described here. A few outstanding cases will, however, be cited.

Among the many notable instances of Public Relations activities initiated and maintained by divisions of General Motors is the Craftsmans' Guild of Fisher Body Corporation, which thereby comes close to many thousands of ambitious youths and, indirectly, to their families and friends in the United States and Canada, as described in Chapters.

## THE BANKING CRISIS IN MICHIGAN

Perhaps the most unusual step that the Corporation has taken to meet a pressing public need was its action in participating with the United States Government in the formation of the new National Bank of Detroit in early 1933, to relieve the paralysis which overtook that great industrial city following the Michigan banking moratorium which began on February 14, 1933. This was the first moratorium of the period effective over the entire area of a great industrial state, and banking troubles spread far and wide over the country until the national moratorium of March 6th closed all American banks. When the national moratorium ended, Detroit was still without the services of its two largest banks, and remained without a major banking institution to serve the needs of that great industrial city until March 24th, when the General Motors Corporation and the Reconstruction Finance Corporation brought relief through opening by their joint efforts the National Bank of Detroit, a unique financial institution, in that it represented a great industrial Corporation and the United States government in partnership to alleviate a distressing situation.

As one of Detroit's large manufacturers and employers, that city being the seat of divisional administration as well as of some of its large plants, the Corporation realized how demoralizing the abnormal situation had become to the public and the business interests of the nation's fourth largest city. While adequate banking facilities were lacking, trade stagnated, enterprise languished, thrift was discouraged, industrial morale declined. In this emergency, and only for the emergency, General Motors entered the banking business in Detroit as an equal partner with the United States Government, itself keenly aware that what Detroit needed more than anything else at the moment was a large commercial bank commanding enough capital and confidence to reassure the community and provide large-scale credit facilities for industry and trade. After various plans had been abandoned, General Motors arranged with the Reconstruction Finance Corporation for the creation of the National Bank of Detroit, with $25,000,000 capital, to which General Motors Corporation and the Reconstruction Finance Corporation each subscribed $12,500,000.

President Sloan expressed the Corporation's motives behind this unusual move in an open letter to the people of Detroit dated March 24, 1933, in which he said, in part:

> I am sincerely happy that after nearly six weeks the people of Detroit again are provided with banking facilities. I feel that it is a privilege to be able to take a small part in a matter so vital to the people of this city.
>
> In the business with which I am so intimately associated I have seen many big things done but never in my experience have I seen so many difficult obstacles overcome in so short a time as has been the case in the past few days in making banking facilities available to the capital of motordom. At ten o'clock on Friday, March 24th, Detroit will again have banking facilities which indeed is a credit to all of those people in government offices, those in responsible capacities in Detroit, and those of responsible position in my own organization.
>
> In bringing about the opening of the National Bank of Detroit, my principal concern has been to accomplish this with a minimum delay. The people of Detroit have needed a depository for their currency. They have needed the usual checking facilities afforded by a bank. This to my mind has been the first need in Detroit. These facilities will be provided on the opening of the National Bank of Detroit which for the present will be limited to commercial deposits including personal checking accounts. Whether savings deposits will be accepted will be determined by the permanent organization later.
>
> I would like to make clear several important facts:
>
> 1. This bank is an entirely new institution.
> 2. It will be the most liquid bank in the United States.
> 3. It will be a Detroit institution owned, directed, and managed by Detroit people.

It may well be that some of the Detroit people will raise the question as to why General Motors has interested itself in the Detroit banking situation. I am sincerely anxious that the people of Detroit appreciate that the only interest which General Motors has in this connection is to be of service to the people of Detroit. In the joint statement which I issued in conjunction with the Reconstruction Finance Corporation, I endeavored to make clear to the public just exactly what position General Motors occupies in relation to the banking situation in Detroit. Perhaps I can do no better than to paraphrase here the statement which I made a few days ago in this connection.

> In underwriting the Common stock of the new bank General Motors Corporation was doing so as a contribution for the settlement of a very serious situation. It had no desire to enter in any way the banking business in Detroit or elsewhere. It was entitled to and had every reason to expect the support of the depositors and the stockholders by subscribing to the Common stock. An offer will be made by General Motors Corporation to all depositors and stockholders of the First National Bank and the Guardian National Bank of Commerce for subscription to the Common stock of the new bank, at the same price as paid by General Motors Corporation, that is, fifty dollars per share.
>
> It was hoped that as soon as the situation was stabilized it would be possible for General Motors Corporation to withdraw entirely, transferring its investment to others to carry on this particular responsibility and duty to the community.
>
> We were happy to cooperate with the government, the President of the United States, the Secretary of the Treasury, the Reconstruction Finance Corporation, and other sincere and able representatives of the government in helping to establish sound banking facilities in Detroit.
>
> In my various contacts with you people of Detroit I feel that you understand the position which General Motors has taken, and that you will lend your support to those men who will be designated to carry on for you a sound type of banking which has been made possible through the sympathetic and financial cooperation of the United States Government.

This position was reaffirmed by Mr. Donaldson Brown, chairman of the Finance Committee of General Motors Corporation, in a statement published on July 6, 1933, giving the background of the negotiations resulting in General Motors entering the Detroit banking situation. Mr. Brown made it clear that General Motors stood willing from the first to cooperate proportionately in any plan to reopen the two large closed banks; and, after the decision had been made by the United States Government not to reopen them, to subscribe to other sound plans which would permit the liquidation of those banks and the relief of their depositors. When other efforts had reached the stage of protracted delay, with the Detroit situation growing more difficult daily, it was decided to confer with Washington to learn whether General Motors could do anything constructive toward a solution of the problem. Of those negotiations, and the considerations involved in them, Mr. Brown comments, in part, as follows:

> During the time prior to the expiration of the originally stipulated period of the Michigan bank holiday we attended meetings of various depositors, directors, and officers of the two large Detroit banks; at which meetings it was stated by government authorities that the old banks would not be allowed to reopen. Following these meetings as well as previously we made known our position of readiness and anxiety to cooperate with the officers and directors of the old banks in any plans deemed to be constructive, and insisted at all times that it was not within our province to take any initiative in the direction of determining the course that should be taken. In course of time we were informed that in the case of each of the old banks the decision had been reached, carrying approval of their boards of directors respectively, to attempt the formation of new banks through the procurement of the necessary capital from important depositors, looking forward to each of such new banks respectively undertaking to purchase certain liquid and sound assets from the old banks together with a suitable assumption of deposit liabilities. After plans in this direction were duly approved and authorized by the respective boards of directors, General Motors Corporation, together with numerous other important depositors, undertook to subscribe to capital stock of proposed new banks.
>
> The solution of the problem in the interest of gaining the greatest possible benefit to the old depositors, together with the supplying of suitable augmentation of banking facilities in Detroit, seemed to the government to call for a new bank or banks having aggregate capital of at least $25,000,000. Government authorities took the position that R. F. C. could not subscribe capital beyond matching the amount of capital that could be derived by private or local contribution, and thus it was made apparent that $12,500,000 of private capital must be forthcoming.

> Recognizing the serious importance of avoiding undue further delay in dealing with the situation, General Motors Corporation undertook to supply initially the full $12,500,000 required, feeling that under the circumstances other depositors who had been cooperative, and who had expressed willingness to supply capital incident to the creation of new banks, should be relieved of any feeling of obligation to supply capital under the new program. General Motors Corporation therefore was in a position to consummate plans in collaboration with R. F. C., and to expedite the culmination thereof, by reason of being free from the necessity of collaboration with other depositors in arriving at agreement upon various important details as to the constitution of the articles of association and understanding as to the principles upon which the project should be carried forward.

Subsequent events amply prove that Detroit benefited by this bold step. From its opening day the National Bank of Detroit was thronged with depositors, many of them returning hoarded gold and currency. Even the most timorous were reassured by the stability of a bank jointly owned by the Federal Government and General Motors. The Common stock representing the Corporation's interest was offered to the public at cost in accordance with the Corporation's promise, soon after the bank opened. About 25 percent of the stock was sold to the public in 1933.

## THE CENTURY OF PROGRESS EXPOSITION

Another instance of General Motors' attitude toward public events of importance was its early decision to participate to the extent of upwards of a million dollars in the Century of Progress Exposition at Chicago, with results which are described in Appendix II. Upon the occasion of the dedication of the General Motors Building at the Exposition, the president of the Exposition, Mr. Rufus C. Dawes, told his hearers that, but for General Motors' encouragement at a critical juncture, the World's Fair of 1933 could hardly have gone forward to a successful conclusion.

## CONSUMER RESEARCH

In the chapter on "Marketing the Motor Car," the trend away from "high-pressure" salesmanship, and toward consumer research, has been noted. Through investigations by trained staffs General Motors endeavors to learn as nearly as may be discovered the needs and wishes of its customers, to the end that those wants can be filled as promptly and fully as possible. The small merchant learns these needs directly from his customers day by day; a large merchant has to take specific steps in order to secure the essential information in all parts of the vast territory served.

General Motors has been conducting consumer research for many years, as part of its settled policy, in acquainting itself with the attitudes of the public, in line with those other essentials of General Motors policy to preserve an "open mind" on all matters, and "find the facts." Under the urge of adverse selling conditions, consumer research lately has been receiving more and more attention. Another investigation, with a definite business objective, that of determining the present wants of the American public with regard to motor cars, is now going forward in the United States, along lines of effort which produced excellent results when tried out by the Corporation in the Dominion of Canada.

In a letter to stockholders, dated September 11, 1933 calling their attention to the twenty-fifth anniversary of General Motors, President Sloan described this phase of consumer research as a "Proving Ground of Public Opinion" and part of a definite program of fact-finding. He said, in part:

> Through modern technique, products undreamed of by our forefathers have been brought into being, and placed within reach of everybody. But as a result of large-scale operations and worldwide distribution, producer and consumer have become more and more widely separated, so that the matter of keeping a business sensitively in tune with the requirements of the ultimate consumer becomes a matter of increasing importance.
>
> Through Consumer Research, General Motors aims to bridge this gap and provide guidance not only with reference to details of design but as regards public relations, advertising, sales, ser-

vice in fact, everything affecting our customer relations, directly, and indirectly.

For a number of years past General Motors has maintained a central staff to conduct various types of market surveys, and the findings of such surveys have contributed in no small measure to the progress of the Corporation. During the past two years this activity has been pursued along more exhaustive lines than formerly, constituting what might be termed a "Proving Ground of Public Opinion," devoting itself to the finding of facts as regards the attitudes of the practical motorist toward various aspects of merchandising and service all of which are vitally important as bearing on customer good-will, and continued patronage. The work is concentrated in a central department known as the Customer Research Staff, which operates in close cooperation with General Motors Research Laboratories, Fisher Body's Art and Color section, the General Motors Proving Ground, and the various divisional engineering, sales, and service organizations, supplying them with data which are constantly flowing into the Central Office directly from owners of cars of all makes and ages all over the country.

The activities of the Customer Research Staff involve sending out questionnaires, calling on owners, and digesting customer reactions flowing into the Corporation through miscellaneous channels. During the past year well over 1,000,000 motorists have been invited to "pool their practical experience with the technical skill of General Motors engineers and production experts." But it would be a mistake to think of consumer research as an isolated department. Sending out questionnaires, calling on people and compiling statistics these are only incidents, or tools; very important tools, to be sure, but tools, nevertheless.

To discuss Consumer Research as a functional activity would give an erroneous impression. In its broad implications it is more in the nature of an OPERATING PHILOSOPHY, which, to be fully effective, must extend through all phases of a business weighing every action from the standpoint of how it affects the good-will of the institution, recognizing that the quickest way to profits and the permanent assurance of such profits is to serve the customer in ways in which the customer wants to be served.

Of course there is nothing really new in this. It is the fundamental basis upon which all successful business is founded, but as stated above, modern industry, with its large-scale operations, tends to create a gulf between the customer and those responsible for guiding the destiny of the institution. We can no longer

depend upon casual contacts and personal impressions our business is too big; our operations too far-flung.
Furthermore, we are passing through a kaleidoscopic era characterized by swift movements social as well as economic and such conditions cannot fail to bring more rapid changes in the tastes, desires, and buying habits of the consuming public. So it becomes increasingly important that we provide the means for keeping our products and our policies sensitively attuned to these changing conditions.
And, irrespective of what these changes may be regardless of what the new economic and social order may hold I am confident that a more intimate, detailed, and systematic knowledge of the consumer's desires will afford the Corporation a sound and progressive basis upon which to meet the new conditions as they unfold.

Speaking of the General Motors consumer research program, Printers' Ink for July 13, 1933, indicates with discernment the point at which public relations and sales merge:

The policy of any consumer research department is to take the people just as they are found; without any attempt to reform them, no desire to change them, but to study them instead and give them something they should have.
The General Motors consumer research is based upon the same purpose which actuates the advertising and selling of its product, namely: to stimulate business and broaden good-will, "by bringing to the buyer information that will aid him in spending money wisely."
This method of going ahead on facts instead of hunches might well be made a guiding principle for many a small business today;

The truth is that General Motors, in so far as it is physically possible, tries to conduct its huge business with its large public as the small business does with its public. Just as the essence of courtesy is the ability to put oneself in another's place, so consumer research is the courteous approach to sales. As a buyer of merchandise in enormous quantities, General Motors has learned to appreciate those suppliers who seriously study its needs and bring forward new products meeting its problems. With the Corporation on the selling end and the public on the buying end, the situation is reversed, but the principle still holds good. General Motors is still going to school. It wants to learn, and within the limits of prudence is willing to take mass opinion as a guide in planning its production and sales programs.

In the quarter century of General Motors' existence, the irresistible march of science, revealing itself most effectively in swifter transportation and communication, has had two effects of the most profound import. On the one hand, the planet has shrunk in size as compared to the reach of the human will; on the other hand, the effective groupings of human energy have greatly expanded in size. Corporations have grown tremendously, both by mergers and by expansion from within; and trade associations are now commonplace features of the business landscape, accepted by the public and vested with certain broad responsibilities by government.

In this relatively new industrial order, geared so directly to communications through which mind meets mind over vast distances and in bewildering numbers and complexity, the use of media for all the senses becomes a necessity with any business enterprise that wants its message to reach all the public. Though the message itself may be simple, the means of spreading it are many and complex, including all variations of communication through the printed word, the spoken word, an appeal to the eye through color, form and motion, and to the reason through logical presentation and detailed information.

To discover, test and apply new methods of using the varied communication facilities of the modern age is part of the task of a public relations staff. It exists, in a sense, to do the things which have not yet been reduced to a system, to meet the unexpected, to anticipate where possible the individual query, to cultivate the human side of business contacts, to act constructively on the mass mind.

This picture, then, reveals itself as one of the many placed before the reader in this history of General Motors. It is a picture of men thinking not alone of the immediate exigencies of production and distribution, but also of the enduring values of human relations, standards of living, and the myriad factors of interdependence which are at the root of community life as well as of trade and manufacture. A search, if you please, for the keys to human behavior, for the remedies which assuage discord, for the understandings which, if they can be brought to pass, will be of value in ushering the future serenely out of the past, unpredictable though that future may be. In this search all the hard-won tools of science, all the arts of grace and beauty, all the systems of idea-distribution are thoughtfully marshalled.

That is the method of public relations, but the method is really of less purport than the motive. How well the arteries, veins, and nerves work is chiefly a matter of technical interest for experts to consider. What the public has a right to know is how the head and heart of General Motors work, whether the ideas and policies thus sent on their way into the thought-stream of the nation are sound, valid, and constructive, and to what extent the incentives of both commerce and the public well-being can be served from the same source.

This question has already been answered in the quoted words of others. The primary motive is to win patronage for General Motors by building for it a foundation of public good-will. There remains merely to register the conviction that in pursuing that objective, General Motors already has created by-products of public opinion and understanding worthy of the notice of sociology and that, in the very nature of things, having in mind both the Corporation's past and present, these indirect influences of its public relations policy will grow more apparent in the future.

# CONCLUSION

FROM its adventurous beginning in the dawn of the industrial age, the self-propelled vehicle has swept on until, as the gasoline motor car of the present, it is a factor in all civilized countries and is helping to civilize others. In the United States, where a broad continental area of fertile land is occupied by an ingenious people with high standards of living, automobile transport has found its widest market, being accepted so completely that the very face of the country has been and is being made over for its wider use year by year. Social life in all its aspects family, school, church, business has been accommodated to the automobile, changing day by day as this flexible mode of transport has proved its ability to move human beings and their goods into new areas and relationships. Under the aegis of the automobile and the rapid communications which are its companions on the modern stage, sectionalism is fading away, population is being redistributed, and new groupings are being formed. The future of America is beyond foreseeing, but at least in that future the automobile will be more highly developed and even more commonly used than it is today.

The position already attained by the automotive trade in America is truly remarkable when the speed of its growth is considered. From a few hundred dollars to millions, from shops in back alleys to the largest and best equipped factories, from an infant under suspicion to a giant universally acclaimed, from an occasional sale to a grand total of $4,774,822,000 retail value of new car sales in 1929 or nearly 10 percent of the total retail trade of the United States such is the sweeping saga of thirty-five years in this industry which has so successfully put power behind space-conquering wheels.

In this advance of the American automotive industry, the experience of General Motors has been unique. It was the first automobile merger, and as such had plants in many localities. Other large automobile companies have grown up around individuals in complete authority; this one from the first has had many bases for widely scattered operations, and in the main has functioned through the meeting of many minds rather than at the word of a single person. Its ownership has become widely diffused over the whole country, a testimony to the general acceptance by investors of the fact that conservatism has definitely succeeded the vigorous slap-dash optimism of youth. The contrast between the General Motors of today and the General Motors of its years of storm and stress, reveals the ground won and held by steady and efficient management.

Only a generation ago the descendants of the original settlers who went West in their covered wagons and beheld the wheel-less red man driving his poor travois over the hill toward oblivion were still as dependent upon horses as their grandfathers were. Their trading range remained the distance a team could travel from farm to town and back in a day; now, within half of an average lifetime, that range has been increased gradually to ten times the old distance. The vehicle which passes over the hill toward oblivion is now an outworn and outmoded automobile ready for the junk heap, while the man of destiny, watching its departure, stands beside a new and superior car, which gives him a better command of time and space than has ever before been at the disposal of millions.

Aboriginal America, the land without wheels, with its trackless woods and plains, is in startling contrast to the America of motor cars and concrete highways; yet the most dynamic elements in that picture are the creation of only a few brief years in which the motor car has been the prime mover of progress. A colossal trade, with a romantic past and a driving spirit of enterprise, has developed swiftly from the early efforts to give the public all the cars it wanted. The part which General Motors and its older subsidiaries have played in the early history of the American automobile is full of drama. Later developments of the Corporation's twenty-five-year history are full of meaning for the serious student of American society, whether it be studied from the standpoint of science, finance, or sociology. General Motors' industrial research will probably be remembered long after its extraordinary earnings are forgotten, but perhaps the historian of the future will look most of all at the sense of high responsibility with which this Corporation has for many years managed its vast and growing business under the eye of the public.

# APPENDICES

## APPENDIX I

*A Chronology of Significant Dates and Achievements in the Evolution of Self-propelled Vehicles and in the History of General Motors Corporation*

c. 130 BC. Hero of Alexandria built first steam engine of record.

c. 1250 AD. Roger Bacon prophesied the coming of horseless carriages. Grand Canal of China constructed.

1487. Lock canal constructed at Milan, Italy, on Leonardo da Vinci's plan. His geared wheel belongs to a slightly earlier period.

c. 1560. "Carriages without horses shall go" prophecy of Mother Shipton.

c. 1600. Simon Stevin of Holland built a sailing chariot.

1601. Delia Porta described how a steam engine like Hero's could be used for pumping water.

1615. Solomon de Caus improved on the steam pumping engine.

1619. English patent granted to Ramsay and Wildgoose for "drawing carts without horses."

1629. Giovanni Branca constructed a simple steam turbine. May have put it in a carriage.

1630. A wooden road laid at Newcastle, England.

c. 1630. Fr. Verbiest in China applied steam to propulsion of a carriage, by applying a steam jet to a small windmill mounted on wheels.

1644. Jean Theson granted a French patent on horseless carriage propelled by two seated men.

1648. Wilkins' Mathematical Magic appeared, summing up mechanical transport to date and propounding ideas for its development.

1649. Hutsch in Germany and Richard in France made autocars propelled by passengers.

1663. Potter of England built a mechanical cart, propelled by moving legs, and Dr. Richard Hooke presented to the Royal Society a plan for a machine with which one could walk on water or land with "the swiftness of a crane."

1676. Parallel wooden rails were laid down at Newcastle, England, to speed horse travel.

1678. Jean de Hautefeuville of France suggested a piston-andcylinder engine using gunpowder.

1680. Christian Huyghens described the first explosion engine (gunpowder).

1689. Model steam engine produced on lines suggested by Sir Isaac Newton ; resembled Hero's in principle.

1690. Denis Papin invented earliest piston-and-cylinder steam engine ; applied this to a carriage and made it move.

1695. Sir Humphrey Mackworth applied sails to his wagons on the trainway at his colliery at Neath, South Wales.

1698. Thomas Savery obtained a patent for a steam engine to raise water: safety valve applied to it by J. T. Desaguliers.

1705. Thomas Newcomen constructed his atmospheric steam engine, which came rapidly into use for pumping water from coal mines, making the steam engine a practical success.

1725 Jacob Leupold described a non-condensing engine in his Theatrum Machinarum.

1736 JonHulls received a patent from Parliament for a steam boat to use a Newcomen engine.

1738. Iron rails first used at Whitehaven, England.

1748. Vaucanson drove before Louis XV a carriage propelled by clockwork springs. Benjamin Franklin's successful electrical jump-spark experiment.

# APPENDIX I

1753 Daniel Bournoulli given prize by French Academy of Science for demonstrating the point at which steam power could be applied to navigation.

1765. Dr. Erasmus Darwin penned a remarkable prophecy of steamboats, motor cars, and flying machines.

1757. Cast-iron plate rails used for coal haulage.

1769. Capt. Nicholas Joseph Cugnot built a steam tractor for French artillery use, the first self-propelled vehicle constructed for definite use as contrasted with experimentation. James Watt patented, after experiments beginning in 1764, his separate condenser for steam engines, and other improvements increasing the effectiveness of steam power.

1772. Oliver Evans, "The American Watt," began experiments in steam transport.

1784. William Murdock, associate of Watt, successfully tested a small steam carnage now in the British Museum. James Rumsey began steamboating experiments on the Potomac.

1775 John Fitch began steamboating experiments on the Schuykill.

1786. William Symington of Scotland patented a road for a steam carriage; not followed up.

1787. Oliver Evans secured right to use Maryland roads for a steam carriage.

1789. Nathan Read drove a paddlewheel steamboat at Danvers, Massachusetts. Oliver Evans took out first American patent on a self-propelled road vehicle. Patent issued by State of Maryland.

1794. Street patented an explosion engine in England.

1798. Robert R. Livingston received right to navigate New York waters with steam vessels.

1800. Sir George Medhurst patented an explosive (gunpowder) engine and proposed to apply this to a carriage.

1801-02. Richard Trevithick built and drove successfully a steam carriage (London, England).

1804. Oliver Evans moved through the streets of Philadelphia and on near-by waters a scow with wheels, the first amphibian, and the first American vehicle to move under its own power for a practical purpose.

1804-05. John Stevens petitioned New York Legislature to encourage railroads.

1807. Fulton's Clermont, paddle-wheel steamship, began to ply between New York and Albany.

1811. Blenkinsop of Leeds made the first practical application of the steam locomotive in transporting coal.

1814. George Stephenson's locomotive made successful trial run, July 25th.

1818. Rudolph Ackermann patented in England a tangential steering device originated in Munich by George Lankensperger, the first to work the modern principle of rounding corners.

1819. Notice given in a London magazine of steam carriage invented in Kentucky.

1820. Parish and Cecil built explosion engines using hydrogen gas.

1821. Julius Griffiths of England patented a steam carriage, which was well constructed by the famous mechanic, Bramah.

1824. W. H. James produced his first steam carriage, which aroused wide interest in the mechanical world. Burstall and Hills's steam "drag."

1825. Two American steam carriages appeared, one by T. W. Walker, in Edgar County, Illinois; and one in Springfield, Massachusetts, by Thomas Blanchard, who received the endorsement of the Massachusetts Legislature. First railroad in regular operation for freight and passengers Stockport and Darlington, England. George Stephenson, engineer. London amazed by a carriage drawn by kites.

1826. Samuel Morey patented two-cylinder poppet valve engine, water cooled, with some compression. Gridley Bryant, of Quincy, Massachusetts, ran steam-propelled cars on rails from his quarry to tidewater.

1827. England entered actively into the building of steam carriages, with Walter Hancock and Sir Goldsworthy Gurney leading. Hancock is credited with being the first to use chain transmission and to make tight metallic joints capable of withstanding high steam pressures.

1828. First successful use of planet gearing Paris. Delaware & Hudson built first railroad track in United States.

1829. First locomotives imported into United States.

1830. James said to have applied the rotary principle to road work for the first time.

1831. Select Committee of British Commons held hearings to review the progress of mechanical transport; decided steam carriages were not harmful to roads. Nevertheless, power vehicles were discriminated against for the next sixty years in England. First railway locomotive built in United States. First link of present New York Central Railroad system opened between Albany and Schenectady, New York.

1833. Richard Roberts of Manchester, England, applieddifferential gearing to a steam carriage. Maceroni and Squire used in a steam carriage a high pressure boiler. Heaton Bros, famous "drag" appeared.

1835. Dietz credited with being the first to use solid india rubber tires. Comte d'Asda drove a Maceroni and Squire steam carriage from Brussels to Paris.

1837. Thomas Davenport brought out his rotary armature electric motor.

1838. William Barnet of England invented a double-acting gas engine. Jump coil made by Dr. C. G. Page.

1839. Robert Anderson of Aberdeen, Scotland, drove a carriage powered by a primitive electric motor. The first electric carriage.

1841. First tractor, or steam "cart horse," by William Worby of England, appeared. F. Hills's steam carriage did 25 miles per hour and covered 128 miles in one day over heavy roads.

1844. S. Perry patented air and water cooled engines; tube ignition.

1845. Robert William Thompson of England patented the pneumatic tire. Dr. A. Drake's engine, the result of experiments begun in 1835, placed on exhibition.

1846. The caterpillar tractor appeared.

1848. The sedan chair, one of the oldest forms of transport, last publicly offered for hire in Edinburgh.

1851. W.M. Storm patented an engine in which gas was compressed before ignition and fired by a jump spark.

1853 JK. Fisher of New York built a steam carriage which ran 15 miles per hour and remained in service two years.

1855. Drake (United States) introduced incandescent metal as ignition method for gaseous mixtures.

1856. Single front steering wheel first mentioned Lotz. Richard Dudgeon of New York City built his first steam carriage; destroyed in the Crystal Palace fire, London, England. Duplicate, built in 1860, carried owner many miles over the space of several years. Still run occasionally by his son.

1860. Lenoir of Paris perfected the gas engine, using electric spark ignition. Placing one in a carriage, he made a three-mile trip.

1862. Beau de Rochas suggested the four-cycle motor.

1863. Important patents granted to Mackenzie in England covering intermediate clutch or "disconnecting device," etc. Brothier in Paris demonstrated a compression type motor. Great Britain, after a long series of restrictive laws, placed on the statute books a law requiring a self-propelled vehicle to be preceded by a man carrying a red flag.

1866. Otto and Langen brought out their improved gas engine using the four-cycle sequence ; patented in United States the next year.

1867. Thompson placed pneumatic tires on steam carriages. Robert McLaughlin, founder of McLaughlin Carriage Co., predecessor of General Motors of Canada, Ltd., began carriage manufacture.

1868. Charles Ravel of Paris produced a steam tricycle with direct drive.

1869. Manufacture of vehicles begun in Flint, Michigan, by W.A. Paterson, one of the first vice-presidents of Buick Motor Company.

1870. Julius Hock of Vienna produced a practical but very low-powered petroleum engine no compression. Todd of England brought out a small steam carriage for two persons which might have become a commercial success except for legal restrictions on use of roads.

1871. Dr. J. M. Carhart built at Racine, Wisconsin, a steam buggy.

1872. George B. Bray ton, an Englishman living in Boston, applied for a patent on a two-cylinder gasoline engine, "earliest to use fuel oil instead of gas." (Enc. Brit.)

1873. George B. Selden of Rochester, New York, began experiments on fuels for internal combustion engines. In Paris, Amadee Bollee built the famous steam omnibus, L'Obesiante, which 22 years later entered the Paris-Bordeaux race and covered the route.

1875. The State of Wisconsin offered a bounty of $10,000 to the inventors of a successful steam carriage.

1877. Siegfried Markus of Vienna placed a petrol engine in a carriage. Sometimes dated as early as 1869.

1878. Robert McLaughlin began operations at Oshawa, Ontario.

1879. Pioneer patent of American automobile industry applied for by George B. Selden. Wisconsin paid a $5,000 bounty (see 1875).

1880. Lawson of England invented an engine driven by the explosion of gasoline.

1881-82. In France and England, electric tricycles were brought out.

1883. Delamare-Debouteville in France and Edward Butler in England came forward with motor tricycles. Butler's was driven by the explosions of benzoline vapors; the Frenchman's by explosions of illuminating gas. De Dion, Bouton, and Trepardoux began the manufacture of steam cars, one of which won the Paris-Marseilles race of 1897. By 1889 one of their steam drags had hauled heavy loads at 24 m.p.h.

1884. John and Thomas Clegg built the first self-propelled vehicle in Michigan a four-wheeled steamer. In operation, 1885. The first oil motor car took the road, by Delamare-Debouteville and Malandin, Paris. The car carried a carbureter by Malandin.

1885. Gottlieb Daimler (Germany) invented his famous petrolvapor engine on the Otto cycle, the first motor to be manufactured in quantity. Fernand Forest (France) built a four-cylinder motor and installed improved carburetion. Carl Benz brought out a three-wheeled motor car, using vapor of benzine.

1886. Daimler applied his engine to a bicycle and developed novel features, notably the bubbling carbureter. Bcnz developed a car with improved devices for variable speeds, reaching 15 m.p.h. L. E. McKinnon, founder of McKinnon Industries, Ltd., St. Catharines, Ontario, began manufacturing at Buffalo, New York.

1887. Daimler brought out his first car. Ransom E. Olds's first steamer appeared, steam being generated by burning gasoline. Radcliff Ward's electric cab appeared, followed shortly by his electric omnibus.

1888-89. Leon Serpollet placed his flash or instantaneous generation boiler in a tricycle, then in a carriage. This boiler gave steam cars a new lease on life.

1889. First American electric built by William Morrison of Chicago. New Departure Mfg. Co. organized at New Britain, Connecticut.

1890. Olds Gasoline Engine Works organized at Lansing, Michigan, for $30,000. Leland, Faulconer & Norton Co. organized in Detroit.

1891. Clincher rim patented by Thomas B. Jeffery. Scientific American records the sale of a R. E. Olds steam carriage to the Francis Times Co., Bombay, India. First American self-propelled vehicle sold for export.

1892. Charles E. and Frank Duryea built and ran successfully the first gasoline car made in America. A "horseless buggy." Hyatt Roller Bearing Company organized, Harrison, New Jersey.

1893. Henry Ford built and ran his first car in Detroit. First importation of an automobile into the United States a Benz shown at the World's Fair, Chicago. Pontiac Buggy Company incorporated at Pontiac, Michigan.

1894. First gasoline automobile driven in Detroit by Charles B. King, later with Olds. Olds began work on his first gasoline car at Lansing, Michigan, and Elwood Haynes at Kokomo, Indiana. Panhard and Levassor, Paris, with French rights on the Daimler motor, led the industry in design, placing the motor under the hood and otherwise creating a car along modern lines. First road-race, Paris to Rouen, 78 miles. All of the starters finished. Won by de Dion with an average speed of 12 m.p.h.

1895. Olds brought out its first car powered with an internal combustion motor. Leland & Faulconer Mfg. Co., forerunner of Cadillac, formed in Detroit by Henry M. Leland. First American Automobile race, organized by the TimesHerald, Chicago, won by Duryea. American Motor League, first association of Automobile enthusiasts, completed its organization in Chicago, on eve of above race. Three hundred cars in production in United States. Paris-Bordeaux race, 732 miles, won by Peugeot on a technicality. Panhard finished first. Selden Patent granted.

1897. Olds Motor Vehicle Company organized. Application made to British Patent Office for electric selfstarter by E. J. Clubbe and Alfred W. Southey. First Oldsmobile produced on display in Smithsonian Institution, Washington, D. C. Reliability run from Cleveland to New York made by Alexander Winton.

1898. Chambered spark plug patented by Frank W. Canfield, Michigan lumberman. Aluminum alloy introduced by Haynes. Electric taxicabs placed in service in New York City. First American sale of gasoline car delivered to purchaser Winton.

1899. Olds Motor Works organized and began building, in Detroit, first factory for automobile production. Automobile Club of America founded.

1900 First automobile advertisement placed in the Saturday Evening Post. First automobile show held in Madison Square Garden, New York City. National Automobile Chamber of Commerce formed. First American car mounted power plant in front Columbia. Connecticut enacted the first automobile traffic law. Olds Motor Works begins production of famous curved-dash runabout, first American car to be manufactured in quantity. First long-distance automobile race, New York City to Buffalo, 500 miles. Of 80 cars, 42 reached Rochester, where tour was abandoned. Roy D. Chapin (Secretary of Commerce, 1932) drove Oldsmobile curved-dash runabout from Detroit to New York City first light car to make the trip. First appearance of five-passenger body designs in United States. David D. Buick established Buick Auto-Vim & Power Co., Detroit. 1902. Cadillac Automobile Company organized. Chrome-nickel and tungsten steels introduced. David D. Buick adapted his marine motor to a "horseless buggy."

1903. Buick Motor Company organized; capital, $75,000. One-cylinder Cadillac appeared 1,895 sold the first year. Ford Motor Company organized. Windshields, front radiators, tilted steering columns, and canopy tops appeared. A. O. Smith & Company made the first pressed steel frame. Association of Licensed Manufacturers formed to manufacture under Selden patent A.L.A.M. Oldsmobile production of curved-dash runabouts reached 4,000. Automobile production in U. S. reached $6,250,000, nearly half of it being Olds. Packard moved from Warren, Ohio, to Detroit.

1904. Cadillac Motor Car Company organized by merger of Cadillac Automobile Company and Leland & Faulconer. Oldsmobile production reached 5,508. Reorganized August, I, 1904. First Vanderbilt cup race, October 8th. Innovations: straight-eight engine, shock-absorbers, pressure engine lubrication, automatic carbureters.

1905. Buick production rose to 750 cars. First Glidden endurance contest. Gus Edwards wrote his famous song "In My Merry Oldsmobile." The Cadillac "30" introduced. Total production of this model over several years, nearly 68,000. American Motor Car Manufacturers Association organized. Knight invented sleeve-valve engine. Innovations included magneto ignition, ignition locks, roadster and touring bodies.

1906. Buick produced its first four-cylinder engine and first sliding gear transmission. Weston-Mott Company moved from Utica, New York, to Flint, Michigan. Introduction of front bumpers, vibrating horns, asbestos brake linings, high tension magnetos, drop steel frame, and air brakes.

1907. Fifth Avenue coach service inaugurated, New York City, and metered taxicabs also appeared there. Oakland Motor Car Company incorporated, Pontiac, Michigan. Multiple-disc clutch introduced. Demountable rims appeared; also magnetic drag speedometers. McLaughlin Motor Car Co., Ltd., organized at Oshawa, Ontario.

1908. July 22nd, Fisher Body Company organized. September 16th, General Motors Co. of New Jersey organized by W. C. Durant and incorporated. Capital increased to $12,500,000 on September 29th. General Motors bought Buick and Oldsmobile. In 1908 Buick produced nearly 8,500 cars. Appearance of motor-driven horns, sleeve-valve engines, silent timing gear chains, left-hand steering, unit power plants, and baked enamel finish. Cadillac won the Dewar Trophy, London, for greatest contribution to motoring for the year interchangeability of parts.

1909. General Motors acquired Oakland, January 8th, and Cadillac, July 27th ; also control of AC Spark Plug. General Motors paid first dividend on Preferred stock. Selden patent sustained by Judge Hough, United States District Court. Here entered electric head-lights, electric generator, four-door bodies, oil gauges on instrument board, etc.

1910. Barney Oldfield cut automobile speed record for the mile to 40.53 seconds from a standing start at Daytona, Florida. Buick brought out its first six-cylinder car. General Motors borrowed $15,000,000 from J. & W. Seligman & Company and Lee, Higginson & Company. New officers and directors elected. Harrison Radiator Co. organized at juockport, New York. Cadillac installed battery ignition.

1911. Selden patent declared invalid in United States Court of Appeals, Second Circuit, Judge Noyes writing the decision. Burman set a new speed record at Jacksonville, Florida, in a "Buick bug" 20 miles in 13 minutes, 11.92 seconds. W. C. Durant bought Flint Wagon Works, present site of Chevrolet home plant. Little Motor Car Company formed to build a light car at Flint, Michigan. Chevrolet Motor Company of Michigan organized by W. C. Durant. General

Motors Export Company organized, June 11th. Electric starting device, by C. F. Kettering, installed on Cadillac. Other innovations of the year: detachable rims, worm gear drive for trucks, improved electric horns. First automobile securities listed on New York Stock Exchange, General Motors voting trust certificates, July 31, 1911. Buick produced its first closed car a limousine.

1912 Cadillac pioneered in installing electric self-starter as standard equipment. Won Dewar Trophy, London, for the second time for this contribution, the greatest of the year. General Motors of Canada organized. Chevrolet absorbed Little and concentrated manufacturing at Flint.

1913. National Automobile Chamber of Commerce organized. Instalment paper first used in financing automobile sales San Francisco. November i8th, C. W. Nash became president of General Motors in succession to Thomas Neal. Wire wheels first used on stock cars. Bendix drive. Chevrolet took over Tarrytown, New York, plant.

1914. Cadillac pioneered the v-type, 8-cylinder high-speed engine. One of the early Cadillacs so equipped is on display at the Smithsonian Institution, Washington, D. C. Vacuum fuel tanks made their appearance.

1915 Chevrolet Motor Company organized in Delaware. Buick brought out its first "six" and Packard, a "twelve." Innovations included aluminum pistons, torsional vibration damper, and self-locking differential. September 16th, General Motors declared first dividend on Common stock $50 a share. General Motors Voting Trust wound up.

1916. W.C. Durant became president of General Motors. October 13, General Motors Corporation was organized under Delaware law, to acquire all assets of the General Motors Company, a New Jersey corporation. Fisher Body Corporation organized under New York laws. United Motors Corporation incorporated, May 16th, a merger of Hyatt Roller Bearing Company, Dayton Engineering Laboratories Company, Remy Electric Company, New Departure Manufacturing Company, Perlman Rim Corporation. December 31, the subsidiaries were set up as divisions of the General Motors Corporation.

1917. Cadillac adopted as standard for war use after exhaustive tests at Marfa, Texas, in July. Buick capacity enlarged to 750 cars a day. General Motors Company of New Jersey dissolved, August 13. Steel disc wheels and steel felloes for wooden wheels appeared.

1918. May 2nd, General Motors Corporation bought Chevrolet Motor Company. Capital of General Motors Corporation rose during year to $370,000,000. Ethyl gasoline developed by General Motors Laboratories. General Motors acquired United Motors Corporation, also completed ownership of McLaughlin Motor Car Co., Ltd., and Chevrolet of Canada. E. I. du Pont de Nemours & Company announced acquisition of 27.6 percent interest in Common stock of General Motors Corporation. Stream-line bodies and adjustable three-piece windshields appeared. General Motors of Canada, Ltd., incorporated, Oshawa, Ont.

1919. General Motors Acceptance Corporation organized. Employes Savings and Investment Fund inaugurated by General Motors. Authorized capital stock of General Motors Corporation increased to $1,020,000,000. Construction of General Motors Building begun in Detroit; completed 1922, at cost of more than $19,000,000. Three fifths interest in Fisher Body purchased by General Motors. General Motors Institute of Technology opened with 300 students; has grown since to 2,000 students.

1920. General Motors Research Corporation incorporated. Pierre S. du Pont became president of General Motors.

1921. Executive offices of General Motors moved to New York City. W. C. Durant retired from the directorate of General Motors, of which he had been a member since 1908. Inter-divisional bodies set up by General Motors to integrate policies and practices. Nickel-plated radiators and lamps.

1922. Balloon tires introduced.

1923. Alfred P. Sloan, Jr., became president of General Motors. General Motors completed its ownership of Brown-LipeChapin Co., Syracuse, New York, manufacturers of differentials. Cadillac adopted two-plane compensated crankshaft. Duco lacquer finish became standard for Oakland. Alemite high-pressure chassis lubrication introduced.

1924. General Motors Proving Ground purchased near Milford, Michigan. Four wheel brakes perfected by Buick in 1923, appeared on its 1924 models. Ethyl Gasoline Corporation formed by General Motors and Standard Oil Company of New Jersey to market "knockless" motor fuel component developed by General Motors Research Laboratories. Oil filters introduced. W. S. Knudsen became president of Chevrolet.

1925. Yellow Truck & Coach Manufacturing Company established. General Motors Truck transferred to it, August 13th. Electric transmission used on buses. Fisher acquired Fleetwood.

1926. Ownership of Fisher Body Corporation completed by General Motors. The Pontiac car made its appearance. Safety glass introduced. Capital of General Motors Acceptance Corporation increased to $25,000,000.

1927. General Motors Corporation distributed to Common stockholders $134,836,081, largest sum paid to date in one year by an industrial enterprise. La Salle introduced by Cadillac. Chromium plating used on stock cars.

1928. Synchro-mesh transmission introduced by Cadillac. Mechanical fuel feed pump brought out by AC Spark Plug Co.

1929. General Motors sold 1,899,267 cars and trucks. Pierre S. du Pont retired as chairman of the board of directors of General Motors Corporation after service from November 16, 1915.

1930. Cadillac introduced America's first sixteen-cylinder engine. Oil cooling systems; also increased use of stainless steel. General Motors had outstanding 43,500,000 shares of Common stock.

1931. Buick brought out its first eight-cylinder engine.

1933 No Draft Ventilation built into all Fisher bodies. "Knee Action" individual front wheel suspension by means of soft springs installed on General Motors cars.

# APPENDIX II

*List of Officers and Directors of General Motors Company (New Jersey) and General Motors Corporation (Delaware) from time of commencing business to date. General Motors Corporation succeeded General Motors Company as of August I, 1917. In preparing this list the two companies have been treated as one.*

* indicates still holding office (December 31, 1933.)

| **CHAIRMEN OF THE BOARD** | **From** | **To** |
|---|---|---|
| Thomas Neal | Nov. 19, 1912 | Nov. 16, 1915 |
| Pierre S. du Pont | Nov. 16, 1915 | Feb. 7, 1929 |
| Lammot du Pont | Feb. 7, 1929 | * |

| **DIRECTORS** | | |
|---|---|---|
| George F. Baker[1] | July 15, 1920 | * |
| H. M. Barksdale | June 27, 1918 | Nov., 1918 |
| H. H. Bassett | May 1, 1919 | Oct. 17, 1926 |
| F. L. Belin | Nov. 16, 1915 | Aug. I, 1917 |
| A. M. Bentley | May 12, 1910 | Nov. 15, 1910 |
| A. G. Bishop | Nov. 1 6, 1915 | * |
| Joseph Boyer | May 15, 1911 | Nov. 16, 1915 |
| Albert Bradley | Nov. 9, 1933 | * |
| Anthony N. Brady | Nov. 15, 1910 | Sept., 1913 |
| Arthur W. Britton | Sept. 22, 1908 | Oct. 20, 1908 |
| Donaldson Brown | Dec. 30, 1920 | * |
| Arthur P. Bush | Nov. 15, 1910 | Nov. 23, 1910 |
| E. R. Campbell | April 1, 1910 | Nov. 15, 1910 |
| Herbert L. Carlebach | Nov. 15, 1910 | Nov. 23, 1910 |
| Walter S. Carpenter, Jr. | Feb. 10, 1927 | * |

[1] Son of George F. Baker, Sr. (1840-1931.)

| | | |
|---|---|---|
| Arthur Chamberlain | May 13, 1920 | Jan. I, 1923 (accepted 4-11-23) |
| W. P. Chrysler | June 27, 1916 | Mar. 25, 1920 |
| Emory W. Clark | Nov. 23, 1910 | June 27, 1916 |
| R. H. Collins | July 27, 1916 | June 30, 1921 |
| Curtis C. Cooper | Aug. 11, 1927 | Apr. 30, 1930 |
| George E. Daniels | Sept. 22, 1908 | Dec. 16, 1908 |
| W. L. Day | June 27, 1916 | May 13, 1925 |
| H. F. du Pont | Feb. 21, 1918 | * |
| Irenee du Pont | Feb. 21, 1918 | * |
| Lammot du Pont | Nov. 17, 1918 | * |
| Pierre S. du Pont | Nov. 16, 1915 | * |
| W. C. Durant | Oct. 20, 1908 | Apr. 20, 1921 |
| William M. Eaton | Oct. 20, 1908 | Nov. 23, 1910 |
| Charles T. Fisher | Sept. 25, 1924 | * |
| Fred J. Fisher | June 30, 1921 | * |
| Lawrence P. Fisher | Sept. 25, 1924 | * |
| William A. Fisher | Feb. 10, 1927 | * |
| Andrew H. Green | Nov. 23, 1910 | Nov. 18, 1913 |
| Harry G. Hamilton | Oct. 19, 1909 | March 20, 1910 |
| Geo. H. Hannum | Apr. 19, 1922 | May 13, 1925 |
| A. B. C. Hardy | Apr. 19, 1922 | May 13, 1925 |
| J. A. Haskell | Nov. 16, 1915 | Sept. 9, 1923 |
| C. R. Hatheway | Oct. 20, 1908 | Nov. 15, 1910 |
| Henry Henderson | Oct. 20, 1908 | Nov. 15, 1910 |
| Robert Herrick | Nov. 18, 1913 | Nov. 16, 1915 |
| F. W. Hohensee | June 27, 1918 | Jan. 13, 1921 |
| L. G. Kaufman | Nov. 16, 1915 | * |
| Charles F. Kettering | Dec. 30, 1920 | * |
| Schuyler B. Knox | May, 12, 1910 | Nov. 15, 1910 |
| William S. Knudsen | Apr. 16, 1924 | * |
| W. C. Leland | May 12, 1910 | Nov. 15, 1910 |
| W. C. Leland | June 1, 1916 | July 23, 1917 |
| Richard Lukeman, Jr. | Nov. 15, 1910 | Nov. 23, 1910 |

| | | |
|---|---|---|
| J. H. McClement | Nov. 15 1910 | Apr. 28, 1920 |
| Sir Harry McGowan | May 13, 1920 | * |
| Benjamin F. McGuckin | Nov. 15, 1910 | Nov. 23, 1910 |
| R. S. McLaughlin | May 12, 1910 | Nov. 15, 1910 |
| R. S. McLaughlin | Nov. 7, 1918 | * |
| William McMaster | May 13, 1920 | May 5, 1930 |
| Benjamin Marcuse | Sept. 22, 1908 | Dec. 16, 1908 |
| W. J. Mead | Oct. 20, 1908 | Nov. 15, 1910 |
| Edwin D. Metcalf | Mar. 9, 1911 | Nov. 16, 1915 |
| James D. Mooney | Nov. 8, 1923 | * |
| Junius S. Morgan, Jr | Sept. 30, 1925 | * |
| C. S. Mott | Nov. 18, 1913 | * |
| M. J. Murphy | Nov. 23, 1910 | Nov. 16, 1915 |
| C. W. Nash | Nov. 19, 1912 | Sept. 26, 1916 |
| Thomas Neal | Nov. 23, 1910 | July 27, 1916 |
| Fritz Opel | May 8, 1929 | * |
| DeWitt Page | Apr. 18, 1923 | * |
| John L. Pratt | Apr. 18, 1923 | * |
| Seward Prosser | July 15, 1920 | * |
| S. F. Pryor | Nov. 16, 1915 | June 27, 1916 |
| Arthur B. Purvis | June 2, 1930 | * |
| J. J. Raskob | Nov. 16, 1915 | * |
| George Reichert, Jr. | Nov. 15, 1910 | Nov. 23, 1910 |
| H. H. Rice | Apr. 19, 1922 | May 13, 1925 |
| Henry Russel | Dec. 16, 1908 | Oct. 19, 1909 |
| C. H. Sabin | Nov. 16, 1915 | Jan. 8, 1917 |
| Alfred P. Sloan, Jr. | Nov. 7, 1918 | * |
| Fred L. Smith | Dec. 16, 1908 | Aug. 10, 1909 |
| John T. Smith | Oct. 19, 1909 | Nov. 15, 1910 |
| John T. Smith | Mar. 25, 1920 | * |
| Edward R. Stettinius, Sr. | July 15, 1920 | Sept. 3, 1925 |
| Jas. J. Storrow | Nov. 15, 1910 | June 27, 1916 |
| Albert Strauss | Nov. 15, 1910 | June 1, 1916 |
| A. H. Swayne | Jan. 13, 1921 | * |

| | | |
|---|---|---|
| Nicholas L. Tilney | Nov. 15, 1910 | Nov. 16, 1915 |
| Edward VerLinden | Aug. 22, 1917 | Apr. 19, 1922 |
| James N. Wallace | Nov. 23, 1910 | Nov. 12, 1912 |
| F. W. Warner | June 27, 1916 | Sept. 22, 1921 |
| Jacob Wertheim | Mar. 9, 1911 | Nov. 16, 1915 |
| George Whitney | Sept. 25, 1924 | * |
| A. H. Wiggin | Nov. 16, 1915 | Nov. 21, 1916 |
| Wm. H. Woodin | July 15, 1920 | Aug. 11, 1927 |
| C. M. Woolley | July 15, 1920 | * |
| Owen D. Young | July 15, 1920 | * |
| K. W. Zimmerschied | Dec. 30, 1920 | Apr. 18, 1923 |

**PRESIDENTS**

| | | |
|---|---|---|
| George E. Daniels | Sept. 22, 1908 | Oct. 20, 1908 |
| Wm. M. Eaton | Oct. 20, 1908 | Nov. 23, 1910 |
| Jas. J. Storrow | Nov. 23, 1910 | Jan. 26, 1911 |
| Thomas Neal | Jan. 26, 1911 | Nov. 19, 1912 |
| C. W. Nash | Nov. 19, 1912 | June 1, 1916 |
| W. C. Durant | June 1, 1916 | Nov. 30, 1920 |
| Pierre S. du Pont | Nov. 30, 1920 | May 10, 1923 |
| Alfred P. Sloan, Jr. | May 10, 1923 | * |

**VICE-PRESIDENTS**

| | | |
|---|---|---|
| H. H. Bassett | May 1, 1919 | Oct. 17, 1926 |
| A. G. Bishop | Nov. 16, 1915 | May 1, 1930 |
| Albert Bradley | May 9, 1929 | |
| Donaldson Brown | Dec. 30, 1920 | * |
| W. P. Chrysler | Dec. 14, 1916 | Mar. 25, 1920 |
| Emory W. Clark | Nov. 21, 1911 | June 27, 1916 |
| R. H. Collins | Aug. 3, 1917 | June 30, 1921 |
| W. L. Day | Dec. 14, 1916 | Aug. 13, 1925 |
| W. C. Durant | Oct. 20, 1908 | Nov. 16, 1915 |
| Charles T. Fisher | May 12, 1927 | * |
| Fred J. Fisher | Feb. 9, 1925 | * |
| Lawrence P. Fisher | Feb. 9, 1925 | * |

| | | |
|---|---|---|
| William A. Fisher | Feb. 10, 1927 | * |
| A. R. Clancy | May 9, 1929 | May 6, 1931 |
| R. H. Grant | May 9, 1929 | * |
| George H. Hannum | June 30, 1921 | Aug. 13, 1925 |
| A. B. C. Hardy | June 30, 1921 | Aug. 13, 1925 |
| J. A. Haskell | June 27, 1918 | Sept. 9, 1923 |
| F. W. Hohensee | May 1, 1919 | Jan. 13, 1921 |
| 0. E. Hunt | May 9, 1929 | * |
| Charles F. Kettering | Jan. 13, 1921 | * |
| William S. Knudsen | Apr. 22, 1924 | Oct. 16, 1933 |
| Exec. Vice-Pres. | Oct. 16, 1933 | * |
| W. C. Leland | Dec. 14, 1916 | July 23, 1917 |
| R. S. McLaughlin | Dec. 31, 1918 | * |
| W. J. Mead | Oct. 20, 1908 | Nov. 18, 1913 |
| James D. Mooney | Nov. 16, 1922 | |
| C. S. Mott | Nov. 21, 1916 | * |
| C. W. Nash | July 12, 1912 | Nov. 19, 1912 |
| Thomas Neal | June 27, 1916 | July 27, 1916 |
| DeWitt Page | Apr. 19, 1923 | * |
| John L. Pratt | Nov. 16, 1922 | * |
| J. J. Raskob | Dec. 31, 1918 | May 9, 1929 |
| 1. J. Reuter | May 9, 1929 | Oct. 21, 1933 |
| H. H. Rice | July 10, 1919 | Aug. 13, 1925 |
| Alfred P. Sloan, Jr. | Dec. 31, 1918 | May 10, 1923 |
| Fred L. Smith | Dec. 16, 1908 | Aug. 10, 1909 |
| John T. Smith | Mar. 25, 1920 | * |
| E. R. Stettinius, Jr.[1] | 2 May 6, 1931 | Dec. 31, 1933 |
| E. T. Strong | Feb. 10, 1927 | May 4, 1932 |
| Alfred H. Swayne | Jan. 13, 1921 | * |

[1] Son of E. R. Stettinius. (1865-1925) See p.450

# APPENDIX III General Motors at the Century of Progress Exposition

AS THE Century of Progress International Exposition at Chicago coincided with the twenty-fifth anniversary of General Motors, the Corporation decided upon active participation in the Exposition, where it became the largest exhibitor. The official Exposition folder, issued in advance of the opening, contained a picture of the General Motors Building erected there, and this description:

> General Motors' quarter of a century of important participation in the industrial and economic progress of America will be dramatized and typified by the Corporation's exhibit at the Exposition. The General Motors Building, designed by Albert Kahn, noted Detroit architect, and decorated by artists and sculptors of international fame, dominates the brilliant skyline of the Exposition with its 177-foot tower. The building will be filled with displays which adequately picture the extent to which the products of General Motors reflect its life-long adherence to the ideal of progress.

This building, boldly modern in architecture, was the tallest in the Exposition grounds, the tower reaching an elevation of 177 feet, while the ground dimensions of the steel-and-armor ply structure were 454 by 306 feet. It contained more than 950 tons of steel. The armor ply was composed of five thicknesses of wood covered on both sides by metal sheets. The walls were almost entirely glass, the semi-circular front of the building consisting of huge plate glass show windows. Each of these bays was n feet high and 48 feet long, and was formed by three sections of plate glass. Over each bay appeared the name of a General Motors division.

The interior was divided into seven main rooms: the entrance hall, two large display rooms on either side, the Chevrolet assembly room, the truck display room, the accessory display room, and the Little Theatre.

In the central entrance hall, which was 118 feet long, 55 feet wide, and 50 feet high, art objects were exhibited in a dignified setting on a scale worthy of the largest art museum. There stood the heroic statue of a workman by Carl Milles, the famous Swedish sculptor, who caught in this impressive work the spirit of the precision worker. Scenes depicting various manufacturing processes occupied the spaces on the side walls between the huge columns, serving as backgrounds for life-sized figures of workmen in action. Carl Hallstammer is the famous artist who carved these figures out of huge blocks of laminated pine. Nine mural marquetries in wood inlay by Matt Faussner, of Munich, decorated the display room. Each of these marquetries measures six by twelve feet, the largest area ever covered by workers in this delicate art.

Around the gallery of the Chevrolet assembly room were hung painted panels on which were depicted scenes from the industrial activities of various states. In this assembly room, factory operations proceeded as practised by Chevrolet in its assembly plants in many cities. Onlookers saw here, for the first time in any exposition, the elaborate synchronization of effort through which the automobile takes its finished shape, until from a multitude of parts a complete automobile is ready to run on its own power. This exhibit contained an automobile and body assembling plant, the latter revealing the exquisite work of the Fisher organization.

The opening of the General Motors Building at the Exposition on May 24, 1933, struck a new note in the adaptation of modern ideas of form and decoration to the requirements of industrial construction. The architect, Albert Kahn of Detroit, will be long remembered for his work on the structures of that city, including the golden Fisher Tower in Detroit and those which he did for the University of Michigan group at Ann Arbor, Michigan. At Chicago Mr. Kahn had to meet new problems, providing for the utilitarian housing of operating exhibits as well as the display of General Motors products, ranging from automobiles and Diesel engines to the smallest ball bearings. Among the special installations was a Little Theater with a capacity of 235 persons, where visitors were entertained with General Motors films. Another unusual feature to be accommodated was a research laboratory exhibit where General Motors engineers revealed some of the scientific aspects of automobile production. Here the most vivid display was an artificial rainbow, illustrating on a grand scale the use of the spectroscope whose sensitivity in minute measurements contributes to the long life and easy operation of the modern motor car.

In designing a building for these requirements, the architect had to remain within the general architectural and landscaping plan specified for all the structures in the Exposition, under which all the buildings were kept in harmony with one another. The location, between Lake Michigan and Leif Erikson Drive, was one that loaned itself exceptionally well to the elaborate landscaping undertaken.

The color scheme for the whole Transportation Group, of which the General Motors Building was a member, followed a daring use of gray, bright orange, red, blue, and silver. Flood-lighting in changing colors made the Joseph Urban color scheme as vivid by night as by day.

The Craftsman's Guild awards of the Fisher Body Company took place at the Exposition in August.

When the General Motors Exposition Building was dedicated, Rufus M. Dawes, president of the Century of Progress, told the assembled notables that but for General Motors encouragement the work could hardly have gone forward. Mr. Dawes said, in part:

> I cannot tell you what encouragement we received when we came into contact with the General Motors Corporation through W. S. Knudsen, president of Chevrolet. Instead of surrendering in the face of difficulties, General Motors saw in them only opportunity. Instead of extending a casual and careless support, it announced in July, 1931, its intention to erect a great building, and to establish a modern assembly line on our exposition grounds. From the moment of this announcement, there has never been a doubt of the success of the Century of Progress Exposition.

Mr. Dawes added another high compliment when he stated that

> So far as I can learn, we hold this meeting in the greatest temporary building ever erected by a private corporation in any world's fair, and shall witness exhibits more expensive and more informative than have heretofore been made by any company as part of a great exposition.

# APPENDIX IV

*Here are grouped short histories of a number of General Motors subsidiaries and affiliates not treated in detail elsewhere. In Appendix V are brief accounts of certain companies now inactive or discontinued.*

### AC SPARK PLUG COMPANY, Flint, Michigan

IN THE Gay 'Nineties the question, "What is a spark plug?" had no answer, for the internal combustion engine was still being fired by means of a hot tube, a bolt of metal, one end extending into the gas chamber of the cylinder, the other, outward, so that it could be heated. One can well imagine the difficulties of heating the tube and keeping it hot. The first spark plugs were intended of course to deliver an igniting spark to the fuel at just the right time, but they fell short of their purpose, so far as to compel the sardonic query, "Why is a spark plug?" The querist, no doubt, was one of those who had tasted experience with these early contraptions, "bought by the peck and thrown away by the bushel."

The AC Spark Plug Company was founded by W. C. Durant and Albert Champion at Flint, Michigan, in 1908, with Mr. Champion as president. It became a subsidiary of General Motors when Mr. Durant sold his interest in it to the Corporation in 1910-11. Mr. Champion was a tireless worker, constantly improving his product and adding new ones. From boyhood he had been connected with transport. He had raced on bicycles, and other foot-driven cycles, motorcycles, and automobiles, having also acquired a mechanic's training in the case of the automobile. In bicycling he stood as world's champion, contesting both in Europe, and the United States, while in automobile racing he ranked among the foremost, his last feat being nothing less than the winning of 18 races in France. After this he returned to America and began making spark plugs at Boston in 1905. Three years later he removed to Flint. His greatest pride was that his product had played an important part in making safe the flights of Lindbergh, Chamberlin, Byrd, and other long-distance flyers.

The company started its operations in a corner of the Buick building with 15 employees turning out a few hundred spark plugs per day. The original quarters in Flint were soon outgrown, and a two-story building was erected, the first on the present site. In 1912 the company had a production of 2,000 spark plugs a day and was working at capacity. Car and truck production in the United States that year was only 378,000, and the total number in the country but 902,000. As the business grew, new buildings were added and increased effort made to meet changes in engine design. The research department kept pace with requirements and went ahead to develop new products. The AC aircraft plug was one of these products, ready when the need came. It anticipated the entry of the United States into the World War in 1917, was soon available, and subsequently was furnished the government at the rate of 50,000 a day for

Liberty, Wright, and Hispano-Suiza engines. The Flint Faience & Tile Company was organized in 1921 by AC because of the relations of its products to the porcelain industry.

ALBERT CHAMPION

In 1922, after previous trial and error, followed by two years of research, the model B speedometer was put on production. Within three years AC became the world's largest makers of speedometers.

The growth of the plant continued as new products were made available until the buildings covered sixteen acres of the twenty-acre site. Employees increased to 4,000, the output of spark plugs to more than a quarter million daily, and of other units to more than seventy thousand representing nineteen newer products. AC became the largest business of its kind anywhere.

AC spark plugs are made also at St. Catharines, Ontario; abroad at Clichy, in Paris; and at Birmingham, England. The start of the overseas business was made in France in 1922 by the acquisition of a spark plug company operating a plant near Clichy. Within six months AC spark plugs were being made there with machinery and clay shipped from Flint. A modern plant was then built at Clichy. Here the clay from Flint is made into insulators by the same processes as used in the home plant; shell and electrode operations, assembly and inspection are similar.

In 1928 there were about 200 employees, with M. Edouard de May as chief engineer in charge of manufacturing. This French company, the Societe des Bougies AC Titan, now a unit of General Motors (France), supplies spark plugs and several other AC products to automobile owners in a score or more of European and French-controlled countries, in addition to supplying AC products as original equipment for over two dozen foreign makes of automobiles. In the countries being served, there were above 1,640,000 car registrations.

The organization in Birmingham is known as the AC Sphinx Sparking Plug Company, Ltd. Whether spark plug, bougie or sparking plug, it is all the same to AC. The English company, also acquired in 1921, was immediately reorganized and the plant was rebuilt. In 1928 there were about 200 employees under the management of Mr. Norman F. Stockbridge. From this plant AC products are supplied to wholesalers and dealers in the British Isles, Germany, the buffer states on the east, the Scandinavian countries, Holland, Egypt, Iraq, Turkey, Persia, and Australasia. Some thirty-odd manufacturing companies also specify AC equipment. There were over 2,560,000 car registrations in the countries being served.

On the death of Mr. Champion in 1927, Mr. Basil W. de Guichard, long associated with him, succeeded to the presidency and Mr. Harlow H. Curtice to the vice-presidency. Two years later Mr. Curtice became president and general manager. He was then only thirty-six years old, but his service with the company dated from 1914, shortly after which he had become an executive believed to have been the youngest in the industry. In October, 1933, Mr. Curtice became general manager of Buick Motor Company, being succeeded as president and general manager of AC by T. S. Kimmerling, formerly general manager of Guide Lamp and more recently assistant to C. E. Wilson, vice-president of General Motors Corporation, in Detroit. In 1927, Mr. Taine G. McDougal, ceramic engineer of the company, was made a vice-president, with supervision over the AC plants in France and England.

Manufacture of all AC products, aside from plugs themselves, is now concentrated at the company's East Side plant in Flint (formerly the Dort Motor Car Company plant), resulting in increased manufacturing efficiency and production economies. From a production as well as an engineering basis, therefore, AC is now virtually divided into two units, the spark plug division, and the equipment group.

During the past decade as new products were added, manufacture was carried on in what at the time was the location least likely to interfere with other products in process. On an over-all plant basis, however, the arrangement left much to be desired as production rates increased.

The complete re-arrangement of AC's facilities recently, under the direction of Mr. Harlow H. Curtice, has corrected these conditions. One of the most important results is that manufacturing operations, which formerly were scattered through three plants with a total floor space of some 424,000 square feet, are now occupying only 251,050 square feet, a net saving in floor space alone of better than 40 percent. At the same time production capacity of individual product set-ups has actually been increased by the addition of new equipment, better handling methods, etc., and production capacity in most cases has virtually been doubled, as a result.

In developing the spark plug and keeping abreast of development of the engine, the company has made many noteworthy contributions to automotivation. Among these are listed:

One-piece patented spark plug and patented machinery used in its construction. A big advantage of the one-piece spark plug is that it is permanently sealed against compression leakage.

Welded side electrode, a patented process, which ensures perfect heat conductivity.

Isovolt electrode, made of a special alloy developed by the company. Contributes to easier starting, and longer electrode life. Manufacturers of radio appliances find use for this material in radio tubes.

Unglazed insulator tip. An exclusive AC provision against encrusting a common cause of spark plug trouble.

A "heat range" line of spark plugs, in which each plug has its individual heat characteristic. This makes it possible to choose the precise type spark plug that will function best for any particular driving or engine condition. One of the most important developments.

A spark plug of only 1 4-millimeter thread size.

A spark plug tester for mechanics the only "shop" tool that will accurately indicate the condition of a spark plug.

Surface combustion principle in firing electric insulators in tunnel kilns, an invention which permits equal and controllable heat to every unit.

Only spark plugs were made until the year 1919. Besides the spark plug development recited, the company is credited with other contributions to the industry, among these, the fuel pump, first introduced by AC about 1927, and now found as standard equipment on most cars. It has improved fuel feed systems from the standpoint of better operating performance and has enabled car manufacturers to effect a material saving in the cost of their product. Some eight million of the AC pumps are now in use.

Among later developments is the AC carbureter intake silencer, also standard equipment, developed in 1930 and since brought to a high state of perfection. It is built in combination with the air cleaner.

In the manufacture of instruments and instrument board panels for the automobile, AC is a leading factor.

In the research and control laboratories both manufacturing and control processes are worked out. Chemical, metallurgical, and ceramic research go on continually. The ceramic laboratory has direct control of every stage in the manufacture of porcelain insulators, from the receiving of the clay and other raw materials, through the stages of coremaking to the removal of the finished product from the tunnel kiln. The materials first pass through machines for mixing and grinding and are hydrated to a pulp called slip. The slip is pumped to filter presses, where a cake is made under high pressure in the course of three hours or so and then laid aside to age. In time the cakes are cut up and worked with a wedging machine, or kneading table, into the form of a circular clay column. A drying process then turns the column into a loaf of clay, which is fed to a press and packed into metal cylinders. Another type of press, the throwing machine, receives the cylinders, and through a die forces them out, giving them their first appearance as an insulator, the bottle shape. The "bottles" move along in trays to a dryer, come out, and are loaded into metal boxes, either for storage or for transfer to the turning lathes upstairs. The lathes make all the different styles of spark-plug cores. After drying and reaming, the plugs go to the glazing machines to receive not only the glaze but to be stenciled with the label AC. Special trays then come into use, holding the glazed cores en route to the firing. Mention has been made of the tunnel kilns as an AC development. They were designed by Mr. McDougal and have various heating and cooling zones. The cores from the kilns are conveyed to the proper departments to receive the work necessary to complete and assemble the spark plug, which includes the insertion and cementing of contact wires, fitting in a metal casing, packing, etc.

Products: AC spark plugs, AC Miko aviation plugs, radio plugs, spark plug testers, spark plug cleaning machines, speedometers, oil pressure gauges, ammeters, thermo gauges, gasoline gauges, techometers, instrument board panels, locker doors, carbureter intake silencers, oil filters, air cleaners, crankcase breather air cleaners, mechanical windshield wipers, fuel pumps, vacuum pumps, combination fuel and vacuum pumps, gasoline strainers, air-pressure gauges, ride-regulator controls, lubrometers, die castings, die-cast machines, flexible shafts and cables, Remo gum solvent injectors and fluid, Reflex warning signals, Flint Faience, Flintcraft, and Vitocraft tiles.

## ALLISON ENGINEERING COMPANY, Indianapolis, Indiana

Before the purchase of the Allison Engineering Company by General Motors Corporation early in 1929, the company had been owned and operated by Mr. James A. Allison, with Carl G. Fisher, a founder and original owner of the Prest-O-Lite Company. They also built the Indianapolis Motor Speedway in 1913 and for the purpose of encouraging automobile racing, organized the Indianapolis Speedway Race Team. Out of whose activities grew the present company.

Allison Engineering Company was organized in 1915 as Allison Experimental Company for the principal purpose of manufacturing fine racing automobiles and to experiment in automotive design and construction.

The first president of the company, James A. Allison, has been succeeded by N. H. Gilman, who also serves in the capacity of general manager and chief engineer. The name was changed to Allison Engineering Company in 1920.

With the entry of the United States into the war in 1917 the facilities of the Allison Experimental Company were offered for war purposes. The first Liberty Motor was constructed in this plant for one of the principal contractors on the production of these engines after which a complete complement of tools was supplied them.

Several contracts were entered into with the War Department on experimental tractor work with a drafting force of two hundred employees under the supervision of Army engineers. At the close of the war, the Navy Department desiring to carry on some of the projects which had been contemplated for the Bureau of Aeronautics, this company designed and built a number of special reduction gears for aircraft engines. Several modifications of the Liberty engine were also made in the years following the war, among them the inverted engines which were used on the "Good-Will" flight through South America. More than two thousand motors were re-worked in this plant for the Army Air Corps and some of the earlier transport companies.

For the past fifteen years the company has confined itself entirely to the aeronautical field with the exception of development work now being conducted in connection with some of the automobile divisions toward the adoption for car use of Allison steel-back bronze-lined bearings. This type of bearing was originated by the company several years ago for aircraft engines and constitutes one of the outstanding achievements in the aeronautical field of endeavor.

Various types of aircraft engines have been designed and built by the Allison Engineering Company for both the Army and Navy air services. In the airship field this company designed and built the propeller reduction and reverse gears for the Navy airship Shenandoah, also the drive units for the Akron and Macon.

At present a new series of liquid cooled v-type engines of 750-1000 h.p. is being developed.

Products: Aircraft power plant engineering, aviation engines, bearings, superchargers, gears and mechanical equipment

## DELCO APPLIANCE CORPORATION, Rochester, New York

ON OCTOBER 22, 1908, Mr. Edward A. Halbleib and his brother Mr. Joseph C. Halbleib organized the Rochester Coil Company. They began business with three employees, manufacturing coils and repairing electric motors. The next year they undertook the manufacture of generators for automobile lighting systems, in which the prospects were so bright that the company was reorganized September, 1909, as the North East Electric Company, capitalized at $125,000. The first North East system, consisting of a combined starter and generator, was placed in production in 1911. Growth of the company was rapid. The capital structure was increased from $1,000,000 to $3,000,000 in 1919, and to $6,000,000 the next year. One of the most important of North East's patents was applied for in September, 1910, and issued on July 30, 1912, for an electric pneumatic starting device. A most successful self-starter built by North

East was the single unit starter-generator used on Dodge from 1913 to 1925, and also on Reo, Franklin, and other cars.

In 1926 the company produced more than 2,000,000 units of various kinds. In 1929 its operations occupied 700,000 square feet of floor space, and it had become the third largest employer of labor in Rochester. The service organization under North East Service, Inc., formed in 1921, included eleven branches and 1,800 service stations throughout the world, the company maintaining a service school to train salesmen and service men for promoting the sale and installation of its products. In conjunction with this training it developed the use of films in industrial educational work, using animated technical pictures from 1917 on. The line of products had enlarged to include starters, generators, ignition systems, horns, speedometers, cowl-board attachments and novelties, and an electrically operated typewriter. Besides its Rochester properties, the company operated the manufacturing branches in Toronto and Paris and distributor)' branches in principal centers in the United States and in London and Antwerp.

On October n, 1929, the North East Electric Company was acquired by General Motors Corporation, and a year later was consolidated with Delco-Light Company to form Delco Appliance Corporation. The Delco line of farm power and lighting plants was transferred to Rochester, and certain lines of automotive production were withdrawn from it and transferred to Anderson, Indiana. Mr. Edward A. Halbleib became president and general manager of Delco Appliance.

From that time there has been a new stage of the development of products at Rochester, no less rapid than the preceding. Delco fans, Delco vacuum cleaners, Delco radio, and Delco heat oil burner products followed rapidly; also new lines of Delco-Light plants, and electric pumps. These show how the Corporation has expanded to supply urban dwellers at the same time as it improves on well-established products in the rural field, giving added emphasis to the phase, "Delco Appliances for homes everywhere."

Products: Delco-Light electric power and light plants; Delco water pumps and Delcogas individual gas-producing units for domestic use; Delco electric fans; Delco radios; Delco vacuum cleaners; Delco and North East blowers and small motors for commercial purposes; Delco Heat oil burners and Delco Heat automatic oil furnaces; North East speedometers and automobile heaters; Delco radio speakers and "B" power units; Delco radio dynamotors; Delco steam automobile heaters.

## DELCO PRODUCTS CORPORATION, Dayton, Ohio

THE present Delco Products Corporation is the outgrowth of an organization founded in 1909 by Edward A. Deeds and Charles F. Kettering. These two men, with a small barn on Central Avenue in Dayton, Ohio, for a workshop, constructed the first practical self-starter for use on automobiles.

The first order for the starters was secured before the two men had even thought about a name for the organization; it was then that they decided to call their new shop the Dayton Engineering Laboratories Company. Later in the same year Mr. Kettering decided that the name was too long, so he shortened it into the trade name of DELCO.

In 1911 Delco was moved to the Beaver Power Building, located at Fourth and St. Clair streets. Later the two men moved their growing industry to the present location on First Street.

In May, 1916, the Dayton Engineering Laboratories Company became a subsidiary of the United Motors. By this time Delco had become a well-established trade name and Delco starters and ignition systems were used on nearly all of the leading automobiles. On December 31, 1918, the United Motors group was purchased by General Motors Corporation. Delco has continued as a part of the General Motors family since that date. Mr. B. D. Kunkle is president and general manager.

On the removal of the automotive electrical equipment work to the Remy plant at Anderson, Indiana, in 1926, Delco Products Corporation was formed. Efforts were then directed to the manufacture of a shock absorber; the result has given Delco a preeminent position in that new line, as well as in its older lines of fractional horsepower motors for refrigeration, pump, and washing-machine use.

The plant has 851,000 square feet of floor space and occupies nearly 20 acres of ground. Over 3,000 employees are regularly engaged in the production of electric motors and hydraulic shock absorbers.

Products: Delco-Lovejoy hydraulic shock absorbers for automobiles, trucks, and busses; electric motors for refrigerators, pumps, washing machines, ironers, oil burners, and air-conditioning equipment; coil springs for engine valves, clutches, refrigeration unit mountings; or any applications where coil springs are required.

## DELCO-REMY CORPORATION, Anderson, Indiana

THE above name of this organization was adopted when the starter and ignition work done by the Dayton Engineering Laboratories Company was transferred to the Remy plant at Anderson, Indiana, in 1926. The Remy Electric Company then took the new corporate name. The manufacture of the Delco battery was begun in 1928. The next year, by consolidation of the Butler Manufacturing Company plant at Indianapolis with Delco-Remy plants, the manufacture of Bu-Nite pistons was brought to Anderson. Delco-Remy for a time made Delco-Lovejoy shock absorbers, since taken over by Delco Products Corporation. Its lamp business was taken over by Guide Lamp Corporation. Delco-Remy operates a battery plant in Muncie, Indiana. Mr. F. C. Kroeger is president and general manager of Delco Remy and its allied companies.

The Dayton Engineering Laboratories Company was organized in Dayton, Ohio, in 1909 to build ignition and generating equipment for automotive and marine uses but one of the most important dates in the history of Delco-Remy Corporation is 1912 for in that year the first cars were placed on the market equipped with the Delco Starter, an electric starting motor which was to do away, for all time, with hand cranking. During the World War, Delco did notable work on ignition for the Liberty Motor and other government projects. From the first, Delco had established an enviable reputation for building quality products. And out of the Delco and Remy organizations have come men to take high position in other General Motors undertakings.

Products: Accelerator pedal-starter controls; coincidental locking devices; dash choke, spark and throttle controls; generators; ignition systems; lock coils; starting motors; starter drives; switches; valve tappets; vacuum controlled ignition systems; wiring harness; Delco batteries; Klaxon horns; Bu-Nite pistons.

## Remy Electric Company, Anderson, Indiana

THE ignition systems of several pioneer automobile manufacturers were made by the founders of this company, B. P. and Frank Remy. In 1895, one was a young man doing an electrical wiring and shop business at Anderson, the other helping his brother after school. That year the older brother produced his first practical electric dynamo. The early dynamos were used as ignition equipment for stationary and marine engines. By 1899 the brothers had a working force of fifteen and occupied the first building in the old plant site. A little later, they started making the permanent magneto type, and then the oscillator. During this time Mr. I. J. Reuter went with the Remy brothers. They incorporated as Remy Electric Company in 1901. They brought out the high-tension magneto about 1904-05, and then the HB machine, which was used on several of the early makes of cars, including the Winton, Marmon, Pope-Hartford, Haynes, Apperson, and Stearns. The plant was moved to its present site in 1906.

Mr. B. P. Remy had previously worked on the inductor and, in 1908, made contracts with Overland and Buick for it. The magneto business expanded in 1909-10 and then gave way to the first of the combination lighting and ignition generators.

From 1911 to 1916 the company was owned by Stoughton Fletcher, banker of Indianapolis. The magneto business was good, with Overland and Buick contracts. The "O" generator came out in 1912, and starting and lighting were taken up in 1913. A contract with Reo was secured by the organization in 1914.

Remy Electric Company was purchased by United Motors in 1916, and through it entered General Motors in 1918. In 1919 it built 163,000 generators, 163,000 motors, and 315,000 ignition systems. In 1924 the plant consisted of 61 buildings, with floor space of 425,000 square feet, on a site of 26.71 acres. Employees numbered 3,672. The manufacture of Klaxon products was transferred from New Jersey to this plant in 1924. Delco starter and ignition work was brought from Dayton in 1926, at which time the organization took the name Delco-Remy Corporation. F. C. Kroeger is president.

### Klaxon Company, Anderson, Indiana

IN 1905 Mr. Lovell and Mr. R. A. McConnell organized the Lovell McConnell Manufacturing Company, making marine electrical specialties and fittings at Newark, New Jersey. In 1908 patents were purchased from Mr. M. R. Hutchinson covering an electrically operated sound-signal device. Until 1913 the company occupied the field alone with these devices of the diaphragm type but on the entrance of others was not able to sustain its patents against competitors. Despite this, its business increased with rapid strides.

The Lovell-McConnell Company was bought by the United Motors Corporation in August, 1916, and the name changed to Klaxon Company. The trade name "Klaxon" has proved a valuable business asset. It was coined by Mr. F. Hallett Lovell, Jr., from the Greek verb klaxo, meaning to roar or to shriek, and was registered at the United States Patent Office, July 14, 1908.

In 1920 the plant was moved from Newark to Bloomfield, New Jersey. By 1923 production totaled 1,317,547, having tripled in two years. Floor space amounted to 175,410 square feet. There were about 500 employees. Under General Motors operations the manufacture of Klaxon products was undertaken by Remy Electric Company (now Delco-Remy Corporation) in 1924, and machinery and equipment transferred to Anderson, Indiana. F. C. Kroeger is president and general manager.

Products: Electric and hand-operated automobile and industrial warning horns and loud-speaking telephones for limousine and industrial use.

### Delco-Remy & Hyatt, Ltd., London

THE business was begun in 1907 by the Walter H. Johnson Company, an automobile-selling company, principally retail. In 1914 it secured rights from Remy Electric Company for the British Isles. When the latter and Delco went into United Motors in 1916, it represented both in London. Mr. Johnson took in Mr. W. O. Kennington, an electrical engineer, formerly with Remy. These two, and General Motors Corporation, became the original incorporators of Delco-Remy, Ltd., July 23, 1920. When General Motors acquired the stock of Messrs. Johnson and Kennington, in 1921, they were made managing directors, respectively, of General Motors, Ltd., and Delco-Remy, Ltd.

The present company is an amalgamation of Delco-Remy, Ltd., and Hyatt, Ltd., formerly a subsidiary of Hyatt Roller Bearing Company.

Functions: Sales and service on all Corporation accessory products in the British Isles; technical and service headquarters at London.

## GENERAL MOTORS BUILDING CORPORATION, Detroit, Michigan

THE Fleming Realty Company was incorporated in Michigan on March 26, 1919, to purchase all parts and parcels of real estate now covered by the General Motors Building in Detroit. This is the block bounded north by West Grand Boulevard, south by Milwaukee Avenue, east by Cass Avenue and west by Second Boulevard.

The name was changed from Fleming Realty Company to Durant Building Corporation on April 30, 1919. Preliminary sketches had been drawn by Mr. Albert Kahn, architect of Detroit, and the Thompson-Starrett Company had been selected to erect the building. Ground was broken on June 2, 1919. The building was to be completed and ready for occupancy in one year, but a change of program prevented this. Purchases of additional parcels were made for the power house and railroad siding and tunnel sites. In the course of operations forty-eight entire buildings and part of another were removed or wrecked, including a three-story brick office building transferred to a new site without interruption to the business of its tenants, even telephone and electric light service being maintained. The first tenant of the new building was Harrison Radiator Corporation, which moved in on November 18, 1920.

The name of the corporation was changed on February 24, 1921, to General Motors Building Corporation and the building under construction received its present name. The original contract was finished by the Thompson-Starrett Company in November, 1921, and a new contract with the same firm for finishing the west half of the building was undertaken in February, 1922, and completed the following January, with the exception of certain items taken over by General Motors Building Corporation. At the time of its erection, this structure was the largest office building in the world. Mr. C. E. Wilson heads the Building Corporation.

## GUIDE LAMP CORPORATION, Anderson, Indiana

WHEN the automobile first began to go out at night, it was provided with oil or gas lamps. Of the latter type was an acetylene gas lamp made at Kenosha, Wisconsin, by the Badger Brass Company. Three of the Badger employees, Messrs. Hugh J. Monson, William F. Persons (1868-1926), and William Bunce, decided to strike out for themselves. They chose Cleveland, Ohio, as the place for their business, and in 1906, with $100 apiece, formed the Guide Motor Lamp Company. Cleveland was then the center of the automobile accessory trade and the location of Royal, Winton, and Stearns car manufacturing, but was without a lamp-making concern.

The partners did a good business in the repair of lamps and radiators. In 1910 they were employing twenty-five persons, and in this year made their first electric lamps for automobiles. Mr. T. A. Willard, the battery man, by the development of his product, furthered the new activity of the firm. A patent was taken out on a reflector arrangement fitted onto the old lamps. The firm made its screw-type sockets by hand. In 1911 the partners secured orders from the Rauch & Lang and Baker companies, and the next year from Reo and from the Detroit Electric for electric side lamps, which were also made by hand.

Guide Motor Lamp Company was incorporated in 1913 with a capital of $40,000. Mr. Monson was its president until 1928. A plant was built on

West Madison Street, Cleveland. Guide Motor Lamp made its first automobile contract in 1913, to furnish the Glide Automobile Company with headlights complete, and batteries, switches, and wiring, to the amount of 500 sets.

Tungsten filament dates from 1914. Mr. Persons introduced the bulb, frosted in the lower part, in 1916. In 1917 Guide took over Emerson Clark's patent, obtained four years earlier on the parabolic reflector of two separate foci. Mr. C. A. Michel, vice-president in charge of engineering of the present Guide organization, developed its Guide Ray reflector. In 1922 the company brought out a bifocal reflector in which the vertical angle of light is controlled in the reflector.

The two-filament bulb came out in 1924, and Guide then developed the Tilt Ray head lamp with which Buick, Pontiac, Oldsmobile, and Reo were equipped.

General Motors Corporation bought out the company in 1928 and formed the Guide Lamp Corporation. Mr. F. S. Kimmerling became president and general manager, Mr. Monson, vice-president in charge of sales. At the end of 1929, Guide Lamp Corporation was located in Anderson, Indiana. To its own lamp operations were added those of Delco-Remy Corporation. Mr. F. H. Prescott is now president and general manager.

Products: Automobile-lighting equipment.

## HARRISON RADIATOR CORPORATION, Lockport, New York

ONE of the chief difficulties encountered in popularizing the gasoline automobile was the tendency of the motor to overheat. Following the primitive and awkward attempts to increase the radiating surface, a steady advance set in which resulted in the cellular or "honeycomb" type of radiator. An outstanding step in this advance was the designing and patenting of the "Harrison Hexagon" by the late Herbert Champion Harrison, founder of Harrison Radiator Company, which he established at Lockport, New York, in 1910. Harrison Radiator Corporation succeeded the company in 1916.

Mr. Harrison was of English stock. Born in Calcutta, India, he was educated at Rugby School and Oxford University, England, being graduated as an electrical engineer. He came to the United States in 1907, organized the Susquehanna Smelting Company, and later became consulting engineer of the Union Carbide Company. From 1910 to 1927 he served as president and general manager of Harrison Radiator.

The other officers associated with Mr. Harrison in the founding of Harrison Radiator Company were B. V. Covert, vice-president and treasurer, and Fred D. Moyer and Oscar A. Loosen, directors. Mr.

Covert deserves particular notice. His Covert Gear Company made bicycle gears, and Mr. Covert himself experimented in motor-car design. An early contact was established between Mr. A. P. Sloan, Jr., and the Covert Motor Vehicle Company, since Hyatt made bearings for the latter. Six years after the founding of Harrison Radiator Company, Mr. Sloan brought Harrison into the newly organized United Motors, and in 1918 it entered the General Motors group. Mr. Frank M. Hardiman is president and general manager.

HERBERT C. HARRISON

During the interval of independent operation the Harrison company had grown from small beginnings, moving from narrow quarters in an inadequate building on a main street of Lockport to a commodious plant built by the American District Steam Company, which was judged ample for many years to come. Expanding business, however, called for more room, and in 1917 a two-story building was added which covered more than half a city block. This was later increased to three stories. By 1923 the original plant of the steam company had been demolished to make way for another building completing the present modern plant. It now contains more than 420,000 feet of floor space, with a potential capacity in excess of 3,000,000 radiators a year. In 1929 the Corporation produced 2,310,440 radiators, since which time various layout changes have been made to accommodate a larger production. The record for 1929 was accomplished by a force of about 2,000 persons.

On the technical side, some of the more important of the Corporation's innovations are: automatic temperature control of the soldering process (1917); dipping of tanks and cores in one operation (1923); a special machine for multiple core strip soldering (1924); the folder header which gives an air and water-tight joint in the core without the use of solder (1914); rotary crimping machine (1915); the removable anchorage (1916), and the intermediate spacing strip (1913), a radical departure in cores from all previous cellular cores. This intermediate spacing strip has been the subject of further development within the past two years, with the result that the hexagon form of radiator, in production from 1914 to 1932, has been changed in appearance, made stronger and more efficient, and radiator leaks are now very rare. Practically all radiators now made by Harrison Radiator Corporation are of the new design.

Harrison also promoted materially the evolution of the vapor cooling system.

From rolling mill to loading docks, the Harrison plant manufactures radiators completely, with a beautifully coordinated system of accurate manufacture and efficient handling. Its research enters three important fields of physics air, water, and metals, since an automobile radiator is a metal container in which water is cooled by air.

Products: Automobile radiators, radiator shutters, oil temperature regulators, thermostats, and hot water car heaters.

## HYATT ROLLER BEARING COMPANY, Harrison, New Jersey

THE name Hyatt occupies a place of importance in transportation, agricultural, industrial, and production equipment wherever shafts and wheels are turning, to make their work more efficient, to reduce their maintenance, and to prolong their operating life.

JOHN WESLEY HYATT

Back of the name Hyatt is the name of its inventor, Mr. John Wesley Hyatt, pioneer in more fields than one, who in the year 1914 received the Perkins Medal of the New York Chemical Society for distinguished services in the field of applied chemistry and engineering. That service included experiments which led to the invention of celluloid. He spent much time on experimenting, and secured over 250 patents, the first (1861) was for a knife grinder with a solid emery wheel. By trade he was a printer. He settled in Albany, New York, in 1863, where he was led into numberless experiments with compositions which might make a billiard ball as a substitute for the ivory one. It was not until 1865 that he found the right composition and took out a basic patent on it. During the course of this work he accidentally discovered the fundamentals on which the celluloid or pyroxylin plastic industry was later established.

A fuller account of this discovery is given in News and Views, September, 1929, and in The Celluloid Mirror, February, 1930, as derived from an interview with Charles S. Lockwood at the experimental laboratories of the Hyatt plant, Harrison, New Jersey. Mr. Lockwood had come into contact with Mr. Hyatt through the Celluloid Manufacturing Company in 1874. In 1888 he was again at work with Mr. Hyatt, who was perfecting a mill for crushing sugar-cane. The difficulty Mr. Hyatt had to contend with in the disintegrators was that he could not find any bearings which would stand up under the loads imposed on the journals. There was trouble with a slanting bearing. He needed something better than a solid bearing to make a line contact with the journal, something touching at every point. Rollers of hardwood, paper, and pressed wood had been found to work on a countershaft, but no one wanted wooden rollers. His inventive genius was straightway brought into action, and in time he developed a roller bearing, the rollers made of flat strips of steel, formed into a coil, or helix. By this means he increased the load-carrying ability, effected a considerable saving in power, and minimized lubrication.

The anti-friction journal bearing was patented that year, and the roller bearing was entered at the patent office May 20, 1890. The first form consisted of a series of loose rollers in a brass cage, from which the roller bearing evolved as a complete assembly in a self-contained cage. Later, raceways were added, and the bearing at present is composed of three parts: the inner race or shell, the cage containing an assembly of rollers, and the outer race. The distinctive feature is the type of roller used, which is made up of hollow cylinders formed by helically winding strips of alloy steel.

The successful results achieved in the sugar-cane mills persuaded Mr. Hyatt that there was a large field for his invention. Accordingly, the Hyatt Roller Bearing Company was founded, in 1892. Its products already bore a high reputation when the automobile industry was born. Here was a new mechanism, subject to heavy loading at certain points, and calling for the least possible loss of power from engine to wheel. Hyatt Roller Bearing Company began production of automobile bearings in 1896, and its product was used in Mr. Haynes' first car, the Haynes-Apperson, also in Mr. Olds's first cars. Mr. Ford became one of the best customers. The automobile of today uses roller bearings on wheels, rear axle, transmission, pinion shaft, power take-off, valve rocker arms, fan, brake shaft (Buick), steering shaft, propeller shaft.

Mr. Hyatt died in 1920, having passed his eighty-second year. With him for many years was Mr. Lockwood, now well past eighty years of age. He took out over eighty patents, and assisted Mr. Hyatt in experimenting. Besides Mr. Lockwood, also early associated with Mr. Hyatt in the manufacture of roller bearings, were Messrs. A. P. Sloan, Sr., and A. P. Sloan, Jr. The latter came into the Hyatt drafting room in 1897 and was made general manager two years later. He received patents for shafting hangers, hanger boxes, and improvements thereon.

In 1917 Hyatt Roller Bearing Company passed into United Motors at $13,000,000, or $3,300 per share of stock, later becoming a division of General Motors along with all other United Motors units.

During the World War the company extended its regular production; it made small shells and Holt tractor parts and employed as many as 5,000 men. The plant at Harrison, across the Passaic from Newark, New Jersey, consisting of 21 buildings, covers some 900,000 square feet of floor space and normally employs about 2,000 men.

The Hyatt Company from its very beginning has always prided itself on the quality of its product. A typical example of how deep-rooted this is may be found in this interesting side light. A few years ago, an installation of line-shaft bearings of almost unknown origin aroused considerable curiosity because they had run so long and so well. Hyatt engineers were called in, and it finally proved to be boxes equipped with Hyatt bearings and set up by Mr. Hyatt himself in the early 'nineties. Nearly two-score years had passed, and the bearings were still performing their work satisfactorily.

With quality and precision production the keynote of Hyatt manufacture, naturally the equipment which produces, and the methods employed, are of the highest caliber. Likewise, the men who operate the machines and supervise the work are specially selected and trained for their respective functions.

A trip through the Hyatt plant will reveal the most modern grinding, heat-treating, assembly, and other operating equipment. Many of the manufacturing devices employed are of Hyatt's own exclusive design. They are the pioneers of centerless internal and external grinding and the exponents of many unusual devices in heat-treating.

The Hyatt method of production control assures a uniform product from the raw steel to finished bearing. Every step of manufacture is carefully checked and rechecked. Special analysis steels are made by leading mills to Hyatt's exact specifications. At the Hyatt physical and chemical laboratories the raw products are carefully analyzed before fabrication and are tested during every manufacturing operation to be sure of their finished quality. The precision machines all along the production line turn out a perfect product, but even here nothing is taken for granted, and a corps of inspectors, with the very latest precision gauging equipment, pass on all work in process as well as finished bearings before they are passed on for final cleaning, wrapping, and shipping. Among Hyatt's gauging instruments are many modern developments in optical measuring devices, one of which is used for checking inspection gauges and tools down to 1/200,000 of an inch.

Hyatt Roller Bearings are extensively used in mining, oil field, textile, steel mill, road building, power transmission, and all other kinds of industrial equipment; likewise in farm tractors, farm machinery, railroad cars, and, as previously stated, they were the first anti-friction bearings used, and continued to be used, in large quantity by the automotive industry.

Thus, on every genuine Hyatt Roller Bearing the die stamp "HYATT U.S.A." is more than just a maker's mark. It means a guarantee of dependable bearing performance, correct design, quality materials, precision manufacturing, and engineering knowledge.

Mr. H. J. Forsythe is president and general manager.

Products: Hyatt anti-friction roller bearings.

## INLAND MANUFACTURING COMPANY, Dayton, Ohio

THE Inland Manufacturing Company was organized in 1923 to take over the steering-wheel activities of the Dayton-Wright Company. During the years 1920 and 1921, experiments had been conducted by Dayton-Wright Company on the construction of a steering wheel with a rim manufactured from a continuous strip of veneer wrapped upon itself. This new type of rim had the advantage of increased strength and lower manufacturing cost. Actual production was commenced on this product at the Dayton-Wright Company in 1921, and by 1923 the steering-wheel business had developed so rapidly that it was found advisable to separate it from the other activities of the Dayton-Wright Company.

The Inland Manufacturing Company was organized and occupied one of the former plants of the Dayton-Wright Company. Mr. Whittaker succeeded Mr. Talbot as head of the enterprise.

During the first year of its operation, Inland Manufacturing Company produced 479,000 steering wheels, and within a few years became the largest producer of wood steering wheels in the world.

Anticipating the gradual obsolescence of the wood-rim steering wheel, the Inland Manufacturing Company, within the years 1929, 1930, and 1931 was completely transformed from a wood-working plant to a rubber plant. During this period a well-equipped rubber laboratory was established with a highly skilled technical staff, and modern equipment for the handling and manufacturing of rubber products was installed.

Constantly increasing appreciation on the part of the automotive industry of the importance of the use of rubber in the refinement of the modern automobile rapidly broadened the field of service of the Inland Manufacturing Company, so that at the present time steering wheels constitute but 25 percent of the volume of this plant.

Products: Steering wheels, metal-rubber running boards, Inlox and other motor supports, Inlox spring eye bushings ; Quickube, Du-flex, Flexotray and Flexogrid, rubber ice trays, and grids for automatic refrigerators, battery containers, and battery container covers, hard and soft rubber, and molded products.

## THE McKINNON INDUSTRIES, LTD., St. Catharines, Ontario

THE McKinnon Industries, Ltd., now a subsidiary of General Motors Corporation, is the outgrowth of a hardware store started at St. Catharines in 1878. The business has since undergone many changes in name, ownership, location, and lines of activity. At the start, Mr. L. E. McKinnon and his partner, Mr. F. F. Mitchell, had $2,000 cash invested. Gradually they took on the manufacture of hardware, including saddlery and wagon hardware and hames. Mr. McKinnon received patents on carriage dashes and fenders, which were made at St. Catharines, and at a branch established in Buffalo. On the expiration of the partnership in 1888, he took charge of the manufacturing end known as McKinnon Dash & Hardware Company. The business at Buffalo grew through a merger of like concerns in 1898. There was also increased business at the Canadian plant, resulting in a move to a new location in 1901 and the construction of a malleable iron foundry.

The advent of the automobile then made advisable the addition of other lines of manufacture. A drop-forging department was opened, and the making of chains begun in 1905. The new line was incorporated the year following as the McKinnon Chain Company, with plants at Tonawanda, New York, and at St. Catharines. Merger in 1917 produced two companies, Canadian and United States, the latter at Columbus, Ohio. Control of these two companies passed from the Dash & Hardware Company in 1922 to the United States concern, the Columbus-McKinnon Chain Company.

The St. Catharines plant, equipped to make saddlery and harness, did a thriving business during the World War. Large additions were made to the plant in 1916 for the purpose of increasing facilities to handle contracts from the British government for shrapnel bullets, time fuses, and high-explosive shells. The end of the war made it necessary to find new lines to replace war contracts. Selecting the automobile field as offering the greatest promise, the company began the manufacture of automobile radiators at the St. Catharines and Buffalo plants, undertaking also, for the first time in Canada, the manufacture of auto differential and transmission gears. The venture exacted years of development work, which ultimately crowned it with success.

Mr. McKinnon died in 1923. Two years later the company was disposed of and reorganized as The McKinnon Industries, Ltd. In 1925 it acquired the Canadian division of J. H. Williams & Company of Buffalo, makers of hammers, wrenches, and other small tools. In 1929 the radiator business of the St. Catharines plant was sold to make room for a more profitable activity, a branch of the gear division of General Motors, which corporation assumed control of the two plants March I, 1929. Considerable expenditures were made for improvements including a new building for the purpose of manufacturing Delco-Remy automotive electrical equipment, theretofore brought into Canada from the Anderson, Indiana, plant. Canadian activities of North East Electric Company of Canada, subsidiary of the Rochester, New York, Company were also taken over late the same year. The Buffalo plant has since been sold. McKinnon Industries manufactures a wide variety of goods and sells 40 percent of its output outside of the General Motors Corporation. Mr. H. J. Carmichael is general manager. Products: Automobile rear axles and differentials;

steering gears; axle shafts; Delco-Lovejoy shock absorbers; Delco fractional commercial motors; AC spark plugs; Delco-Remy starting, lighting, and ignition systems; tool kits; malleable castings; stampings; drop forgings and saddlery hardware.

## THE MORAINE PRODUCTS COMPANY, Dayton, Ohio

THE Moraine Products Company is the largest producer of self-lubricating bearings in the country. These bearings are sold under the trade name of Durex, and rolled bronze bearings, also made in quantity, under the name of Moraine.

Durex bearings are the result of three years of research. They are made in large variety for use in automobiles, electric refrigerators, electric motors, electric clocks, musical instruments, food mixers, washing machines, ironing machines, casters, electric fans, adding machines, counters, cash registers, moving-picture cameras, airplane pulleys, textile machinery, conveyors, farm implements, toys, and shoe machinery.

Mr. James H. Davis is president and general manager.

Products: Durex oil impregnated metal bearings; Moraine rolled bronze bearings.

## THE NEW DEPARTURE MANUFACTURING COMPANY Bristol, Connecticut

THE New Departure is one of the largest accessory units of the General Motors Corporation. It makes nearly three fourths of all the ball bearings used in the United States, and more than one half of the requirements of the world. Yet, perhaps no unit had more modest beginnings.

Two boys, Edward D. and Albert F. Rockwell, of New England ancestry, left their home in Illinois in the late 'seventies and started a hardware store at Jacksonville, Florida. Albert, the younger, conceived the idea of a doorbell with a clockwork mechanism and operated by a push button. This, he claimed, would give "electrical results without the use of a battery."

Manufacture of the new device was let to the H. C. Thompson Clock Company, Bristol, Connecticut. It was so revolutionary in type that it was called the "New Departure Bell." The boys and their store janitor peddled these bells in Florida and later elsewhere: Edward, in Binghamton, New York, and Albert, in Michigan. Success led them to form the New Departure Bell Company in 1888, which began assembling, packing, and shipping in a single room at Bristol. The next year they were incorporated for $50,000 and took larger quarters in what is the No. I building of the present company.

The bicycle craze, then at its height, led Edward Rockwell to invent a bell with a rotary mechanism actuated by a thumb lever. This venture was successful, and the mechanism was adapted to various types of bells for trolley cars and fire apparatus, as well as for bicycles. Other products such as bicycle lamps, plumber supplies, and trolley harps were added to manufacture, but later dropped. Between 1891 and 1895 the business grew rapidly, during which time Edward left to establish the Liberty Bell Company in East Bristol, specializing on bicycle bells.

In 1898 the New Departure Bell Company made its controller, a free-wheel mechanism which permitted coasting without movement of the pedals. It necessitated a separate hand brake applied against the front tire.

Later, an employee of the company developed an improvement of this device, permitting the rider to brake as well as to drive and coast. This device, when perfected, became the original New Departure Coaster Brake, the broad basic patents for which were assigned to the New Departure Bell Company. Bicycle manufacturers were quick to see the value of the coaster brake, and it was not long before several companies were manufacturing under the New Departure patents. The brake gave such an impetus to bicycle riding that the New Departure product became known as "the brake that brought the bike back."

In 1901 the state approved a change of name of the company, substituting the word "Manufacturing" for "Bell." Sales of the brake were extended overseas. The company began marketing its product through its own sales organization in 1906. From 1904 to 1907 the company operated a plant in a suburb of Berlin. In 1907 the Rockwell brothers united their interests, New Departure absorbing Liberty Bell. Edward Rockwell, the former owner of Liberty Bell, was placed in charge of the East Bristol plant and in time took over all bell operations. Sale of bells grew apace and reached 1,000,000 a year. About 1921 the East Bristol plant was vacated and operations were transferred to the New Departure plant.

The new automobile industry was a natural field to engage the attention of the bicycle and parts manufacturer. New Departure officials formed the Bristol Engineering Company, and in a plant across the street made the original Allen-Kingston and Houpt-Rockwell automobiles; also the Rockwell Cab, the original "Yellow Taxi" which was first operated in New York City.

From his experience with coaster brakes and automobiles, Albert Rockwell conceived the idea of the first two-row angular contact ball bearing which in one integral unit combined the functions of a strictly radial bearing and two pure thrust bearings. New Departure sponsored this bearing type and in 1907-08 started its development of ball bearings, later extending the scope to Single Row radial and Single Row angular contact or one-direction thrust bearings. They were first used by the Bristol Engineering Company and Apperson Brothers Company as well as by Weston-Mott, makers of complete rear axles. Driggs-Seabury Ordnance Company, of which Mr. F. G. Hughes was the chief engineer, was among the first to use New Departure Double Row ball bearings.

These bearings were first made of carburized steels, case hardened. Today they are made of electric furnace, high carbon, chrome alloy steel. Bearing manufacture calls for precision workmanship, and until recently one tenth of a thousandth of an inch was the extreme limit. This was made possible by a system of gauging wherein, by amplification, the dimension is readily visible. Today, by perfection of the gauges, accuracy down to one hundredth of a thousandth of an inch is obtained.

Ball bearings are now found in many points in the automobile, commonly as follows: front wheel, generator, fan and water pump, steering gear, clutch, pilot and throw-out, transmission, free wheel unit, pinion, differential, and rear wheels.

In 1933 there were 32 makes of American passenger cars and 62 makes of American commercial cars using New Departure Ball Bearings as regular equipment. Some thirty manufacturers of farm machinery also applied these bearings to their products. Airplanes, motor boats, construction equipment, industrial machinery use them in fact they are applicable wherever a shaft revolves or power is transmitted.

The company has two plants, situated at Bristol and Meriden, Connecticut, with the main office at Bristol, and branch engineering and sales offices at Detroit, Chicago, and San Francisco. The plants include the world's largest steel ball manufactory, the largest grinding department housed in a single building, and the largest upset forging plant in the East. Floor space exceeds two million square feet or approximately fifty acres. Normally there are 7,000 employees. Its plants rank first in the world in consumption of quality steel and of grinding wheels, and also perhaps in the number of specially designed machines in use. Equipment includes 26,000 precision gauges. Daily capacity of the plants amounts to 225,000 ball bearings, 9,000 coaster brakes and front hubs for bicycles, 20 tons of precision steel balls, and 300,000 high carbon chrome steel forgings.

The Single Row ball bearing has four principal parts inner race, outer race, separator and balls, requiring 106 major operations in its manufacture. The Double Row, with two separators and two rows of balls, requires 149 operations.

The company, starting in 1889 with $18,000 cash and $32,000 goodwill, increasing its capital by $25,000 cash in 1893, $25,000 in 1902, $100,000 in 1903, $300,000 in 1904, $1,000,000 in 1909, $500,000 in 1912, $500,000 in 1913, and $500,000 in 1915, brought the capital to $3,000,000, represented by $500,000 Preferred and $2,500,000 Common stock. Mr. DeWitt Page, president, has been connected with the company since his entrance as an office boy in 1894. Mr. Hughes, vice-president and general manager, has been with New Departure since 1911.

The company has taken much interest in the welfare of its employees. Its institutions for their benefit includes Endee Inn, Endee Club, and Endee Manor. The last is a residential section developed by the company and handled by the New Departure Realty Company, organized June 20, 1916. There were 102 houses built by 1924, which were offered to employees practically at cost, with a 10 percent payment, half of balance on first mortgage with the bank, and the other half on second mortgage with the company. In the later established Bristol Realty Company, New Departure owns a majority of the stock.

Products: Ball bearings for machinery and automobiles; coaster brakes, front hubs, and bells for bicycles.

Bristol Realty Company, Bristol, Connecticut

THE New Departure also owns 63 percent of the stock in the Bristol Realty Company, which was founded in May, 1913. Like the New Departure Realty Company, this organization assists people in the purchase of homes. Nearly two hundred houses, mostly of the two-family type, have been purchased through this plan.

## PACKARD ELECTRIC CORPORATION, Warren, Ohio

THE Packard Electric Company, now Packard Electric Corporation, was originally incorporated in 1890 by J. W. and W. D. Packard, for the purpose of manufacturing incandescent lamps and transformers. In 1898 the company designed and brought out the Packard automobile. Because they were unable to purchase satisfactory cable in this country or in Europe, they took up the manufacture of automotive cable as a matter of necessity.

In 1903, the automobile division was sold to the Packard Motor Car Company of Detroit and the lamp department was sold to the General Electric Company.

Automotive cable production, however, was continued at the Warren factory, supplying the many car manufacturers. About 1910 the replacement end had its start as a natural outgrowth of the demand for such service on the part of the car owners. The manufacturing facilities for automotive cable were added to from time to time to take care of the increasing demand for replacement service as well as for original equipment.

In the meantime, transformers had also been made by the Packard Electric Company. However, in 1919 the Small Transformer division, comprising bell-ringing, toy and radio transformers, was sold to another Warren concern. Large central-station transformers were continued until in 1929 when this division was sold outright and the Packard Electric Company devoted its efforts exclusively to automotive cable and automotive cable products.

The Packard Cable Company of Canada was established in 1931. An assembly plant is maintained in Toronto. Early in 1932 negotiations were begun which resulted in the sale of the Packard Electric Company, effective March 1st, 1932, to the General Motors Corporation.

The basic product of the company is automotive ignition, lighting, and starting cable. There are many sizes, with hundreds of variations as to the number of strands of fine wire used in the core, rubber thickness over the core, and the various kinds of insulation used, including rubber, tape, etc.

Cable is supplied directly to the car manufacturers and this original equipment runs into many millions of feet annually. The maximum capacity of the Warren plant is about forty million feet per month. Automotive cable is made complete from the copper rod to the finished product. A quarter-inch rolled copper rod as received from the copper mills is passed through several drawing processes until it reaches the required sizes. These range from number thirty-six to number eleven gauge. The wire progresses to the stranding department where it is bunched or stranded on twisting machines. Thus a core is formed over which the rubber insulation is placed, after which it is cured and then sent to the braiding department for covering of the various colored braids as seen on the finished wire.

Other departments include the lacquer department where the secret Lac-kard finish is put on high tension cable and where the lighting cable is given a protective coat of lacquer. There is also a fabricating department where battery cables and ignition cable sets are prepared to fit any of the hundreds of makes and models of cars on the road.

Mr. F. C. Kroeger is president of the organization.

Products: Automotive ignition, lighting and starting cable.

## THE SAGINAW MALLEABLE IRON DIVISION, Saginaw, Michigan

THE Saginaw Malleable Iron Company was organized in January, 1917, with a capital of $250,000 Common and $150,000 Preferred stock. Mr. C. F. Drozeski was president, and Mr. G. H. Hannum, treasurer. Its purpose was to supply malleable castings, which were demanded by the large automotive concerns in the vicinity of Saginaw. A plant was erected, and production begun before the end of 1917. Initial melting capacity was about 80 tons daily, doubled the next year, capital being increased by $250,000 on that account. Additions and expansions called for a further increase of $500,000 in capital stock in 1919. In July that same year the board considered the offer of General Motors to buy the Common at $125 per share and accepted it. The company thereupon became affiliated with General Motors and was placed under the management of Saginaw Products division. It was at the company's plant that the Dressier kiln or tunnel-over method of annealing was worked out and perfected, having been installed in 1920. Four tunnel kilns are in use at the present time. In 1926 the plant was entirely revamped, molding conveyors were installed to replace the individual floor method and the duplex method of melting, consisting of cupolas and electric furnaces, was installed to replace the hand-fired air furnaces. As a result the daily melting capacity is in excess of 600 tons as compared with 80 tons in 1917. All of the above tonnage is consumed by the various divisions of General Motors Corporation. 1500 persons are now employed. Mr. D. O. Thomas is general manager. Products: Malleable iron castings for passenger cars and trucks.

## THE SAGINAW STEERING GEAR DIVISION, Saginaw, Michigan

THE parent company was the Jackson-Church-Wilcox Company, organized April 21, 1908, by John L. Jackson, E. D. Church, and Melvin L. Wilcox, with a capital of $25,000 of which $15,000 was paid in. The purpose of the organization was to manufacture automobile parts, but for the first two years it was engaged principally on machining parts for the Buick Motor Company and others. Then it obtained control of a British patent covering the double-thread screw and half-nut type of gear, which was made and sold to Buick. This gear was improved and put on the market under the trade name of "Jacox."

The history of the company is one of rapidly increasing production. Under constant pressure from the Buick Motor Company for greater output, the original capacity of twenty gears per week was rapidly expanded until 1909, when it was planned to produce twenty thousand. Before the year of 1909 was far advanced, it became evident to the Buick Company that their steering gear source must be increased in capacity. Their demands were backed by an order for 60,000 gears.

The views of the owners of the Saginaw Company were not in accord with the Buick Company as to the future of the motor car and they declined to expend the money for the additional plant and equipment; the result was that negotiations were opened for the purchase of all the capital stock of the Jackson-Church-Wilcox Company by the Buick Motor Company and this was accomplished in January, 1910.

The Jacox plant continued to grow along with other General Motors units, and in 1919 all the General Motors operations in Saginaw were consolidated under one management known as the Saginaw Products division, General Motors Corporation. This larger unit was composed of the original Jacox plant, the Motor plant, the Grey Iron Foundry, the allied plant of the Saginaw Malleable Iron Company, and the Crankshaft plant, which was an outgrowth of the National Engineering Company.

By 1928 the Saginaw Products division had grown to such an extent that it seemed wise to decentralize the activities. This was done, and the following organizations were set up to function as independent units: Chevrolet Grey Iron Foundry Division of Chevrolet Motor Co. Saginaw Malleable Iron Division of General Motors Corp. Saginaw Steering Gear Division of General Motors Corp. Saginaw Crankshaft Division of General Motors Corp.

During 1931 the Crankshaft division's operations were transferred to the individual car manufacturers, and this plant was closed.

During the years of 1929 and 1930 two new types of steering gears, known as the worm and sector gear and the worm and roller tooth gear, were developed and marketed under the trade name of "Saginaw Gear," and the "Jacox" name was dropped. These designs have been improved upon and today are used on all General Motors cars.

At the present time the Saginaw Steering Gear division has approximately 180,000 square feet of floor space with a capacity of 7,500 gears daily with 800 employees.

In 1932 and 1933 the Saginaw Steering Gear division produced more steering gears than the combined output of all other exclusive steering gear manufacturers. A complete line of steering gears is produced for all sizes of passenger cars, trucks and busses. Mr. Max L. Hillmer is general manager.

## UNITED MOTORS SERVICE, INC., Detroit, Michigan

THIS organization arose from the needs of certain automotive equipment makers for a distribution organization which would carry complete stocks of replacement parts and provide adequate service facilities everywhere, in the performance of a nation-wide service of uniform character. The manufacturers represented in it at the outset were Dayton Engineering Laboratories Company, Remy Electric Company, and Klaxon Company. It was originally a subsidiary of United Motors Corporation (1916-19) and since then has been a subsidiary of General Motors Corporation.

Beginning with a small office in Detroit, the company in a few years expanded to twenty-five branches in principal cities, in general each branch consisting of office, stock room, shop, and garage. Supplementary to this, it appointed authorized service stations in each branch territory.

The company became the national field organization of Harrison Radiator Company, in 1918, and has since added to its service and distribution until today it represents the following units of General Motors:

AC Spark Plug Co.
Delco Appliance Corp.
Delco Products Corp.
Delco-Remy Corp.
Guide Lamp Co.
Harrison Radiator Corp.
Hyatt Roller Bearing Co.
Klaxon Co.
McKinnon Industries, Ltd.
New Departure Mfg. Co.

Its authorized service stations and dealers exceed five thousand.

United Motors Service deals with all classes of the automotive trade except car manufacturers, whose equipment sales are handled directly by the factory organization. From a service standpoint, the activities of the company cover not only cars of General Motors production but, with few exceptions, all domestic makes of cars, trucks, and tractors. Mr. F. A. Oberheu is president and general manager.

Services: Through its twenty-five branches in the United States and Canada, United Motors Service provides authorized national service for Delco-Remy and North East starting, lighting, and ignition systems; North East and Harrison hot water heaters; AC and North East speedometers; Delco batteries; Delco-Lovejoy hydraulic shock absorbers; Delco commercial motors; Automotive radio; Klaxon horns; Harrison radiators; New Departure ball bearings; Hyatt roller bearings; AC air cleaners, oil filters, gasoline strainers, fuel pumps, gauges, spark plugs, guide lamps. It also provides authorized national distribution and service for Delco-Light electric plants, and Delcogas systems for rural homes; Delco electric pumps and water systems; Delco fans, vacuum cleaners, and household appliances.

## WINTON ENGINE CORPORATION, Cleveland, Ohio

TWENTY-ONE years ago, Alexander Winton, the famous pioneer automobile manufacturer, was in the market for the best marine gasoline engine power plant obtainable, to be used in his motor yacht La Belle, then under construction. He visited various yachts equipped with gasoline engines, but was disappointed in their performance, so after exhaustive investigation of different types of marine gasoline engines then in use, he decided to design and build the La Belle's power plant himself. This decision was the foundation upon which has been developed the Winton Engine Corporation of Cleveland. When the La Belle was launched, her power plant consisted of three six-cylinder type Winton marine gasoline engines, of 9" bore, 12¼" stroke, each rated at 150 h. p.

ALEXANDER WINTON

In its uniformly satisfactory performance, its smoothness, quietness, cleanliness, and dependability, this new Winton power plant proved a revelation to marine architects and engineers. Many of them pointed out to Mr. Winton the broad field for such an engine, and, consequently, he began seriously to consider designing and building marine gasoline engines for the market.

Early in 1912, the Winton Gas Engine & Manufacturing Company was organized and incorporated under the laws of the State of Ohio, and in July of that year began to develop the first Winton marine gasoline engines for the market. In the meantime, prior to the formation of the new company, Mr. Winton had devoted considerable attention to the possibilities of the Diesel type engine, which was then coming to the front rapidly in Europe. During this time he also developed a stationary gas engine electric generator set which was used successfully as an industrial power plant for many years, until, in fact, the cost of fuel, natural gas, became prohibitive. This was a six-cylinder engine, with 9" bore, 14" stroke, developing 150 h. p. at 350 r. p. m. The high cost of fuel, of course, prevented further development of this type of power plant.

Shortly after the formation of the company, late in 1912 and early in 1913, three Winton marine gasoline engines of remarkably advanced design for their time were brought out. These were Models W5 and W6. These three models, without major change in design, are still being built, sold, and used successfully, more than twenty years after they were designed and built for the first time. They are today modern high-class power plants, and afford convincing proof of the advanced design characteristic of the Winton line of engines. All of them are six-cylinder engines of the four-cycle type. The Model II was a development of the La Belle's engines, with an increase in the bore to 9 1/2 and stroke to 14", and such other refinements as experience in the La Belle and additional research work suggested. This engine develops 200 h. p. at 450 r. p. m.

The Models W5, and W6, which are smaller engines, came along about the same time, the former with a bore of 8" and stroke of n" developing 125 h. p. at 450 r. p. m. and the latter with 6 1/2" bore and 9" stroke, developing 80 h. p. at 500 r. p. m. While these engines were in process of construction, the company developed a six-cylinder, 7 1/2 k. w. generator set, known as Model W2, which was the forerunner of the present-day line of Winton auxiliaries.

In the fall of 1913 the first Winton Diesel engine was completed. This engine was the first all-American Diesel ever built, and its successful use as an industrial power plant for many years has fully justified Mr. Winton's early decision to strike out along original lines in developing Diesel engines for the American market, instead of following practices then prevailing in Europe.

During the war, two new Winton gasoline engines made their appearance. These were the Models W28 and W29. The W28 was a six-cylinder unit, with 6 1/2" bore, 9" stroke, developing 150 h. p. The W29 was an eight cylinder, with the same bore and stroke as the W28, and developing 200 h. p. A large number of these two models were sold to the American and Russian governments for use in mine-layers and river vessels. The company at this time also developed a Diesel engine known as Model W35. This was a six-cylinder unit, with 11" bore, 14" stroke, developing 225 h. p. A number of these engines were sold to the Italian government for war service. In 1915, the company produced what was probably the first V-type twelve-cylinder Diesel engine ever built. This was known as the Model W30. Two of these engines were installed in the La Belle, replacing the three six-cylinder gasoline engines with which she was originally powered. In 1916, the Model W24, a six-cylinder Winton Full Diesel type marine engine was produced. This engine had a bore of 12-13/16" and a stroke of 18", and developed 225 h.p. This was the first Winton engine designed strictly for workboat service, and was installed in such ships as the auxiliary freight schooners Esperanca, Pecheney, Adrien Badin, Charles Gawthrop, Sherewog, and Erris, ranging from 230 to 255 feet in length. Shortly thereafter followed the Model W40, an eight-cylinder Diesel, with the same bore and stroke as the Model W24, and developing 500 h. p. Engines of this type were installed in the Mt. Baker, Mt. Hood, Mt. Shasta, and James Simpson, all freighters, ranging from 286 to 393 feet in length.

Since these early installations, the company has enjoyed a steady, substantial growth and today is a recognized leader in its, field. From the beginning of 1918, the Diesel end of the business has increased greatly from year to year. Today, many of America's finest yachts and workboats are propelled by Winton Diesel engines, and their range of usefulness is constantly broadening. The gasoline engines, too, have kept pace, not only in the development of the units themselves, but also in their highly successful application to the diversified classes of service for which they are designed and built. One of the most important factors is a liberal service policy. Users of Winton engines are assured of competent, willing assistance when needed, to the end that their Winton power plants may be, in every sense, a complete success.

The company has also found it desirable to engage in the manufacture and sale of auxiliary equipment, such as generator sets, air, fire, and bilge pump sets, and air compressors, as well as rail car engines, tender engines, and industrial engines of various kinds.

Winton Engine Company and the Electro-Motive Company, also of Cleveland, the latter manufacturing power plants for railway cars, etc., were consolidated into the Winton Engine Corporation on January I, J 933General Motors acquired both companies in 1930.

This consolidation followed upon close relations maintained between the two companies since 1923, when Electro-Motive was incorporated, becoming a pioneer in gas-electric transportation and one of the chief outlets for Winton power plants. Four hundred and fifty-six Winton powered cars now operate on forty-two railroads in the United States, Canada, Mexico, and Australia, having been accepted as dependable by the leaders of the railway world under all conditions. On some of the most difficult of its runs the Santa Fe operates huge passsenger and mail cars, seventy-five feet in length, driven by Winton engines. The types of Winton engines used in Electro-Motive installations run from 220 to 800 h. p. In addition to use in transportation on sea and land, Winton Diesel engines drive the latest Marion steam shovels, developing 150 h. p. at 150 r. p. m.

George W. Codrington was president of both Winton and ElectroMotive at the time of the merger. He joined the Winton organization in 1917 and became general manager in 1918 and president in 1928.

Products: Marine and stationary gasoline engines, marine and stationary oil engines, rail-car oil and gasoline engines, and locomotive oil and gasoline engines.

# APPENDIX V

## THE ARMSTRONG SPRING DIVISION

IN 1888 Mr. J. B. Armstrong of Guelph, Ontario, founded the J. B. Armstrong Manufacturing Company at Flint for the manufacture of a single-leaf spring for carriages and wagons. Later a full line of carriage and wagon springs was made. The company had already made some automobile and truck springs before 1904 and thereafter turned all its equipment to their manufacture. The plant had an output of about 200 tons per month in 1909, when the company was merged with the Western Spring and Axle Company of Cincinnati, and this was gradually increased to 550 tons per month. On the merger of Western Spring and Axle with Standard Parts Company of Cleveland in 1909, the Flint plant operated as the Armstrong division. Ten years later Standard Parts erected the modern spring plant at the request of Buick Motor Company.

Before the plant was finished, Mr. R. T. Armstrong, son of J. B. Armstrong, organized the Armstrong Spring Company, in March, 1920. They purchased the plant and began operating it. The volume of business in 1923 exceeded 30,000 tons and some 300 men were employed.

The company was purchased by General Motors, December 31, 1923, for $1,623,186 and became a division, Mr. Armstrong remaining for a time as general manager and being succeeded by Mr. James Parkhill, formerly plant manager.

The division became part of Buick Motor Company in 1932.

Recently production has been as high as 70,000 tons, consisting of 4,500,000 automobile springs or about one third of all the country's production in that time.

## THE BROWN-LIPE-CHAPIN DIVISION

GENERAL MOTORS first became interested in this company in 1910 through purchase of $170,000 of its Common stock from the Buick, Cadillac, and Olds companies. Earlier the same year it had contracted for differentials. Ownership of Brown-Lipe-Chapin was completed in 1923. Of the three men whose names formed the company's name, Mr. Alexander T. Brown and Mr. Charles E. Lipe are dead, and Mr. H. W. Chapin retired in 1931. There were several Lipes connected with the business, including Mr. Willard C. Lipe (1861-1924). Both Mr. Brown (1854-1929) and Mr. Charles E. Lipe (1851-1895) were well known engineers and inventors as well as manufacturers. The former had to his credit the L. C. Smith shotgun, the Smith-Premier typewriter, the clincher tire, and the automatic switchboard for telephones. He was a founder of the Franklin Company, Syracuse, makers of automobiles. Mr. Lipe was interested in broom machinery and was the first to perfect the invention of a machine for sewing brooms. He also invented a milling machine applicable to general machine-shop work. The success of several companies was due to his mechanical and inventive genius.

Messrs. Brown and Lipe formed a partnership in Syracuse in 1895 to manufacture a two-speed gear for bicycles, the invention of Mr. Brown. This was not a commercial success, and Mr. Brown turned to designing a spur gear type of differential for an automobile he was using. While the differential was not patentable, the application to the automobile was new. The differential was put in production in 1900 and scored an immediate success, being adopted by Winton, Mobile, Locomobile, and other pioneer makes of cars. The first large order, however, came from Cadillac, 3,000 differentials for the one-cylinder cars. Growth and expansion were paid out of earnings. The business was incorporated as Brown-Lipe Gear Company.

Mr. H. W. Chapin (born 1867), who had been associated with the sales department of the L. C. Smith Company, joined Messrs. Brown and Lipe in the gear company in 1900. He had become its secretary when he was suddenly called to Flint by Mr. C. S. Mott of the Weston-Mott Company, their largest customer. Mr. Mott showed him a bevel gear type differential of a new design which would replace the gear company's goods in Buick, Cadillac, Olds, Oakland, and probably other cars as well. To produce the new gear meant retooling at an expense of at least a half-million dollars. Mr. Chapin had to make the decision quickly, and he did, placing an order by telegram for new machinery. To him is given the credit for saving one of the largest industries developed in Syracuse. As a result of the need for new capital, the Brown-Lipe-Chapin Company was formed in 1910 to take over the differential gear department of the gear company, and Mr. Chapin became manager of its plant at the beginning of 1911.

The new company had a five-year contract with General Motors. It sold $150,000 of treasury stock. Its business progressed with success. Mr. Chapin became president and general manager late in 1922, and as of January I, 1923, Brown-Lipe-Chapin affiliated with General Motors as a division.

The Syracuse plant occupied 450,000 square feet of floor space on five acres and had about 1,800 employees. Much of the special equipment used had been designed and built in the plant. Operations were transferred elsewhere in 1933, the Brown-Lipe-Chapin division being discontinued as an operating unit.

## THE ELMORE MANUFACTURING COMPANY

THE Automobile Trade Journal for December 1, 1924 gives this historical note on one of General Motors' early purchases, to which reference has been made in the text:

The Becker Brothers, who organized the Elmore Manufacturing Co., Clyde, Ohio, in 1901, gave the industry the two-cycle automobile, which they continued manufacturing until the company was sold to General Motors in 1910, and manufacture of this car discontinued two years later. The car had a twin-cylinder, two-cycle motor, rated at 4 h. p., which was mounted in a Stanhope design, with curved dash, and lever steering. The radiating coil, later to be known as the radiator, was concealed under the floor-boards. As the Becker Brothers had been in the bicycle business, it was natural that bicycle type tires should be used. A dynamo furnished ignition current. The car weighed 775 pounds and was very similar in design to the popular steam vehicles of that day.

## GENERAL MOTORS RADIO CORPORATION

GENERAL MOTORS entered the radio field primarily to develop and produce radios for automobile use in the expectation that radio would become standard equipment in all its cars. The Corporation bought into an already existing Dayton plant, licensed under RCA patents, and in that way entered the wider radio market, the plan being to have General Motors radios sold by its car dealers. This arrangement did not prove commercially successful, and the above company quit radio production in 1931.

In the meantime, after many difficulties had been overcome, the research engineers had gone a long way toward developing successful automobile radios. Automobile radios manufactured by other producers are sold by United Motors Service, Inc.

## JAXON STEEL PRODUCTS COMPANY

THIS company, first known as the Jackson Rim Company, was organized in 1911-12 and headed by Mr. O. W. Mott, who went from the Mott Wheel Works at Utica, New York, to Jackson, Michigan, and established this industry as an overflow from the Utica works. Operations began in 1913, but the real expansion dates from 1917 when the Perlman Rim Corporation of New York took over the company. Later the organization operated as a division of United Motors and was taken over by General Motors January I, 1919.

It manufactured automobile rims, felloes, wheels, tire carriers, battery box hangers, and stampings. In 1923 it produced 1,050,000 sets of rims. There were normally 1,050 to 1,150 employees. Floor space totaled 201,000 square feet, about eight acres being comprised in buildings, sidings, etc. Beside General Motors some ten other car manufacturers were supplied with equipment, and the Hayes Wheel Company, with certain materials.

All the assets of the Jaxon Steel Products Company were sold to and taken over by the Kelsey-Hayes Wheel Corporation of Jackson, June 19, 1930.

## LANCASTER STEEL PRODUCTS CORPORATION

THE immediate predecessor of this corporation was the Lancaster Steel Products Company, to which the name had been changed May 16, 1916, from New Process Steel Corporation. Mr. H. B. Cochran was president from 1915 to late in 1920.

There had been two earlier incorporations: New Process Steel Wire Manufacturing Company, incorporated in 1908, and the New Process Steel Company, incorporated in 1909 to operate the plant at Lancaster, Pa. In 1911 New Process Steel Corporation was organized, and the earlier companies were later dissolved.

First production consisted of drill rod and high carbon cold-drawn wire. In 1911 began the production of cold-drawn roller-steel for the Hyatt Roller Bearing Company. Following the change of name in 1916, operations were extended, and the plant was enlarged to a capacity of 4,000 tons per month.

Lancaster Steel Products Company passed into the hands of General Motors January I, 1919, and was dissolved the next year, being succeeded by the above corporation.

In 1923 production totaled 32,133 tons, consisting of cold rolling of strips, band, hoop, diaphragm, and steel for stamping purposes, and cold drawing of wires, bars, and special shapes. A large percentage of the output went to General Motors. The plant, occupying about nine acres, was sold to the Armstrong Cork Company in 1927. Lancaster Steel Products Corporation was subsequently dissolved.

## THE MUNCIE PRODUCTS DIVISION

THIS activity was not separately incorporated but was operated as a division of General Motors Corporation until 1932. In September, 1909, the T. W. Warner Company was organized for the purpose of specializing in the manufacture of steering gears, transmissions, and control levers. The original plant at Muncie, Indiana, was moved to a larger building in 1910. Additions and extensions were made rapidly. The company furnished one or more of its products to many of the earlier cars, including the Moon, Norwalk, Davis, Kissel, Columbia, Willys Overland, Marion, Chandler, Haynes, Studebaker, Wescott, Stoddard-Dayton, American, Stutz, and Cole. It had also established relations with the General Motors group through furnishing transmissions for the Scripps-Booth car, and in 1915 began supplying Chevrolet F. B. transmissions, and steering gears for all Chevrolet models. It then added transmissions for General Motors Truck, Oakland, and Olds, and also steering gears for the Sheridan, so that the Corporation eventually absorbed most of the company's output.

The result was the absorption of the company by General Motors in October, 1929, with Muncie Products taking its place as a division of the Corporation. At the time a large plant was erected for the manufacturing of Chevrolet chassis parts. In 1924 all the plants of Muncie Products had 377,500 square feet of floor space, occupying seven buildings in an area of eight acres. All operations were built on a progressive line-up of machinery. There were 2,200 employees. Annual production in units consisted of 250,000 transmissions, 800,000 steering gears, 750, 000 chassis parts sets, and 500,000 sets of valves. The annual volume of business amounted to from $12,000,000 to $15,000,000.

In 1929 another building was erected for the manufacture of gear forgings for use in transmissions and automobile engine valves for Buick, Olds, Oakland, and Pontiac.

Muncie Products plants were closed down in August, 1932. The transmission equipment was shipped to Buick Motor Company, Flint, Michigan, and to General Motors Truck Company, Pontiac, Michigan.

## NORTH WAY MOTOR AND MANUFACTURING COMPANY

THE original incorporators were Messrs. R. E. Northway, A. F. Knoblock and M. McMillan. The company was organized for the purpose of building, in Detroit, motors and parts for passenger cars and trucks. In 1913 the Michigan Auto Parts Company was consolidated with it.

Motors were built for Oakland, Oldsmobile Company, and various other buyers. Clutches and transmission were also manufactured. After the World War, with Oakland and Oldsmobile equipped to build their own motors, production of motors for Samson Tractor was increased. On the completion of that program in 1920, activities were transferred from the Maybury plant to the new plant, occupying a site of five acres on Holbrook Avenue, and the manufacture of motors for General Motors Trucks was begun, the General Motors Truck Company later becoming the exclusive customer. These motors had removable cylinder walls known as sleeves, thus obviating the need of pulling the motor to replace any that were scored or worn, also the need of replacing with a whole set. Northway division has been discontinued, its property being turned over chiefly to Chevrolet.

## RAINIER MOTOR CAR COMPANY

RAINIER began manufacturing in New York City in 1905. Its product was rather well known and ranked in the high-priced field. After removing to Saginaw, Michigan, Rainier Motor Car Company failed in 1908, as the result of conditions caused by the panic of 1907.

Mr. J. T. Rainier, president of that company, bought in the company at the receiver's sale and in turn sold it to General Motors Corporation. After Mr. Durant relinquished control, the banker management ceased manufacturing the Rainier car.

"In 1917 Rainier Motor Car Company was wound up by Mr. Durant, who had regained the presidency of General Motors, for the purpose of permitting J. T. Rainier to manufacture trucks and call them Rainier trucks, doing business as Rainier Motor Corporation. This company in turn was succeeded in October, 1924, by Rainier Trucks, Inc."

## THE SAGINAW PRODUCTS DIVISION

In 1919 the name Saginaw Products division had been given to the Saginaw activities of General Motors Corporation. These comprised the

Jacox steering gear plant, Motor plant, Grey Iron Foundry, and the allied plant of the Saginaw Malleable Iron Company (q.v.)

The motor plant was the Rainier Motor Company's property acquired by General Motors in 1908. There, for about three years, the Corporation built the Marquette, a high-priced car, and subsequently the plant lay idle until 1917. It was then devoted to the assembling of trench-mortar shells for the government, with a production peak as high as 25,000 shells assembled in nine hours. In 1919 the plant was equipped to manufacture overhead valve motors of the four-cylinder type, with Chevrolet and Oldsmobile taking the product. Chevrolet discontinued its FB model in 1922, and Oldsmobile became a "light six" in 1923, thus ending the demand on the motor plant and bringing about its close. Over 130,000 motors had been made there. The 25-acre property was turned over by the division to General Motors in November, 1923.

Grey Iron Foundry was erected under the management of Saginaw Products division in 1919 to furnish an adequate supply of grey iron castings for motors. The site, 100 acres of low ground near the river, was chosen with a view to expansion and filling. Original capacity of 100 tons of finished castings was augmented to 225 tons by the installation of continuous mold conveyors and other labor-saving and material handling devices. Floor space of the foundry building itself, core room, pattern shop, etc., amounted to 225,720 square feet in 1924. The plant was turned over to Chevrolet in September, 1927.

Before 1921, when it was absorbed by Saginaw Products, there existed for a while a Crankshaft division, the outgrowth of one of Saginaw's long-standing industrial concerns, the National Engineering Company. The company produced crankshafts first in 1907, its exclusive customer being Reo. However, it was not until 1913, when it received orders from the Northway Motor Company, that quantity was obtained. In 1917 the concern passed to Lansing capital, already interested in crankshaft manufacture in that city, and the name became Michigan Crankshaft Company. The Saginaw plant had run up to a production of 300 per day, and the Lansing plant to 200 in 1919, when they became affiliated with General Motors. The two were then eliminated in favor of a plant at Saginaw to take care of their previous productions and increased needs. This was the Crankshaft division of General Motors, continuing about two years as such, when its proximity to Saginaw Products division resulted in its absorption by the latter. The plant in 1924 had a floor space of 121,520 square feet on a 44-acre site. Production was about 2,000 crankshafts daily, employing 500 to 550 men. Some 65 to 90 operations were required in making the product. Close grinding, especially on the bearings and pins, was exacted down to one quarter of one thousandth of an inch. The plant supplied General Motors and several other car manufacturers.

## APPENDIX V

# THE WESTON-MOTT COMPANY

THIS company originated with the manufacture of bicycles, hubs, and wheels by Messrs. I. A. and G. B. Weston at Jamesville, New York, in 1886. Some years later the brothers J. C. Mott and F. G. Mott of Bouckville, New York, became interested in the concern and, in 1898, built a new factory at Utica. The bicycle business dropped away swiftly, but the company also made wire wheels for buggies, improved pushcarts, and axles. As early as 1897, inquiries regarding wire wheels for automobiles began to come in. The next year the company received an order for 500 sets of wheels for an automotive quadricycle. There were orders in 1899 from the Grout Motor Company of Orange, Mass., and the Autocar Company of Philadelphia. Mr. C. S. Mott became superintendent at Utica in 1900. Mr. Mott had received a technical training at Stevens Institute and in 1897 had engaged in the manufacture of carbonators with his father, Mr. John C. Mott, who died in 1899. Their plant was moved to Utica and merged with that of the Weston-Mott Company.

In 1900 orders began to come in from the Olds Motor Works for 500 sets of wire wheels at a time for the famous curved-dash runabout. Two years later Oldsmobile went to wood wheels and Weston-Mott had to adjust itself accordingly. Mr. C. S. Mott became president and general manager in 1903.

The company was pressed by its customers to supply axles, which it did in the spring of 1903 furnishing 1,500 sets, chiefly to Cadillac, but also to Elmore, Blomstrom, and others. These were crude affairs, of the chain-drive type, for nearly two years. Buick began buying in 1903, a few at a time and then by the dozen, for the two-cylinder car. The bevel drive axle of Weston-Mott found its first customer in Stevens Duryea.

Through the efforts of Messrs. Durant and Dort, the company was moved to Flint, a factory site having been granted in 1905, and the Weston-Mott Company of Michigan was incorporated with $500,000 capital, Buick interests taking a part. The new factory began operations late in 1906. Mr. H. H. Bassett, later president and general manager of Buick, came to Flint as part of the Weston-Mott organization. After the death of Mr. Doolittle of Weston-Mott, Mr. C. S. Mott, as agreed upon, purchased the decedent's 40 percent interest, which had equaled his own.

Following its organization in 1908, General Motors Company bought into Weston-Mott until it owned 49 percent against Mr. Mott's 51 percent. Weston-Mott at the time made a contract to furnish Buick axles for ten years. General Motors completed its ownership of Mr. Mott's stock in 1913, at which time he was serving as Mayor of Flint for the first of three terms. Weston-Mott is now part of Buick Motor division. Mr. Mott has been a director of General Motors since 1913 and a vice-president since 1916.

# Bibliography

ADAMS, Charles Francis. Railroads, Their Origin and Problems. G. P. Putnam's Sons, New York, 1878-86.

ADAMS, Walker Poynter. Motor-Car Mechanism and Management.C. Griffin & Co., London, 1907.

ALLEN, James Titus. Digest of U. S. Automobile Patents from 1789to July 1, 1899. By Authority, Commissioner of Patents. H. B. Russell & Co., Washington, 1900.

Americana Biographies. Americana Corporation, New Yorkand Boston, 1923-30.

ANDRE, H. L'Industrie Automobile en 1911. G. Mathiere, Paris,1911.

ARAGO, Dominic F. J. Life of James Watt. F. Didot, Paris, 1834. Tr.pub. in Edinburgh, 1839.

ARENE, Georges d', Vicomte. L'Evolution des Moyens de Transport.E. Flammarion, Paris, 1919.

ASTON, Wilfred Gordon. The Book of Motors. E. and F. N. Spon,Ltd., London, 1924.

AUCAMUS, E., and GALINE, I. Tramways and Automobiles. H. Donot & E. Pinat, Paris, 1909.

Automobil und Automobilsport. W. Isendahl, Ed. R. C. Schmidt &Co., Berlin, 1908.

Automobile and the Village Merchant, The. Business Research Bull. No. 19., pp. 9-42, Univ. of 111., Urbana, 111.

Automobile Biographies. Lyman H. Weeks, Ed. Monograph Press, New York, 1904.

Automobile History, A Bit of Early. J. I. Case Manufacturing Company, Racine, Wis., 1914.

Automobile in Munsey's Magazine, The. F. A. Munsey Co., New York, 1910.

Automobile, The: Its Past, Present and Future. Annals of the

Academy of Political and Social Science. Philadelphia, November, 1924.

Automobile Trade Journal, Silver Anniversary number. David Beecroft, Ed. The Chilton Co., Philadelphia, December I, 1924. BARBER, Herbert Lee. Story of the Automobile. Its History and Development from 1760 to 1917. A. I. Munson & Co., Chicago, 1917.

BARRON, Clarence W., Notes of. They Told Barron, 1930; MoreThey Told Barron, 1931. Arthur Pound and Samuel T. Moore, Ed. Harper & Bros., New York.

BAUDRY DE SAUNIER, Louis. L 'Automobile , Theorique et Pratique. Paris, 1899L'Art de Bien Condulre une Automobile. Paris, 1907.

BEAUMONT, W. Worby, Cantor Lectures on Mechanical Road Carriages. Wm. Trounce, London, 1896. Motor Vehicles and Motors. Archibald, Constable & Co., London, 1900; revised edition, 1906.

BEECROFT, David. History of the American Automobile Industry.Published serially in The Automobile, Vols. 33-35. Class Journal Publishing Company, New York, 1915-16.

BELL, Sir I. L. Development of the Manufacture and Use of Rails inGreat Britain. Inst. Civil Engineers, London, 1900.

BERRIMAN, Algernon E. Motoring: An Introduction to the Car andthe Art of Driving It. Methuen & Co., London, 1914.

BERRY, William H. Modern Motor Car Practice. H. Frowde & Hodder & Stoughton, London, 1921.

BOULTON, W. H. The Pageant of Transport through the Ages. S. Low, Marston & Co., Ltd., London, 1931.

BOYD, T. A. Gasoline: What Everyone Should Know about It. Frederick A. Stokes, New York, 1926.

BOYER-GUILLON, A. Les Vehicules Automobiles. J. B. Balliere et-Fils, Paris, 1926.

BRERETON, Frederick S. Travel: Past and Present. B. T. Batsford,Ltd., London, 1931.

BRIGHT, Charles. The Locomotion Problem. P. S. King & Co., Ltd.,London, 1905.

BRISCOE, Benjamin, Jr. The Entire Story of General Motors. DetroitSaturday Night, Detroit. Vol. XV.

BROWN, Edward T. Book of the Light Car. Chapman & Hall, Ltd.,London, 1926. Motors and Motoring. Williams & Norgate, London, 1927.

BUCK, Max. Die Automobiltechnik. J. A. Barth, Leipzig, 1908.

BURTON, Charles M. History of Detroit, Michigan, 1701-1922. 4Vols. S. J. Clarke Publishing Co., Cleveland, 1922.

BUTLER, Edward. Carburetors, Vaporizer s t etc. C. Griffin & Co., Ltd., London, 1909. Cadillac Participation in the World War. Cadillac Motor Car Co., Detroit, 1919.

CARTER, Charles F. When Railroads Were New. Henry Holt & Co., New York, 1909.

CLERK, Sir Dugald. The Gas and Oil Engine. J. Wiley & Sons, New York, 1902.

CONRADI, Charles A. Mechanical Road Transport. Macdonald & Evans, London, 1923.

COOKE, Stenson. This Motoring: The Story of the Automobile Association. Cassell & Co., Ltd., London, 1931.

COURTOIS, C. Traite: Theorique et Pratique des Moteurs. 2 vols. L. Mathias, Paris, 1846-50.

CROMPTON, Rookes E. B. Modern Motor Vehicles. Inst. of Civil Engineers, London, Proceedings: Vol. V. London, 1907.

DERHAM, W. Experiments and Observations of Dr. Robert Hooke, W. J. Innys, London, 1726.

DESMONS, Robert. Les Derniers Progres de VAutomobile, 1920.Paris, 1920. Development of Motor-vehicles Trade Abroad. U. S. Bureau of Foreign and Domestic Commerce, Govt. Printing Office, 1913.

DOOLITTLE, James Rood. The Romance of the Automobile Industry. Klebold Press, New York, 1916.

DUNBAR, Seymour. A History of Travel in America. Bobbs-Merrill Co., Indianapolis, 1915.

DUNCAN, Herbert Obaldeston. The World on Wheels. Paris, 1926. DUNCAN, W. Galloway. The Modern Motor Car. C. Lockwood & Son, London, 1911.

DUNN, Robert W. Labor and Automobiles. International Publishers, New York, 1929.

DURANT, Margery. My Father. G. P. Putnam's Sons, New York and London, 1929.

DURYEA, Charles E. Handbook of the Automobile. American Motor League, New York, 1906.

DURYEA, Charles E., and HOMANS, J. E. The Automobile Book.Sturgis & Walton, New York, 1916.

EPSTEIN, Ralph C. The Automobile Industry: Its Economic and-Commercial Development. A. W. Shaw Co., Chicago and New York, 1928.

EVANS, Oliver. The 'Young Steam Engineer's Guide. 1st ed. Philadelphia, 1804.

FARMAN, D. Auto-Cars, Cars, Tram Cars and Small Cars. L. Serallier, tr. Whittaker & Co., London, 1896. Les Automobiles. Paris, 1898.

FARMAN, M. Manuel du Conducteur --Chauffeur d 'Automobiles a Petrol. B. Tignil, Paris, 1897.

FAUROTE, Fay Leone. The How and Why of the Automobile. Motor Talk Publishing Co., Detroit, 1907.

FLETCHER, William. The History and Development of Steam Locomotion on Common Roads. E. & F. N. Spon, Ltd., London, 1891. English and American Steam Carriages and Traction Engines.Longmans, Green & Co., New York and London, 1904. FORBES, B. C., and FOSTER, O. D. Automotive Giants of America.B. C. Forbes Pub. Co., New York, 1926.

FORESTIER, A. Essai ... Tractions Mecaniques. Le Genie Civil, Paris, 1900.

GALLOWAY, Elijah. History and Progress of the Steam Engine. B. Steill, London, 1829, 1831.

GIBSON, Charles R. Proceedings, LV, Philosophical Society of

Glasgow, Scotland, 1927. The Motor Car and Its Story. J. B. Lippincott Co., Philadelphia, 1927.

GORDON, Alexander. An Historical and Practical Treatise upon Elemental Locomotion, etc. 2nd ed., B. Steuart, London, 1832; reprinted 1836.

GRAFFIGNY, H. E. Manuel du Constructeur et du Conducteur de Cycles et d 'Automobiles. J. Hetzel et Cie. Paris, 1897.

GRAND-CARTERET, John. La Voiture de Demain; Histoire de L'automobilisme. Paris, 1898.

GREAT BRITAIN, House of Commons. Report on Steam Carriages-by Select Committee. Reprinted as Document 101, 22d Congress, First Session, House of Representatives, Washington, D. C., 1832.

GRIFFIN, Clare E. The Life History of Automobiles. Bureau of Business Research, Univ. of Mich., Ann Arbor, Mich., 1928.

GRIFFITHS, Joseph P. Transport. A. Philip & Son, London, 1919.

GUEDON, P. and Y. Manuel Pratique du Conducteur d' Automobiles.J. Fritsch, Paris, 1897.

HADLEY, Arthur Twining. Railroad Transportation: Its History and Laws. G. P. Putnam's Sons, New York, 1885, 1903, 1906.

HALE-SHAW, H. S. Road Locomotion. Inst. of Mechanical Engineers, London, 1900.

HANCOCK, Walter. Narrative of Twelve Years' Experiments (1824-36). Steam Carriages on Common Roads, London, 1838.

HASLUCK, Paul N. The Automobile. Based upon Lavergne's L'automobile sur Routes. Cassell & Co., Ltd., London, 1904.

HAWKS, Ellison. The Romance of Transport. T. Y. Crowell Co., New York, 1931.

HEIRMAN, Edmund. L' Automobile a I'Essence. C. Beranger, Paris, 1908.

HELDT, P. M. The Gasoline Automobile. 2 vols., Horseless Age Co., 1911-13; Nyack, N. Y., 1916; 7th ed., 1920.

HERBERT, Luke. Practical Treatise on Railroads and Locomotive Engines. London, 1837. Highway Education Board Proceedings, Univ. of Tenn. Conference, 1921, Washington, D. C., 1922.

HISCOCK, Gardner D. Gas, Gasoline and Oil Vapor Engines. N. W. Henley & Co., New York, 1898. Horseless Vehicles, etc. N. W. Henley & Co., New York, 1900.

History of Automobiles in St. Louis, A. St. Louis Society Automobile Pioneers, St. Louis, Mo., 1930.

HODGINS, Eric, and MAGOUN, F. A. Behemoth: The Story of Power.Doubleday, Doran & Co., New York, 1932. HOLLAND, Rupert Sargent. Historic Railroads. Macrae Smith Co., Philadelphia, 1927.

HOMANS, James E. Self-Propelled Vehicles. T. Audel & Co., New York, 1902; ran to 7 eds., 1902-10.

HOOKE, see DERHAM.

HOWE, Henry. Memoirs of the Most Eminent American Mechanics. J. C. Derby, New York, 1856.

HUNT, Rockwell, and AMENT, William S. Oxcart to Airplane, Powell Pub. Co., Los Angeles and San Francisco, 1929.

HUTTON, Frederic. The Gas Engine. J. Wiley & Sons, New York, 3d ed., rev., 1908.

JACKMAN, William T. The Development of Transportation in Modern England. University Press, Cambridge, Eng., 1916. JARROTT, Charles. Ten Years of Motors and Motor Racing. E. A. Richards, Ltd., London, 1912. JENKINS, Rhys. Power Locomotion on the Highway. W. Cate, London, 1896.

JUDGE, Arthur William, Ed. Modern Motor Cars. Caxton Pub. Co., London, 1924. Automobile Engines. Chapman & Hall, London, 1925.

KENNEDY, Rankin. The Book of Modern Engines. Caxton Pub. Co., London, rev. ed., 1912.

KETTERING, Charles F. and ORTH, Allen. The New Necessity. Williams & Wilkins Co., Baltimore, 1932.

KIRKALDY, Adam W. and EVANS, A. D. The History and Economics of Transport. Sir. I. Pitman & Sons, Ltd., London, 1924.

KLIMA, Anton. Das Auto in des Karikatur. O. Stollberg, Verlag, Berlin, 1928.

KNIGHT, John H. Notes on Motor Carriages. Hazell, Watson. &. Viney, Ltd., London, 1896.

LAIDLER, Harry Wellington. Concentration of Control in American Industry. Thomas Y. Crowell Co., New York, 1931.

LAVERGNE, G. see HASLUCK, Paul N.

LAVILLE, Ch., and GATOUX, A. Voiturettes et Voitures Legeres. H. Donot et E. Pinat, Paris, 1910.

LE GRAND, Georges. L'Automobile et la Guerre. H. Donot et E. Pinat, Paris, 1916.

LEROUX, A., and REVEL, A. La Traction Mechanique et les Voitures Automobiles. J. B. Balliere et Fils, 1900.

LEVITT, Dorothy. The Woman and the Car. John Lane Co., London, 1909. LOEW, Von and Zu STEINFURTH, Ludwig, Freiherr. Das Automobil. C. W. Kriedel, Wiesbaden, 1909.

LOUGHEED, Victor. How to Drive an Automobile. J. C. Chase, Ed., Motor, New York, 1908.

LOUGHEED, Victor, and HALL, M. A. The Gasoline Automobile. American School of Correspondence, Chicago, 1912.

LYND, Robert S. and Helen M. Middletown: A Study in Contemporary American Culture. Harcourt, Brace & Co., New York, 1929.

MACERONI, Francis. Expositions and Illustrations of Steam Power. Wilson & Hebert, London, 1835.

MACMANUS, Theodore F. and BEASLEY, Norman. Men, Money and Motors. Harper & Bros., New York, 1930.

MAKEPEACE, Gordon. Capetown to Stockholm: A Chevrolet Adventure. General Motors South African, Ltd., Cape Province, S. A., 1929.

MARTIN, R. N. Railways, Past, Present and Prospective.

W. H. Smith & Son, London, 1849.

MASON, Otis Tufton. The Human Beast of Burden. Govt. Printing Office, Washington, D. C., 1889. Primitive Travel and Transportation. Govt. Printing Office, Washington, D. C., 1896.

Memorial Pamphlet on Richard Trevithick. Institution of Civil Engineers, London, 1883.

MERZ, Charles." and then came Ford". Doubleday, Doran & Co., New York, 1929.

MEYER, Balthasar Henry. History of Transportation in the United States before 1860. Carnegie Institution, Washington, D. C., 1917.

MILANDRE, Charles, and BOUQUET, R. P. Traite de la Construction, etc. E. Bernard et Cie, Paris, 1898-99.

MITMAN, Carl W. The Beginning of the Mechanical Transport Era in America. Smithsonian Institution, Washington, D. C., 1930.

MONTAGU, John W. E. D. S., 2d baron, Ed. Cars and How to Drive Them. Car Illustrated, London, 1905.

Motor Machines. Manufacturers' Bureau. Govt. Printing Office, Washington, D. C., 1907.

Motor Transport Corp. W. S. Report, 1918-19; 1919-20. Govt. Printing Office, Washington, D. C., 1919-20.

Motor Vehicle Regulation and Taxation in Foreign Countries. Govt. Printing Office, Washington, D. C., 1930.

MUIRHEAD, James Patrick. The Life of James Watt. Arago's Life, with notes by Muirhead. 2d ed., John Murray, London, 1859.

National Parks. Report of R. B. Marshall, director. Govt. Printing Office, Washington, D. C., 1917.

National Parks. Report of H. N. Albright, director. Govt. Printing Office, Washington, D. C., 1933

NORTHCLIFFE, Alfred C. W. H., ist baron, and others. Motors and Motor Driving. Badminton Library. Longmans, Green & Co., 1904.

PAGE, Victor W. The Modern Gasoline Automobile. N. W. Henley Publishing Co., New York, 1912-18. PANGBORN, J. G. The World's Railway. Winchell Pub. Co., New York, 1894

PARKER, J. W. The Roads and Railroads, Vehicles and Modes of Traveling of Ancient and" Modern Countries. London, 1839. Passenger Car Industry, The. Curtis Publishing Co., Philadelphia, 1932.

PATTERSON, James. History and Development of Road Transport. Sir I. Pitman & Sons, Ltd., 1927.

PEMBERTON, Max. The Amateur Motorist. Hutchinson & Co., London, 1907.

PERISSE, Lucine. Automobiles sur Routes. Masson et Cie, Paris,1898.

PETER, M. Das Moderne Automobil. R. C. Schmidt & Co., Berlin, 1912, 1920.

PETIT, Henri. Traite Elementaire d 'Automobile, etc. H. Donot et E. Pinat, Paris, 1919.

PHELPS, D. M. Effect of the Foreign Market on the Growth and Stability of the American Automobile Industry. Business Studies, Vol. Ill, No. 5., pp. 553, 728. University of Michigan, Ann Arbor, Michigan, 1931.

POUND, Arthur. The Iron Man in Industry. Atlantic Monthly Press, Boston, 1921.

READ, David. Nathan Read, Life of. Hurd & Hough ton, New York, 1870. Recent Social Trends. Reports of the President's Research Committee on Social Trends. McGraw-Hill Book Co., Inc., New York, 1933.

Report of U. S. Commissioners. International Exhibition, Vienna, 1873. R. H. Thurston, Ed., Washington, D. C., 1876.

Report on Highway Transport. International Chamber of Commerce, American Section, Washington, D. C., 1925.

ROLL, Erich. An Early Experiment in Industrial Organization, being a History of the firm of Boulton & Watt, 1775-1805. Longmans, Green & Co., London and New York, 1930.

SELIGMAN, E. R. A. The Economics of Instalment Selling. 2 vols., Harper & Bros., New York, 1927.

SELLEW, William H. Steel Rails. D. Van Nostrand Co., New York,

SELTZER, Lawrence Howard. A Financial History of the American Automobile Industry. Hough ton Mifflin Co., Boston and New York, 1928.

SENNETT, A. R. "Carriages Without Horses Shall Go/' Whittaker & Co., London, 1896. SERAFON, E. Let Tramways, les Chemins de Per sur Routes, les Automobiles, etc. E. Bernard et Cic, Paris, 1895.

SLOSS, Robert T. The Book of the Automobile. Introduction by Dave Herman Morris. D. Appleton & Co., New York, 1905.

SMILES, Samuel. The Life of George Stephenson. Lives of the Engineers. J. Murray, London, 1862-68. Lives of Boulton and Watt. Also a history of the steam engine. J. B. Lippincott & Co., Philadelphia, 1865.

SMITH, F. L. Motoring Down a Quarter of a Century. Detroit Saturday Night Company, Detroit, 1928.

SOUVESTRE, Pierre. Histoire de I'Automobile. H. Donot et E. Pinat, Paris, 1907.

SPOONER, Henry J. Motors and Motoring. Dodd, Mead & Co., New York. Story of the Automobile, The. Sport of the Times. New York, 1904.

STRATTON, Ezra M. The World on Wheels. 400 illustrations. New York, 1878.

TALBOT, Frederick A. A. Motor Cars and Their Story. Cassell & Co., London, 1912.

TREDGOLD, T. The Steam Engine. J. Weale, London, 1838.

TREVITHICK, Francis. Life of Richard Trevithick. W. J. Welch, London, 1872.

WAGNER, Adolphe. Sozialekonomische Theorie des Kommunikations und Transportwesens. C. F. Winter, Leipzig.

WALLIS-TAYLER, Alfred J. Motor Cars or Power Carriages for Common Roads. C. Lockwood & Sons, London, 1897. Motor Vehicles for Business Purposes. D. Van Nostrand Co., New York, 1905.

WETMORE, John C. Surviving Pioneers of Automobile Building. Evening Mail Supp. N. Y. Evening Mail, New York. January 14, 1914. Motoring Memories. N. Y. Evening Mail, New York, September ber 26, 1922.

WHITMAN, Roger B. Gas Engine Principles. D. Appleton & Co., New York, 1907.

Wonder Book of Motors. The. Harry Golding, Ed. Ward, Lock & Co., London, 1926.

WOOD, Nicholas. Practical Treatise on Railroads. Knight & Lacey, London, 1825, revised eds., 1831, 1838.

WRIGHT, Harold E. The Financing of Automobile Instalment Sales. A. W. Shaw & Co., Chicago and New York, 1927.

YOUNG, A. B. Filson. The Complete Motorist. McClurc, Phillips & Co., New York, 1904. YOUNG, Charles F. T. The Economy of Steam Power on Common Roads. Atchley & Co., London, 1861.

Automobile bibliographies are obtainable from the Division of Bibliography, Library of Congress, Washington, D. C., as follows:

List of Bibliographies on Automobiles: 28 items; dated November 1 6, 1922.

Brief List of References on the Automobile Industry: 21 items; dated November 10, 1916.

List of References on the Taxation of Automobiles: 68 items; dated September 28, 1922.

List of References on the Automobile: 36 items and the names of periodicals, etc.; undated.

Card records of 1,500 publications on the automobile, as listed by the Library of Congress, may be purchased from the Card Division of that institution.

Detroit Public Library has a twenty-page list of automobile titles, dated 1916.

# Index

B

C

E



F

G

**Life of an American Workman**
**Walter P. Chrysler**
Oxford City Press, 2011
156 pages
ISBN: 978-1-84902-395-5

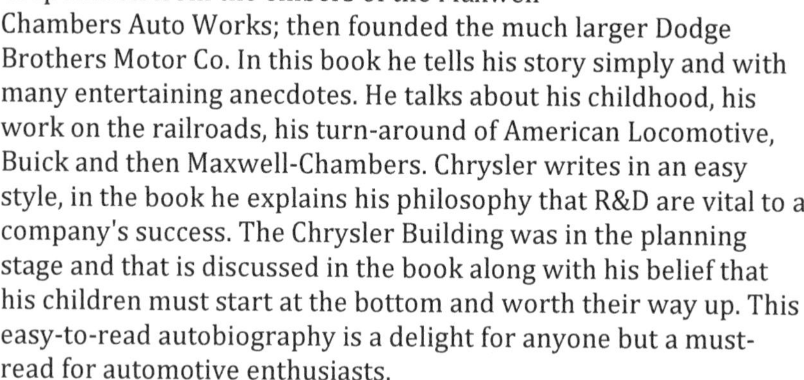


Available from www.amazon.com, www.amazon.co.uk

Walter Chrysler established the Chrysler Corporation from the embers of the Maxwell-Chambers Auto Works; then founded the much larger Dodge Brothers Motor Co. In this book he tells his story simply and with many entertaining anecdotes. He talks about his childhood, his work on the railroads, his turn-around of American Locomotive, Buick and then Maxwell-Chambers. Chrysler writes in an easy style, in the book he explains his philosophy that R&D are vital to a company's success. The Chrysler Building was in the planning stage and that is discussed in the book along with his belief that his children must start at the bottom and worth their way up. This easy-to-read autobiography is a delight for anyone but a must-read for automotive enthusiasts.

**Journal of a First Fleet Surgeon (1788)**
**George B. Worgan**
Benediction Classics, 2011
60 pages
ISBN: 978-1-84902-460-0

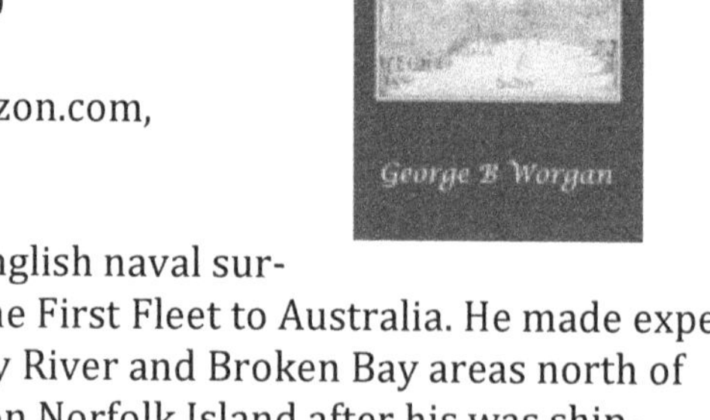


Available from www.amazon.com, www.amazon.co.uk

George Worgan was an English naval surgeon who accompanied the First Fleet to Australia. He made expeditions to the Hawkesbury River and Broken Bay areas north of Sydney and spent a year on Norfolk Island after his was shipwrecked there. Although he kept a journal, it was not published on his return, unlike his contemporary, Watkin Tench. This book consists of letters to his brother in England, written in 1788, the second letter journaling the first six months after the First Fleet's arrival in Sydney Cove.

**The Story of Paul Boyton:**

**Paul Boyton**
Benediction Classics, 2011
340 pages
ISBN: 978-1-84902-390-0

Available from www.amazon.com, www.amazon.co.uk

This book was originally marketed with this words: "A rare tale of travel and Adventure. Thrilling experiences in distant lands, among strange people. A book for boys, old and young." The description of the book is no exaggeration. Paul Boyton (1848-1924) was clearly a remarkable and fearless man and indeed had adventures that can only be described as thrilling. He discovered and started working with a rubber suit, similar to modern drysuits . It allowed the wearer to float on his or her back, using a double-sided paddle to propel themself, feet-forward. Eventually, he was to found the first "amusement park" featuring performing sea lions and water chutes.

**Chapters from My Autobiography**
**Mark Twain**
Benediction Classics, 2011
274 pages
ISBN: 978-1-84902-343-6

Available from www.amazon.com, www.amazon.co.uk

Mark Twain (or Samuel Clemens) intended for his autobiography to be published long after he died. He felt that he couldn't be honest about his experi- en- ces and contemporaries if he was wor- ried about the reaction of others. However, in 1906 he agreed to publish selections from the autobiography in the North American Review, from September 1906 through December 1907. The twenty-five "Chapters from My Autobiography" have been brought together in this book.

Also from Benediction Books ...

**Wandering Between Two Worlds: Essays on Faith and Art**
**Anita Mathias**
Benediction Books, 2007
152 pages
ISBN: 0955373700

Available from www.amazon.com, www.amazon.co.uk

In these wide-ranging lyrical essays, Anita Mathias writes, in lush, lovely prose, of her naughty Catholic childhood in Jamshedpur, India; her large, eccentric family in Mangalore, a sea-coast town converted by the Portuguese in the sixteenth century; her rebellion and atheism as a teenager in her Himalayan boarding school, run by German missionary nuns, St. Mary's Convent, Nainital; and her abrupt religious conversion after which she entered Mother Teresa's convent in Calcutta as a novice. Later rich, elegant essays explore the dualities of her life as a writer, mother, and Christian in the United States-- Domesticity and Art, Writing and Prayer, and the experience of being "an alien and stranger" as an immigrant in America, sensing the need for roots.

**About the Author**

Anita Mathias is the author of *Wandering Between Two Worlds: Essays on Faith and Art*. She has a B.A. and M.A. in English from Somerville College, Oxford University, and an M.A. in Creative Writing from the Ohio State University, USA. Anita won a National Endowment of the Arts fellowship in Creative Nonfiction in 1997. She lives in Oxford, England with her husband, Roy, and her daughters, Zoe and Irene.

Anita's website:
http://www.anitamathias.com, and
Anita's blog Dreaming Beneath the Spires:
http://dreamingbeneaththespires.blogspot.com

**The Church That Had Too Much**
**Anita Mathias**
Benediction Books, 2010
52 pages
ISBN: 9781849026567

Available from www.amazon.com, www.amazon.co.uk

The Church That Had Too Much was very well-intentioned. She wanted to love God, she wanted to love people, but she was both hampered by her muchness and the abundance of her possessions, and beset by ambition, power struggles and snobbery. Read about the surprising way The Church That Had Too Much began to resolve her problems in this deceptively simple and enchanting fable.

**About the Author**

Anita Mathias is the author of *Wandering Between Two Worlds: Essays on Faith and Art*. She has a B.A. and M.A. in English from Somerville College, Oxford University, and an M.A. in Creative Writing from the Ohio State University, USA. Anita won a National Endowment of the Arts fellowship in Creative Nonfiction in 1997. She lives in Oxford, England with her husband, Roy, and her daughters, Zoe and Irene.

Anita's website:
    http://www.anitamathias.com, and
Anita's blog Dreaming Beneath the Spires:
    http://dreamingbeneaththespires.blogspot.com